DNA and

DNA and You

Blog Posts from the Golden Age of the Human Genome Project

DONALD N. YATES and TERESA A. YATES

Panther's Lodge Publishers

Longmont

LIBRARY OF CONGRESS CATALOGUING-IN-PUBLICATION DATA

Yates, Donald N., 1950– author
DNA and You: Blog Posts from the Golden Age of the Human Genome Project
/ Donald N. Yates and Teresa A. Yates
p. cm.
Includes index.

ISBN-10: 0692663606
softcover : acid free paper ∞

1. Science news. 2. Genetic genealogy.
3. Genetics—Social aspects. 4. DNA fingerprinting.
5. Indians of North America.
6. Human population genetics. 7. DNA, Fossil. 8. Prehistoric peoples. I. Title. II.
Series.

DNA Consultants Series on Consumer Genetics 2

Manufactured in the United States of America

*Panther's Lodge Publishers
P.O. Box 2477
Longmont, Colorado 80502
www.pantherslodge.com*

CONTENTS

INTRODUCTION

The Human Genome Project, first sketched out in the 1980s, has been heralded as the "the most ambitious writing endeavor" in human history. A rough draft was announced to the world jointly by President Bill Clinton and British Prime Minister Tony Blair on June 26, 2000. A target of "85 percent completion" had been achieved, covering most of the 3 billion nucleotides in DNA. Phase II, the conquest of genetic disease, has occupied the best minds in medicine for over a decade now. All these exciting developments, and their importance for the field of genetic genealogy, were followed and reported on in the blog started by DNA Consultants in the early 2000's, continuing down to the present.

Our oldest post dates from May 15, 2000 ("Six Generations of Cherokee Blood," pp. 50-53). In the early days, entries often consisted of a bare notice. "A Future without Emperors," for example, picked up on a news report by David Cyranoski in the journal *Nature* on February 22, 2006 and asked simply, "Will Japan have a queen?" (p. 94). Other posts minimally registered the appearance of a newsworthy scientific investigation with the proper citation and full abstract. Increasingly, a précis, summary or critical review might be provided, especially for product news and favorite interests, such as Native American testing, autosomal DNA methods and Jewish markers. Readers' comments accrued, reflecting a wide range of interests, professional affiliations and diverse backgrounds. During the years between 2012 and 2015, I was joined by Teresa, my wife, in writing posts that were gaining greater and greater attention.

In addition to the two indispensable, leading journals *Science* (U.S.)

and *Nature* (U.K.), we have blogged about articles from the *American Journal of Human Genetics, Current Biology, European Journal of Human Genetics, Human Biology, Journal of Human Evolution, Molecular Biology and Evolution* and *PLoS (Public Library of Science).* Occasionally, we have referenced lesser known journals such as *Genome Biology, International Journal of Legal Medicine, Journal of Forensic Science* and *Proceedings of the National Academy of Sciences.* Newspaper sources from which we found material include *The Guardian, The Huffington Post, The New York Times* and *The Wall Street Journal.*

Breaking news in the field of genetics is represented by a host of hot topics cropping up as they crossed over from technical publications and gained resonance in the popular press, some little more than brief notes and others full-blown reviews. Readers will be particularly interested in startling revelations about Neanderthals and Denisovans, the non-Asian roots of Native Americans, the introduction of personal genomics made possible by 23andMe ("Google Wants to Do What?" p. 296) and the skeleton of Shakespeare's villain and England's last York Dynasty monarch exhumed in a Leicester parking lot ("Richard III's New Winter of Discontent," pp. 87-90). One of the last items in our blog, still going strong on the DNA Consultants website, is an update from October 21, 2015 on the discovery of water on Mars (see "Alien DNA," originally dated October 4, 2012, pp. 6-7).

Mention is made in these pages of some of the most famous geneticists of our time. These include James Watson ("DNA Pioneer Rebuked For Racist Remarks," *BBC News*, October 20, 2007, p. 112), Swedish-Estonian biologist Svante Pääbo, whose team sequenced the Neanderthal and Denisovan genomes; Bryan Sykes, author of *The Seven Daughters of Eve* and founder of the commercial firm Oxford Ancestors; Danish geneticist Eske Willerslev, whose work has revolutionized the study of American Indian origins; Stephen Oppenheimer (*The Real Eve* and *The Origins of the British*); Mark Thomas of Cohen Modal Haplotype fame; Stanford's Luigi Luca Cavalli-Sforza; Columbia's David Goldstein; and Harvard's Richard Lewontin.

The classic study, "Population Structure and Genome-wide Patterns of Variation in Ireland and Britain" prompted a considered appraisal after it appeared in *European Journal of Human Genetics* in 2010. See

"Surprises in English and Irish DNA," pp. 170-172. Jon Entine's *Abraham's Children: Race, Identity, and the DNA of the Chosen People* gets major notice in the chapter "DNA and Religion." Who could fail to be fascinated by the chronicles of the royal family of Egypt whose mummies were investigated by Zahi Hawass' team in that same year?

Among the science journalists whose writing is followed are Nicholas Wade and Amy Harmon of the *New York Times*. There are also many lesser known bloggers and tweeters whose work in the trenches of genetic geneaology receives attention.

Exotic themes touched on are Roman soldiers in the DNA record of Britain, Egyptian and Greek DNA among the Cherokee, Basques in Idaho, ancient Sea Peoples detected in the DNA and rock art of the American Southwest, Chinese surname systems, Itza Mayas in north Georgia and India's caste system.

From our series "People behind the Numbers," come interviews and customer stories of special interest. Noteworthy are "Peter McCormick, Cultural Geographer" (pp. 174-175), "Elizabeth Hirschman, Modern Pioneer" (183-185), "Jim Bentley, DNA Frontiersman" (360-367) and "Phyllis Starnes: Designer Genes," (216-19). Starnes, who was an assistant investigator for DNA Consultants for many years, poignantly describes her Melungeon ancestry and long struggle with familial Mediterranean fever in the piece "Tapestries of Illness" (226-232).

Variety and diversity are signaled by the inclusion of unusual personal cases. In this category, we have Elvis Presley's "King-sized Sheets" with the adventures of canny Houston P.I., Bobbi Bacha, featured in the movie *Suburban Madness* (pp. 103-105), "A Red-Hot Tale from the Nation's Capital," by Monica R. Sanowar (130-133) and "Melungeon Revelation," by Kari Carpenter (212-215).

There are numerous excerpts and previews of books we have written over the years on history, culture and DNA. These include the "trilogy" I co-authored with Elizabeth Caldwell Hirschman, begun with *When Scotland Was Jewish* (2007), followed by *Jews and Muslims in British Colonial America* (2012) and concluded with *The Early Jews and Muslims of England and Wales* in 2014, as well as the solo study *Old World Roots of the Cherokee* (2012) and *Cherokee DNA Studies: Real People Who Proved the Geneticists Wrong* (2014). The last-named forms volume one in DNA Consultants Series on Consumer Genetics, in which the present title is the volume two.

Blog posts relating to our publications range from "Customs and Beliefs of the Roma and Sinti" (pp.118-122) and "Shot across the Bow of the Good Ship Mayflower" (204-206) to "Signs of Crypto-Judaism" (322-326) and "Cherokees Spoke Greek and Came from East Mediterranean" (3-5).

Cutting-edge research, often from peripheral disciplines and out-of-the-way sources, is woven into the posts on history and archeology. Instances include contributions to pre-Columbian studies by Carl L. Johannessen, John L. Sorenson, Gunnar Thompson, Cyclone Covey, Barry Fell, Harold S. Gladwin, Gloria Farley and Stephen C. Jett. The last-named, a geographer, began his scholarly activity over 50 years ago with studies of Navajo place-names, Peruvian textiles and the diffusion of blow-gun technology in Pacific Rim archaic cultures. Jett has published over 130 articles and book chapters. At press time, his definitive *Atlantic Crossings* was still eagerly awaited by all who follow his work.

Historians often seem to pay more attention to beginnings than endings. When did the Middle Ages end? When did the last flat-earther draw breath or last white supremacist in a position of academic authority go into retirement? In this regard, it is interesting to see that the obituaries of certain concepts are remarked upon in this collection equally with the birth of new thinking. Browse these pages for the passing of outmoded theories on American Indians, Homo sapiens, ancient giants, Australian Aboriginals and numerous disease origins, including Alzheimer's, diabetes and cancer.

The human genome has turned into a library of information beyond our wildest imagining. "Genes are the teachers" (pp. 370-371), geneticists were the first patrons, but the Internet has opened access to all. May this anthology of news from the golden age of discoveries serve not only to inform people about how far we have come in this genetic landscape but also inspire us to be curious about our own DNA story and the scope, future and possibilities ahead.

D.N.Y.

1 DNA AND WEIRD SCIENCE

Self-Healing DNA Discoveries

Three astonishing experiments with DNA prove that it can heal itself according to the "feelings" of the individual as reported recently by Gregg Braden at the Global Healing Center in Houston, Texas.

March 2, 2005

Ruins of Possible Maya Settlement Uncovered in North Georgia

In what a December 22 report in Raw Story describes as possibly "the most important archeological discovery in recent times," the ruins of a stone city believed to be at least 1,100 years old have come to light in the mountains of North Georgia. Interest in the Kenimer Mound, a large, five-sided pyramid in the foothills of Georgia's tallest mountain, Brasstown Bald, near Blairsville in the Chattahoochee National Forest, goes back to a 1999 University of Georgia dig led by archeologist Mark Williams.

At least 154 stone masonry walls for agricultural terraces were exposed, plus evidence of a sophisticated irrigation system, prompting Richard Thornton to speculate that the site corresponds to Yupaha, a town explorer Hernando de Soto searched for unsuccessfully in 1540. Thornton relates the site to other Maya-like ruins in the Southeast and believes the people who built it were the Itza Maya, a word that carried over into the Cherokee language of the region. Traditional oral histories such as those recorded by Constantine Rafinesque have always traced the origins of the Natchez and other pre-Muskoghean Native Americans in the Southeastern U.S. to Mexico. Rafinesque thus distinguishes between the Iztacans (or Aztec) and the Oguzhians (Algonquians).

The site is called Archaeological zone 9UN367 at Track Rock Gap and is a half mile (800 m) square and rises 700 feet (213 m) in elevation up a steep mountainside.

December 27, 2011

Genetics and Psychiatry

Mood disorders and schizophrenia have traditionally been approached as unrelated extremes on a continuum of psychiatric disorders. But now it seems the genes causing them, and the linkage disequilibrium associated with them, extend from one end of the spectrum to the other. In a report by Nick Craddock and Liz Forty in the *European Journal of Human Genetics* (2006) 14, pp. 660-668, the gene for bipolar disorder has been shown to be implemented in other disorders also, while manifestation depends more than previously believed on the interaction between one's environment and genetic susceptibility. The title: "Genetics of affective (mood) disorders." This entire issue is devoted to genetics and behavioral psychology/psychiatric medicine.

October 15, 2006

History Reburied Daily

The year 2011 has gone down as the year of faked scholarship, but what if sound (although undaring) research is the victim of scientific controversy? The prestigious journal *Human Immunology* first published the article "The Origin of Palestinians and Their Genetic Relatedness with Other Mediterranean Populations," then yanked it, instructing their subscribers to rip out the offending pages because they showed that Middle Eastern Jews and Palestinians are genetically almost identical.

As of today, we still found the article online along with the editor's retraction and protests, but you'd better hurry if you want to read it. The censors who guard the scientific fables about Jewish DNA may discover a way to rewrite World Wide Web history as well as world history. In the meantime, you can read about the whole lamentable mess in *The Guardian* in a story by Robin McKie, "Journal Axes Gene Research on Jews and Palestinians."

January 7, 2012

Cherokees Spoke Greek and Came from East Mediterranean

Possum Creek Stone and Anomalous Cherokee DNA Point to Eastern Mediterranean Origins
Keynote Address for Ancient American History and Archeology Conference, Sandy, Utah, April 2, 2010
Donald N. Yates
In memoriam Gloria Farley

Summary. Three examples of North American rock art are discussed and placed in the context of ancient Greek and Hebrew civilization. The Red Bird Petroglyphs are compared with Greek and Hebrew coins and the Bat Creek Stone. The Possum Creek Stone discovered by Gloria Farley is identified as a Greek athlete's victory pedestal. The Thruston Stone is interpreted as a record of the blending of Greek, Cherokee, Native American, Egyptian and Hebrew civilization. Keetoowah Society traditions, as captured in "The Vision of Eloh'," are adduced to confirm a general outline of the origins of the Cherokee people in a Ptolemaic Greek trans-Pacific expedition joining pre-arriving Greeks, Jews and Phoenicians in the Ohio Valley around 100 C.E. Recent DNA investigations showing Egyptian, Jewish and Phoenician female lineages and the Y chromosome of Old Testament priests among the Cherokee are also touched upon.

A cave entrance overlooking the Redbird River, a tributary of the South Fork of the Kentucky River in Clay County, Kentucky, in the Daniel Boone National Forest, has inscriptions that according to Kenneth B. Tankersley of the University of Cincinnati display a nineteenth-century example of writing in the Cherokee syllabary. A local resident (Burchell) recognizes Greek writing in one inscription (called *Christian Monogram #2*), but his reading is unsatisfactory for a number of reasons.

Evaluation by experts in Greek and Semitic epigraphy identifies two distinct inscriptions, one in Greek and one in Hebrew. They appear to be contemporaneous with the Bat Creek Stone unearthed in the 1889 excavation of a tomb in East Tennessee by Cyrus Thomas of the Smithsonian Institution.

Another record of Greek-speaking people in ancient America is the Possum Creek Stone, discovered by Gloria Farley in Oklahoma in the 1970s. It is discussed by her in Volume 2 of *In Plain Sight* as proof that the man whom history knows as Sequoyah did not invent the Cherokee syllabary. The inscription can be read as Greek, HO-NI-KA-SA or 'o nikasa, i.e. "This is the one who takes the prize of victory," a common inscription for the pedestal upon which victors were crowned at athletic games. The use is Homeric, and the spelling Doric.

A third piece of evidence helps fill in the background of the arrival of Greeks and their intermarriage with Asiatic and other Indians in North America. In 1870, an engraved 19-by 15-inch limestone tablet was uncovered in a mound excavation on Rocky Creek near Castalian Springs in Sumner County, Tennessee (see *Ancient American*, vol. 12, no. 77). Dating to an earlier time than its Mississippian Period context, it commemorates a peace treaty between the Cherokee and Shawnee. The Cherokee chief wears a horse-hair crested helmet and carries the spear and shield of a Greek hoplite. His Shawnee adversary clasps hands in a wedding ceremony with a Cherokee woman who bears wampum belts as a pledge of peace, has her hair in a maidenly bun, wears a Middle Eastern-style plaid kilt, and displays a large star of David. In the Red Record or Walam Olum, we learn that before crossing the Mississippi, somewhere along the south bank of the Missouri, the Algonquians or Lenni Lenape (Delaware Indians), who are later allied with the Cherokee, encounter a foreign tribe they call the Stonys. Cherokee legends about Stone-coat demonstrate that the original Cherokee had metal armor and weapons. DNA studies confirm a mixture of "anomalous" East Mediterranean mitochondrial lineages such as Egyptian T, Greek U and Phoenician X with "standard" American Indian haplogroups A, B, C and D in the Cherokee and certain other Eastern Woodlands Indians.

To sum up, the Red Bird Petroglyph is a Greek inscription from the 2nd to 3rd century C.E., not a crude Cherokee scratching of around 1800 as announced recently by the Archeological Institute of America and *The New York Times*. It occurs above what is, in all likelihood, an inscription in Maccabean-era Hebrew. The Sequoyan syllabary for which these Greek and Hebrew inscriptions were mistaken originated in the Greek world of the Bronze Age along with other syllabaries like Linear A, Linear B and Cypro-Minoan. The Cherokee language, which today is Iroquoian, is the result of a relexification process in the distant past. It contains many relics of words of Greek origin, especially in the area of government, military terminology, mythology, athletics and ritual. Cherokee music also reflects Greek origins. The Cherokee Indians are, quite literally, the Greeks of Native America.

Greek Words and Customs in Cherokee

Greek	Meaning	Cherokee	Meaning
alomenoi	wanderers (in a hopeless sense)	eloh'; elohi	migrants, wanderers; earth
dakos	noxious, devouring beast, whale	dakwa	mythic great fish
dasis	hairy, shaggy like a beast	dachi	hairy water monster
tynchana	things that befall	tikano	history
etheloikeoi*	volunteer settlers	eshelokee	Cherokee; original people
gennadas	noble	kanat(i)	doctor, hunter
huios Dios	Son of Zeus (title of Herakles)	Su-too Jee	mythic strong man
illo, illas*	wrap, twist; rope	kilohi	twisted hair clan (cf. Hawaiian *hilo*)
kakotechneo	base arts, perjury, fraud	kaktunta	taboo regulation
kanon	straight-edge used by athletes	kanuga	scraper used by ballplayers
karanos	a chief	Koranu**	war chief title
kateis*	assembly	cahtiyis	assembly house
kerux	herald	skarirosken**	speaker, herald
mona*	stopping place, way-station	mona	land where the Elohi tarried
neika*	contest	anetcha	ballplay
Ogyges	titan of Greek mythology	Ootschaye	rival of Sutoo Jee (Herakles)
ouktenna	one not killed	Uktena	name of a dragon or serpent
oulountata	declared healthy	oolungtsata	divining crystal for health
skia	ghost, shade	atchina	ghost; cedar
stix	abominable	Stichi	name of dangerous serpent
tanawa*	astronomical instrument	Tchlanua	Great Hawk
(hoi en) telei	those in authority	tilihi	brave, warrior
theatas*	spectator in a play	tetchata	Playful Cherokee fairy
theatron	theater, assembly	tetchanun	ceremonial enclosure
Thrax	Thracian	tchaskiri**	sorcerer, Stoneclad
typho	raise a smoke, make sacrifice	Tathtowe, Tistoe	ceremonial title; firecracker (smoke) bringer (Santa Claus)

*Doric Greek **Lower Cherokee

June 17, 2010

Alien DNA

A few months ago, someone called us and asked me if we had alien DNA. I thought to myself, "Does she think we have an alien from Roswell in the basement?" All I said though was, "No, but it sounds like a bestseller, and I hope we get it first!" The whole office got a huge laugh over that, thinking of all the different certificates we could create for it. It was a splendid joke. Seriously, I don't work for "Warehouse 13." We are a scientific and serious business. But when I thought about it, I realized the joke was on us. She was on to something.

Although it does not look like bacteria can grow in arsenic (*New Scientist*, No. 2849), as scientists have been unable to replicate Felisa Wolfe Simon's findings at NASA Astrobiology Institute in California culturing bacteria in arsenic soup, this does not mean it is the end of our search for alien DNA. Or for DNA that acts in an alien manner. In fact, it is just the beginning.

First, there was a study in *Nature* indicating that some people have ancient DNA from Neanderthals and Denisovans. Though both are related to humans, they are considered another species, so it is alien in that sense. This shook science. Many of us may have some ancient, alien DNA. If that is so, pinpointing where that is on our genetic code might be a next step.

And scientists are now looking for life in Martian soil with the robot rover Curiosity. They are concerned over what they are calling "black, spidery things" that show up every spring wondering if this is life. And if they are dangerous. Nasa says they have been seen during the Martian spring for some time. Some scientists think that these "black, spidery things" could possibly be microbial Martian strains. Strains of life. And strains of DNA. Now, bacteria have a much simpler form of DNA than do humans, a single strand, but the evidence of life would be a monumental step. We are living in the future. And it reads more like a novel by Isaac Asimov.

How could there be life on Mars? Curiosity has already confirmed that there were once waist-deep Martian rivers. In fact, some scientists now believe there may be underground rivers on Mars. Water could mean life. Of course, what kind of life we do not know.

For one thing, we do know that our own underground oceans are teeming with life, and with "a biodiversity greater than the rainforest," according to David Gallo, a pioneer in ocean exploration. And much of it that is unexplored seems alien enough. Bacteria thriving from toxic, sulfur dioxide. Worms thriving in boiling and freezing water at the same time. He compares it to something like putting one hand in a boiling pot of liquid and the other in freezing water at the same time. Fire and ice. How is life possible in such extreme conditions? We still have questions, but Gallo purports that underground volcanic activity "creates and supports" bacterial life. In fact, some scientists think these underground volcanoes might have been where life started. And Mars has volcanoes.

And though I am not ready, just yet, to create alien ancestry certificates, I think the potential for discovery is enormous. Katherine Dudley Hohn in her article, "Nasa Curiosity Rover on the Verge of Potentially Historic Discovery: Water on Mars," says that we are about to find water on Mars. Where there is water, there is life. And where there is life, there is DNA. In this case, alien DNA.

Will we find microbial life and alien DNA on Mars? Will we one day be able to determine on your genetic code other alien DNA—if one of your grandmothers was a Denisovan or a Neanderthal? We are living in exciting times. I am just worried about one thing: the Mars rover bringing some of those "black, spidery things" back to Earth.

October 4, 2012

Update 10/21/15: NASA has discovered water on Mars! What more will they find?

The Werewolf Gene

Did you believe in werewolves as a child? There is a bit of truth to the myth. After school, one of the favorite things my brother and I loved to do was to watch grade B horror movies about monsters like *The Mummy* and *Count Dracula*. We would settle down with popcorn on our orange '50s style sofa and have a blast watching movies like *The Curse of the Werewolf, The Wolfman*, and *I Was a Teenage Werewolf*. I thought it was just fun.

Or perhaps you have read of Julia Pastrana, the "bearded lady" who toured Europe and North America in the 1850s or "Fedor Jepticheff," who toured with P.T. Barnum as "Jo-Jo the Dog-Faced Man," and who became the inspiration for Disney's "Beast." Jo-Jo played along, barking on cue for P.T. Barnum, although he was very intelligent and made quite a good living from his "genetic misfortune," according to Ricki Lewis, a science writer who has a Ph.D. in genetics, in his *PLOS* article, "The Curious Genetics of Werewolves." What these people had has been historically known as the "werewolf syndrome." They actually look like a werewolf.

I think I would have been much more horrified and saddened if I realized there really was a rare, genetically inherited condition called the "werewolf syndrome," and how this "likely contributed to the ancient "werewolf" folklore. And how—from ignorance—these people have been labeled monsters. The IMDB database describes the 1941 movie, *The Wolfman*, starring Lon Chaney Jr., like this: "A practical man returns to his homeland, is attacked by a creature of folklore, and infected with a horrific disease his disciplined mind tells him can not possibly exist." But a horrific disease such as this does exist. Just not exactly as they describe it.

The "werewolf" disease known as hypertrichosis or Ambras syndrome is a very rare genetic condition. According to the Live Science article, "Werewolf Gene May Explain Hair Disorder," there are "fewer than 100 cases documented worldwide." Lewis says there are only 50 cases known.

The name for the disease comes from the case of a "Petrus Gonzales," who "at age 12 in 1556" was "brought as a slave … to the court of France." The French royalty called him "man of the woods" because of his "strikingly hairy face," thinking he "represented a race of hairy people from the Canary Islands" (Lewis). He and his hairy family—three of his children had the trait—sat for family portraits. One of these was displayed at Ambras Castle, which "led to their notoriety" and the name.

The degree of hairiness differs. A man "Shwe-Maong" in Burma had Ambras syndrome, and an observer noted in 1826: "… the whole face, with the exception of the red portion of the lips, were (sic) covered with fine hair. The hairs were 4 to 8 inches long, straight and silky." He had one daughter, Maphoon, who was completely covered "in a pelt of long, silky grey hair" (Lewis).

Lewis says that those with Ambras syndrome are not evolutionary throwbacks:

Charles Darwin mentioned Maphoon in *The Descent of Man and Selection in Relation to Sex* (1859), but it was his readers who brought up that the striking hairiness was atavistic, a turning-on of an ancestral trait silenced through evolutionary time. But the quality of the hair isn't like that of an orangutan or gorilla. It's silky, and covers places, especially on the face, where our ape cousins don't have hair. People with Ambras syndrome aren't evolutionary throwbacks.

Then how do we explain it? It is a genetic mutation handed down in families. In 1995, scientists traced the approximate location of the mutation to a section of the X (female) chromosome ("werewolf gene").

What kind of mutation? Extra DNA. A "man in China with the disease helped researchers break the case." Xue Zhang, a professor of medical genetics at the Peking Union Medical College, tested the man and his family and found an extra chunk of genes on the X chromosome.

But just recently the 1000 Genomes Consortium announced that even healthy people have mutations ("Deleterious and Disease-Allele Prevalence in Healthy Individuals"). I was just getting used to being a living, breathing mutant myself, after reading this article, so how are we different from people with Ambras syndrome? According to Lewis, "apparently redundancies built into our genomes protect us" and we can "have a mutation (a disease-causing genotype) yet not be sick."

So with any DNA test, you have to be careful to not over-interpret results. Lewis says, "They are not a crystal ball." If you have a mutation for diabetes, it doesn't mean you will necessarily get diabetes. There are other factors. It is not the same, but in a similar fashion you can catch a virus and not come down with it. Just because you carry genetic markers from your parents for diabetes does not dictate you will necessarily get diabetes. It may mean you might want to watch your sugar intake though.

If there are environmental triggers, one wonders what the triggers might be for the "werewolf syndrome"? We don't know as we don't even have a cure (Lewis). And no—I did not mean the full moon.

November 29, 2012

Bigfoot DNA

What if Bigfoot—also called Yeti and Sasquatch—was an elusive relative? Has folklore and myth ever met science? Most of us assume that myth—or what is perceived as myth—never collides with the world of reality. Myth is supposed to belong to the realm of the mind, according to most scientists. However, there are a few rare exceptions when a creature has been incorrectly categorized as mythic when it belongs to the world of reality.

Some creatures once relegated to the world of myth or thought to be a hoax have since been redeemed and discovered by science. According to Brian K. Hall in his *Bio Science* article, "The Paradoxical Platypus," when European naturalists first saw sketches of the platypus in the 18th century, they were first thought to be a hoax. George Shaw, an English botanist and zoologist, who was one of the first to describe the animal, went as far as to put scissors to its duck-like bill to see how it was attached, according to "Platypus Facts on File" by the Australian Platypus Conservancy.

And this is but one. According to Australian zoologist, Tim Flannery, as well as other scientists, Boudai and Alexandra Szalay, in their *Mammalia* article, "A New Tree Kangaroo," they found a creature that was only supposed to belong to the world of myth. The elusive, tree kangaroo, the bondegezou, or dingiso, was called the "men of the forest" by the Moni tribespeople. It was not discovered until 2009 according to Flannery.

But no one has found a Yeti. In a PRWeb release in November 2012, it was announced that a team of experts "in genetics, forensics, imaging, and pathology," led by Dr. Melba Ketchum found Bigfoot DNA. According to the article, Ketchum, a participant in another published study mapping the equine genome and a veterinarian with 27 years of research in genetics and forensics, along with her team of experts, sequenced "20 whole mitochondrial genomes" and used "next-generation sequencing "from purported Sasquatch samples" to determine that "the species (Sasquatch) is a human hybrid … that arose approximately 15,000 years ago as a human hybrid cross of modern Homo Sapiens with an unknown primate species." Ketchum says the "data indicate that the

North American Sasquatch is a hybrid species, the result of males of an unknown hominin (all members of the genus Homo including modern humans, Neanderthals, Denisovans, and others) species crossing with female Homo Sapiens."

But modern humans have never interbred with other species, have they? Actually, though this is still in debate, many scientists now believe that modern humans did once interbreed with Neanderthals and Denisovans. According to Matt Kaplan, in the *Nature* article last year, "Human Ancestors Interbred With Related Species," early humans having hot and heavy dates with other hominins "occurred in Africa some 35,000 years ago." Then could the origin of this species—if it is a species—be from this? Genetic testing has ruled out either Neanderthals or Denisovan DNA as "contributors to Sasquatch mtDNA (mitochondrial DNA, or mother's line) or nuDNA (nuclear DNA, found on loci throughout all your cells)." She says that the male line is more "distantly removed than other recently discovered hominins like the Denisovan individual."

To what then does Ketchum compare the purported, Sasquatch DNA? She says she cannot compare it to anything: "Sasquatch nuclear DNA is incredibly novel," and "while it has human nuclear DNA within its genome, there are also distinctly non-human, non-archaic hominim, and non-ape sequences. We describe it as a mosaic of human and non-human sequences." Guy Edwards, in his perhaps-too-quickly-named piece, "Dr. Bryan Sykes Has Not Even Begun Testing Yet," (he has) even quotes Dr. Bryan Sykes, author of the book *The Seven Daughters of Eve* as "agree(ing) with Dr. Melba Ketchum's study that there is indeed a new hominid in North America (though he) disagrees with Ketchum regarding the origin of this new hominid." However, he does not disclose his own verdict.

What other conclusions could there be other than the one Ketchum gives, the one Sykes hints at? Benjamin Radford, in the Live Science article, "Bigfoot is Part Human, DNA Study Claims," concludes that either her theory is correct and half-human beings have been hiding in the hills, or the samples were somehow contaminated: "Whatever the sample originally was—Bigfoot, bear, human, or something else—it's possible that the people who collected and handled the specimens accidentally introduced their DNA into the sample, which can easily occur with something as innocent as a spit, sneeze or cough." (This is an interesting point. What do we know about the quality of the lab/s? Were ISO certified, the highest quality labs and the standard of the industry used?) Radford also argues that "unknown" does not mean "Bigfoot." However, John Thomas Didymus, in his Digital Journal article, "Scientist Melba Ketchum claims Bigfoot DNA Human-Cryptid Hybrid," quotes Ketchum as saying in *The Register,* "Four other university laboratories have tested the samples prior to this study and they all found human DNA, so they all made the assumption that the samples were contaminated. That's why we took a forensic approach, going to two different laboratories and testing two different ways and we were very careful to avoid contamination."

But what evidence is there? So far, we have none, and scientists remain skeptical. According to Guy Edwards in his blog article, "Dr. Jeff Meldrum Responds to Melba Ketchum's TV Interview," Dr. Jeff Meldrum, "the highest profile scientist and academic when it comes to Bigfoot research, and an Associate Professor of Anatomy and Anthropology at Idaho State University" remains skeptical. Edwards quotes Meldrum as saying, "My criticisms stem from the lack of available substantiation of her interpretation of the mtDNA results, and the difficulty I have envisioning a scenario that accounts from what is proposed—a hybridization event 15,000 years ago in Eastern Europe that resulted in a population dispersed across North America." Meldrum adds, "Many people don't understand the purpose of a null (working) hypothesis." He does believe in Bigfoot but claims, "Whatever is out there is likely a relict ape, or a relict early hominin (e.g. Paranthropus)." Meldrum wishes Ketchum the best: "Please don't get me wrong, I truly hope she has the brass ring. I want very much for her study to be legitimate and significant."

Of course, the vast majority of scientists and others do not believe in any version of Bigfoot. *Wired* blogger and science writer Brian Switek writes about an interview of Meldrum on a 2011 NPR interview: "Doug Fabrizio interviewed Idaho State anthropologist and Sasquatch devotee Jeff Meldrum. The stated point of the interview was to see how Meldrum applied scientific reasoning to the search for a creature that, at best, exists on the fringe of scientific investigation." He goes on to say that that did not occur and he felt the interviewer was overly swayed by emotional appeal rather than reason when evidence was lacking. This is not just the concurring opinion of the vast majority of scientists … they ignore the topic since it is viewed as "fringe science."

It does not help that none of this has yet been published in a peer-reviewed journal. Moreover, according to another blog article by Edwards, "Dr. Ketchum's BigFoot DNA Study, Peer-Rejected," her application for publication in a scientific journal (not named) has just been rejected. However, should that alone be a determining factor? It is hard enough to get published, but a peer-reviewed article is very difficult, and sometimes the work has to be polished and reviewed and re-sent. Edwards quotes Katrina Keiner, a *Science* magazine editor, as saying," We have to reject more than 90% of the papers submitted to us."

So, is that the end of hearing about Bigfoot DNA? No. According to the article, "Where Is the Ketchum Bigfoot DNA Study," posted on Doubtful News, "Igor Burtsev, a Russian Bigfoot/Yeti researcher who originally leaked the results of his study on his Facebook page," said that Dr. Ketchum had given him the papers to be published in a Russian journal, and they are currently under review. The author of the article concludes that either Burtsev is a "loose cannon" or Russia will embrace Bigfoot (though he hardly sounds serious). According to the initial PRWeb article Ketchum says she is available to interview or to answer further questions and provides contact information for those interested. And Edwards, in the article, "Dr. Ketchum's BigFoot DNA Study, Peer-Rejected," states that Sykes, who leads another Bigfoot Study, will try "multiple formats" instead of "hinging the research on a peer-reviewed paper." Hints are dropped of a later book and possibly a BBC show.

And so we are left to wonder. Is Sasquatch just a mythical creature? Or sometime in the near future will we learn of another hominim to add to our already very complicated family tree?

December 20, 2012

Hobbits Aren't Just in Hollywood

Hobbits, or something like them, it turns out, were once real. We cannot get enough of hobbits' mythic journeys and adventures in the movies, but who would have thought there really were hobbits? Who would have thought there once was someone like a Bilbo Baggins? Well, someone his size, and he and his companions once walked the Earth some 17,000 years ago according to a recent ZME Science article, "High Resolution Genome Sequence of Ancient Ancestor Released Online."

Going on their own quest, in 2004 a team of anthropologists led by Mike Morwood, a professor of anthropology at the University of New England in Australia, discovered a miniature hominin in Flores, Indonesia, which they nicknamed the "Hobbit." Announcing they had discovered a new species of humans in the Ling Bua Cave, they shook science and the world. According to Morwood and Penny Van Oosterzee, in their book, *A New Human: The Startling Discovery and Strange Story of the 'Hobbits' of Flores, Indonesia*, these were beings that were "no more than three feet tall with brains a little larger than a can of cola."

However, is this a new species of humans? Scientists have battled back and forth about this. Megan Gannon, in her recent Huffington Post Science article, "Hobbit Face: Homo Floresiensis," says, "Scientists have debated whether the specimen actually represents an extinct species in the human family tree, perhaps a diminutive offshoot of Homo erectus."

Guy Gugliotta in his 2008 *Smithsonian* article, "Were 'Hobbits' Human?" says some thought they were a distinct species while others thought they were diseased humans. However, Gannon says a 2007 study has since disproved this supposition, as an examination of the "Hobbit's brain" was "inconsistent" with a diseased brain.

Many writers have now decided the "hobbits" were a kind of human after all. Gugliotta quotes Rick Potts, director of the Smithsonian's Human Origins Program, "who once doubted that the Hobbits were a separate species," as saying he has since changed his mind: "Flores was this wing in the building of human evolution that we didn't know about. There is no reason that 800,000 years of experimentation could not evolve a small but advanced brain."

Hominin brain expert Dr. Dean Falk analyzed the hobbit-like brains and determined that just because their brains were small did not mean they were not capable of thinking. He identified neural patterns that indicated cognitive and reasoning ability despite their smaller size (Smithsonian, "'Hobbits' on Flores, Indonesia").

What else do we know? We cannot sequence their DNA according to the article "High Resolution Genome" because "the hot, humid climate of the Indonesian island of Flores, where their remains were found, impairs the preservation of DNA."

Gannon quotes Dr. Falk as saying, "It's not just that their brains are small; they're differently shaped. It's its own species." Anthropologist, Susan Hayes, who uploaded 3D imaging scans, says about a female specimen, "She's not what you'd call pretty, but she's definitely distinctive" ("Hobbit Face"). The 3D picture is startling with deep set, haunting, and thoughtful eyes. It does look like at least an early ancestor of a hobbit. They were short and heavy—only about a "yard tall" but "squat, " she says.

Morwood, in his book, *A New Human*, says this was a "tool-making, fire-making, cooperatively hunting person." Erin Wayman, in the *Smithsonian* article, "Were the Hobbits' Ancestors Sailors?" postulates that in order to get to Flores, perhaps they were seafarers. If that is the case, perhaps they truly were adventurers like Bilbo.

Hopefully, we will be able to get better preserved "hobbit" bones, so we can do some DNA testing and determine the exact connection to our family tree as well as the connection with other hominins like Neanderthals and the Denisovans. Until then, we want to find out more about this species of miniature humans that walked the Earth with dwarf elephants, giant rodents and Komodo dragons.

"There is nothing quite like looking, if you want to find something. You certainly usually find something, if you look, but it is not always quite the something you were after."—J.R.R. Tolkien, *The Hobbit*

December 23, 2012

Sykes' Sighting

Bigfoot might not be a hoax after all. We have been doing our best to make sense of our world since we lived in caves and first looked up at the stars. It is in our DNA to do so. We have not always been right because we did not have enough information. As a result, we thought the world was flat for a long time, and we created stories about the thunder gods and how the stars affected our everyday lives. And sometimes we have thought something was a hoax, like the platypus, when it actually belonged in the realm of scientific reality.

Though we are not accustomed to using Bigfoot in the same sentence with science, Bryan Sykes, famous Oxford geneticist, has made insightful deductions about Yeti or Bigfoot and its Russian cousin, Zana, from DNA testing and other evidence. Sykes has determined with this evidence that what some of these people did see could be something that actually belongs in the realm of science. They just did not see a monster called Bigfoot.

What did they see? For one thing, Sykes does not think we are discussing one creature. He conducted extensive genetic testing on hairs and tissue samples from the jaw of a purported Yeti mummy and other specimens and declared that the Himalayan "Abominable Snowman" was some type of ancient, Arctic bear though it could be an unknown type of bear or bear hybrid (Ker Than, *National Geographic,* "Is the Abominable Snowman a Bear?").

However, the Russian Bigfoot evidence is even more exciting than a possibly unknown type of bear. Sykes has made a bold theory in this case. He believes that "Zana," who was captured in the Russian forest in the 19th century, might have been a slave that was a "remnant of an earlier human migration" and a representation of a "hitherto unknown human tribe." DNA testing results show that her maternal line is Sub-Saharan African and human; however, Sykes says this is not the end of the story. This does not explain some of the primitive features of the skull of her son, Khwit, according to Sykes. Despite this, she is not a Neanderthal as some thought she might be. DNA testing results show she has as much Neanderthal as the rest of Europeans and Asians which is only 2 percent to 4 percent (shown when the Neanderthal genome was sequenced in 2010 and no doubt from early inbreeding). It is quite a DNA mystery, and we cannot wait for the end result and the published paper on this ("Was Russian 'Bigfoot' Actually an African Slave?").

October 22, 2013

Genetic Genealogy Like Astrology?

Dust off the crystal ball. Scientists consider DNA ancestry services "genetic astrology," according to a recent BBC article by Pallab Ghosh. In "Some DNA Ancestry Services Akin to 'Genetic Astrology'," Ghosh quotes Professor David Balding as maintaining that "such histories are either so general as to be personally meaningless or they are just speculation from thin evidence." One article, "Don't Believe the Guy Who Claims He's Descended from Vikings," quotes evolutionary geneticist Mark Thomas, as saying, "these tests have so little rigor that they are better thought of as genetic astrology."

But not all DNA ancestry tests or companies are created equal, so you may have to find where the astrologers and crystal balls are hidden.

First, not all DNA ancestry companies have an ISO-certified lab—the highest benchmark in the genomics industry. Why not? It is not required in direct-to-the-consumer DNA testing. With noncertified labs, there is a stronger possibility of contamination or lost or swapped samples. I know a person to whom this happened. He jokes that he thought he was someone else for two years!

And there are weaknesses with the most common DNA tests for ancestry—haplotype tests. The mitochondrial genome is small compared with the nuclear genome, according to the article "Mitochondrial Genome Analysis," which means there cannot be as much variation. Live Science writer Stephanie Pappas quotes Maria Avila, a computational biologist at the Center for GeoGenetics at the British Natural History Museum, as saying, "People could share mitochondrial DNA even if they don't share a family tree." And the article, "Doubts Remain," quotes Thomas, who concurs with this, saying, "People can have matching mitochondrial DNA by chance and not be related."

Male line haplotype testing has different limitations. "The Male Y-linked tests have very rapid mutation rates and are very fragile, so you can get a lot of errors with that type of testing," according to Dr. Donald N. Yates of DNA Consultants.

And it seems less stable than we previously thought. Bryan Sykes, professor of genetics at Oxford University and author of *The Seven Daughters of Eve*, makes a strong argument that the Y chromosome is weakening and in trouble in his book, *Adam's Curse*. He says it is "doomed" because of its instability and rapid genetic mutation and is headed toward extinction. (That is a scary thought—*Planet of the Females*.) Before the 1990s paternity testing was based on Y chromosome comparisons and limited to fathers and sons. Sometimes, an uncle would be mistaken as the father. Today, it relies on autosomal DNA comparisons that can be applied to females and is 99.99 percent accurate.

Moreover, there is no standard usage for the term "autosomal." Why does this matter? A conventional autosomal DNA test provides you with a 16 STR profile and is based on forensic and academic studies.

Autosomal DNA testing as it is traditionally understood is based on different science, connecting you to populations, not people, and gives you a comprehensive map of your ancestry.

This type of testing is more stable and has more scientific validity. It uses the same science used by the courts, government and law enforcement. And its validity has been confirmed over and over, using the magic of math and laws of large numbers.

Two research teams independently reached the same groundbreaking results that the DNA mutation rate is slower than previously thought: James X Sun et al., in the article, "A Direct Characterization of Human Mutation Based on Microsatellites," in *Nature Genetics* 44/10 (October 2012):1161-65, and A. Kong et al., in the article, "Rate of de novo Mutations and the Importance of Father's Age to Disease Risk," *Nature* 488 (2012):471-75.

What does this mean? The alleles on your DNA have been relatively unchanged for the past 20,000 years. These are markers that everyone has, the reason why anyone can take an autosomal ancestry test. That's about twice the length of what we call world history, hence a perfect time depth for valid inferences about population patterns and ancestry of individuals. Of course, anything can be over-interpreted. DNA testing is not magic—at least traditional autosomal DNA testing based on forensics isn't. If you want magic, look for someone with a crystal ball.

March 18, 2013

Science Fiction and Fictional Science

The future is here. According to Emma Huang in the article, "Personal Genomics: Where Science Fiction Meets Reality," the Human Genome Project started in 1990, and we are not far off from an individual medical diagnosis based on your personal genetic code with a unique prediction for disease "predict[ing] exactly which treatments will work best for you." This is where science meets science fiction. She says the "similarities with newborn testing in the movie, 'Gattaca,' are clear" though we are not yet where we can "tell you anything about your child's risk for heart disease." But we are ever so close.

In a recent *Smithsonian* article, "Ten Inventions Inspired by Science Fiction," Mark Strauss says the inventor of the modern helicopter, Igor Sikorsky, was inspired by Jules Verne and used to quote him, saying, "Anything that one man can imagine, another man can make real." Art then becomes reality. Goethe, the German poet and philosopher, once said, "Few people have the imagination for reality."

But some do. Scientists and inventors have frequently been so inspired by fiction that they create something new to bend our idea of reality and what is possible. Personal genomics is one such discovery that "has the potential to change medicine and society as a whole" (Huang).

There are others. Robotics is another invention that could potentially transform our society. Could it eventually alter how we think about what it means to be human? An ancient idea, the idea of robots stems from Greek myths. The stirring, golden image of C3PO of "Star Wars" trekking across the desert or chatting in the eclectic, intergalactic bar is reminiscent of the bronze man, Talos, and the talking, golden handmaidens fashioned by Hephaistos, the Greek god of fire, in Greek mythology (Encyclopedia Mythica).

A recent *Smithsonian* article notes that Japan has of late created talking and very human-looking androids and even babies ("In Long History of Creepy Robot Babies, This One Really Takes the Cake"). They remind one of the robotic androids in the *I Robot* book series by science fiction writer Issac Asimov.

Other innovations that have shaped our society were written about long before they were invented. Martin Cooper, the director of research and development at Motorola credited the "Star Trek" communicator as his inspiration for the design of the first mobile-phone device in the early 1970s.

Indeed, H.G. Wells depicted a video Ipod (moving picture player) and a DVD/VCR (entertainment player) in his 1899 novel *When the Sleeper Wakes* and a wireless wrist intercom (like a cellphone later used in "Star Trek") in his 1936 novel *Things to Come*. And according to a National Geographic Daily News article, "8 Jules Verne Inventions That Came True," we now have several things he wrote about: Electric submarines, rocketships, and television newscasts are but a few.

The "father of the modern submarine, American inventor Simon Lake," was "captivated" after reading Jules Verne's novel *Twenty Thousand Leagues under the Sea* in 1870 and built the first submarine—the Argonaut in 1898. Robert Goddard, "the American scientist who built the first fuel-rocket in 1926 became fascinated with spaceflight after reading *War of the Worlds*."

What future inventions may come from fiction? Harry Potter's cloak may soon no longer be the stuff of fantasy. According to Carri Sieczkowski in a recent article of hers in Huffington Post Science, physicists at Duke University successfully "cloaked an object perfectly" with a "microwave cloak." And Robert Lamb and Patrick Kiger, in a How Stuff Works article, "10 Future Inventions Everyone's Been Waiting For," claim we are on the brink of realizing cars that drive themselves. "Google scientists and engineers" are actually working on creating "autonomous vehicles that utilize intelligence software and Google maps to navigate. These have been depicted in several science-fiction movies, like *I Robot*.

But getting a personal answer to your child's genomic history in under 50 hours? That could be vital. Especially when you have an ill child and "time is of the essence." And that is just what Matthew Herper in a recent article in *Forbes*, "DNA Sequencing: Is Science Fiction Becoming Medical Fact?," says may soon be realized.

Getting a flying car or a robotic maid like the one on "The Jetsons" is not happening that quickly though Lamb and Kiger do say you can go buy Roomba, a robotic maid that vacuums, although she doesn't have the capability or the personality of Rosey. And you will have to wait a while to acquire that invisibility cloak.

But the car that drives itself might be on the roads soon; however, it will be too pricey for most.

According to Herper, DNA sequencing could become a standard first choice for infants in neonatal intensive-care units and routine use in the clinic. He quotes Han Brunner of the Radboud University Nijmegen Medical Center as saying, "This is a paradigm shift to genome-first medicine" for those patients where conventional strategies do not work. We are truly living in the future.

I just hope Rosey doesn't take the automatic smart car for a test drive.

August 22, 2012

Customizing Our Offspring

Has *Gattaca* almost arrived? Remember the scene in the film when the parents are in a fertility clinic ordering their perfect bundle of joy by going over a shopping list of possible genetic traits they do or do not want their baby to inherit? We are nowhere near that capability of using genetic testing in this capacity, but some scientists are concerned about us opening that door from several fronts. Should we be concerned or not?

Actually, no one has proposed trying to use DNA testing to do this. All that has been proposed is using genetic testing to cure disease and help parents understand diseases that would be passed down to their children. That sounds like a positive benefit for society, so why all the hype? The argument sounds like this: We don't know where this science will lead, but it sounds as if it will lead to designer babies. That is actually a slippery slope argument as there is no reason to believe that one event must inevitably follow another. If X has occurred, then Y must happen. This type of argument is especially illogical if, as in this case, there are many other steps that must lead to Y.

According to Rob Stein's recent NPR article, "Proposed Treatments to Fix Genetic Diseases Raises Ethical Issues," some are concerned about the science of introducing the healthy mitochondrial DNA of a second mother into an egg of a woman who lacks healthy mitochondrial DNA. This has just been developed in order to prevent the passing of genetically inherited diseases; however, some are concerned this could lead to using DNA science to create designer babies. Stein quotes Mark Sauer of the Columbia University Medical Center, as saying his team of scientists and another are both hoping to "cure disease and to help women deliver healthy normal children."

And others are concerned about DNA testing companies going down this path though none has said it is. According to Alexandra Sifferlin in the *Time* article, "Company Patents First Designer Baby Maker, Now What?" one DNA testing company has just been granted a patent that could allow prospective parents to look at the genetic traits their future child could have with "drop-down menus." That is a huge breakthrough, and we have no idea where that science will eventually lead. However, the company denies it is in the business of making designer babies and says this is solely for the knowledge of genetic traits parents might pass down to their child. So why are some people concerned?

Since Galileo there has been a fear of science about how it could change society. DNA testing is a marvelous tool. We have discovered diseases, tracked criminals and opened doors for people to discover their past. And we are just beginning. We are at the beginning of an exciting era with genomics. Science is not the problem, nor is DNA testing. But like money, it is what you do with it that matters. There are, perhaps, some things we should not do. Like recreate *Gattaca*. But no one has suggested that, so maybe we need to take a deep breath and calm down. At least for now.

October 14, 2013

DNA and the Future

We live in a world that once only science fiction authors tread. But is this "Brave New World" of genetic testing so alarming after all? Many things have seemed frightening before the technology was fully understood or developed. Many thought air flight impossible before the Wright brothers. And we could not see how much computers and the Internet would change our world.

Of course, we can only imagine how much the genomics field will change our world. But we do hold the keys. And as Mark Henderson points out in his article, "Human Genome Sequencing," "this advent of personal and medical genomics" calls for "broad(er) personal engagement" beyond the world of science. It also calls for a more positive imagination than the "Frankensteinish" ones envisioned in films *like The 6th Day* with Arnold Schwarzenegger.

DNA testing is no longer science fiction. It is part of our world. And we need to understand the science properly. And decide not just what we can do, but what we should do.

For instance, Britain is now set to create babies soon from the DNA of three people by using the healthy mitochondrial DNA of a second mother ("UK Takes Step Toward 'Three-Parent' Babies"). Some say this will help prevent disease.

But others are concerned about the ethical and social challenges for this and a range of issues within the field of genetics. Some are just ideas—like cloning and designer babies. But the ideas of today may be science tomorrow. There are questions the scientific community has to resolve. But will science alone be enough?

DNA testing is now entrenched in the argument of rights versus security or freedom of expression. Most know that genetic testing can also be used to identify someone accurately. Think CSI. Forensics and the FBI. However, now cops can swab anyone they arrest—before they are charged ("Supreme Court Says Police Can Take DNA Swabs"). What do they do with the samples? Some feel this is an infringement of our constitutional rights.

Some think the busts of faces an artist created from DNA found in gum and hair found in public is questionable (Megan Gambino, "Creepy or Cool?" *Smithsonian*). Actually, scientists have declared specific depictions like this impossible despite the advances made in genetic testing. However, you could get a glimpse of facial possibilities. But where is this headed?

What we have to solve is not just what we do with DNA. It is how we understand how DNA is used. Henderson says the perception that genetic testing is a "simple and deterministic science" is wrong—it is complex. Other factors have to be considered. Otherwise, isn't this a belief in a kind of scientific pre-determinism? Having a gene for cancer does not automatically mean you get cancer any more than carrying a virus means you get sick.

Such genetic testing could and probably will go way beyond these fields. Perhaps it could be used for identity scanners at airports, hospitals and government offices. Perhaps everyone will be required in the future to carry their DNA with them to be scanned for identification much as we carry around a driver's license today. It could even be used for online dating services. Identity theft of one's DNA could be a serious future crime.

However it is used, it is up to us to envision a brighter future with DNA testing—not just a braver one.

June 28, 2013

Three-Parent Babies

As DNA sequencing gets cheaper, it seems that we live in a world that once only science fiction authors tread like *Gattaca* and *Brave New World*. Both depict a frightening future with genetics testing playing a key role. But will it be such a frightening world after all when DNA testing is the norm?

Many things have seemed frightening before the technology was developed or when it was new. Air flight seemed impossible before the Wright brothers. Not a few were afraid to fly. And we could not envision how much computers and the Internet would change our world. We are just at the beginning of the Genomics Age and genetics testing. We can only conjecture what it will look like. But we hold the keys to creating a positive era both in our imagination and in the communication of the possibilities we wish to bring into fruition. As Mark Henderson points out in his article, "Human Genome Sequencing," this advent of personal and medical genomics" calls for "broad(er) personal engagement" beyond the world of science.

So, we have to begin talking more about how genetics and genetic testing affect our world and how we want it to affect our world. What is the basis for such unfounded fears? Henderson says one problem lies in thinking genetic testing is a "simple and deterministic science" when it is complex. Other factors have to be considered. Otherwise, isn't this a belief in scientific pre-determinism? Having a gene for obesity or cancer does not mean you have to be obese, or you are destined to get that type of cancer. Other factors have to be considered. This type of thinking is depicted in *Gattaca*, but the lead character, considered genetically deficient with a bad heart, out swims and outpaces his genetically enhanced brother and reaches his dreams. He is chosen for a prestigious space-flight mission despite all odds. Genetic testing that indicates certain health conditions is not usually, for most health conditions, a life sentence. Many have missed that final point of the movie. But genetics testing is a useful tool we can use to make better lifestyle choices. It is knowledge.

But that doesn't mean the path does not have some difficult spots and all fears are unfounded. As Henderson notes, there are other "social and ethical challenges" to DNA testing that we have to resolve. And there will likely be more. The Supreme Court just decided police can do a DNA swab test of anyone arrested—not charged. Some see this as protecting us as citizens in a world fraught with terrorism and others see this as an infringement of our constitutional rights ("Supreme Court Says

Police Can Take DNA Swabs"). It is up to us as a society to uphold this law or overturn it.

There will be more issues to solve as well. Should we allow the creation of babies from three parents? Britain is set to do this by also using the healthy mitochondrial DNA of a second mother when the biological mother's mitochondrial DNA is diseased ("UK Takes Step Toward Three-Parent Babies"). Some people see this as a great medical advance while others think it unethical.

And what of an artist who has been busy painting portraits of people from salvaged DNA from stranger's hair and cigarette butts found in public places? Is this an invasion of privacy? Or avant-garde? Matthew Herper, in his article, "Artist Creates Portraits From People's DNA," quotes the artist, a doctoral student, Heather Dewey-Hagbo, as saying these were only "speculative interpretations" and her key idea was to point to the future possibilities of DNA testing.

The possibilities of genetic testing in the future are only limited by our imagination. We just have to stay engaged.

June 30, 2013

The Retrovirus That Came from Outer Space

In the future, a doctor might pull some of her patients aside to inform them that she has determined with genetic testing that they have a rare Neanderthal virus. She needs more DNA tests run just to be sure. It sounds like a bad mix of a movie somewhere in between *The Flintstones* and *Contagion*. How could this be a likely scenario? Scientists have just proven that DNA testing results of some people in modern times show that they still carry Neanderthal retroviruses in their DNA, and there is some question whether or not this could be linked to diseases (Colin Barras, "Neanderthal Virus DNA Spotted Hiding in Modern Humans").

It is almost as if Neanderthals are getting some revenge for us having wiped them out. How could modern humans have signs of Stone Age viruses? First of all, it seems that many of us share a small percentage of our DNA with Neanderthals and Denisovans. This was discovered through advanced genetic testing of the genomes of both (Barras).

How did we get DNA and viruses from Neanderthals? It seems rather likely that some of our ancient ancestors may have had some swinging hot dates with Neanderthals. That is not the picture most desire of their original ancestors though, and the debates over this have been just as heated ("Hot for Hominids," *Smithsonian*).

These hidden ancient retroviruses found in some modern DNA could be a clue for cancers and other diseases. It is found in people with cancer and has been shown as a link to disease for mice, but it could be another five years to make a definitive link with more genetic testing to see how widespread the virus is in the human population and whether or not there are definitive links to it causing diseases in humans (Barras).

We need to do more genetic research and more genetic testing for more definitive answers. Perhaps we can then use this information to help wipe out cancer and other diseases like HIV. Before a super Neanderthal virus has real revenge and wipes us out. That is, hopefully, a joke.

November 20, 2013

The Persistence of (Genetic) Memory

How many times have you looked at a painting by Salvador Dali and seen something you did not see before? Our DNA contains a wealth of information, and we have learned to access that information for many purposes. DNA testing for ancestry as well as genetic testing for forensics and medical purposes. But scientists have just discovered that our DNA contains more genetic instructions than we even previously knew. It has memory. How is that possible?

It seems that a second DNA code was playing hide-and-seek behind another one. Per Liljas, in *The Times* article, "Second Code Uncovered Inside the DNA," quotes the genetic research team leader, Dr. John Stamatoyannopoulos, as saying, "Now we know that this basic assumption about reading the human genome missed half the picture." That is like catching half a plane. How can you take off? Imagine what we have yet to discover if this is true.

But how did they discover this second code? It was part of a genomics project called "Encyclopedia of DNA Elements Project" or ENCODE which was funded by the National Human Genome Research

Project (Heather Tooley, Examiner, "Second DNA Code"). It writes out a new language much as John Lennon would compose and play a new tune with the same piano. We just weren't listening.

And what is it telling us? DNA writes in code like an undercover spy in *Mission Impossible*. This new DNA language to decipher is a code about gene regulation that is written over the first code about making proteins (UW Health Science and UW Medicine, "Scientists Discover Double Meaning in Genetic Code").

How will this impact genetic testing and genetic research? A new language means new information. Scientists can use this extra information to take someone's DNA testing results from a genetic test and better diagnose and treat for disease. ("Scientists Discover"). And scientists may be better able to understand the role of epigenetic, or lifestyle, factors and be able to reverse triggers that cause disease and even aging. The complete picture of our DNA is coming into focus. What will we see?

December 13, 2013

2 DNA AND ANCESTRY

Basque Argosy

Isolated Populations as Treasures-Troves in Genetic Epidemiology: The Case of the Basques, by Paolo Garagnani et al., *European Journal of Human Genetics* 17 (2009): 1490-1494.

The Basques living on the western border between Spain and France are a unique population, one often considered as an exotic isolate. "Basques" often comes up as a match in people's DNA Fingerprint results, often because (as is widely believed, at least) a people resembling Basques helped repopulate the British Isles after the last Ice Age.

But Basques are not an isolate. This article proves they blend gradually together with and into their closest neighboring populations in Spain and France and, so, are not a candidate population, as the Finns are, for the study of disease associations.

"Basques do not show the genetic properties expected in population isolates," according to the authors. On the contrary, as many previous studies suggest, the Basques have so much diversity among themselves they were probably the source of population diffusions in prehistory, not a backwater trap for inbreeding.

October 22, 2009

Population Turnover and the Etruscans

In a recent research article published in the journal *Molecular Biology and Evolution*, a team headed by Silvia Guimaraes of the University of

Florence documents how the Tuscans of the Middle Ages preserved Etruscan bloodlines while the contemporary inhabitants of the Italian state of Tuscany seem to have little or no connection with those mysterious antecedents from the Bronze Age.

It is an example of discontinuity in the mitochondrial DNA record. The paper is titled "Genealogical Discontinuities among Etruscan, Medieval and Contemporary Tuscans" (published online on July 1, 2009: you must have a subscription or pay to read the full text). The authors are on sure ground with their findings since they had access to ancient, medieval and modern DNA for comparisons.

It is often assumed that whoever lives in a place belongs to a population whose ancestors settled there thousands of years ago, and who created a sort of genetic bedrock beneath the present-day DNA landscape. The Italian study, however, disproves the applicability of this theory in a country famous for suffering many invasions but enduring and retaining its native population structure and composition. It was to be expected that the same mitochondrial lineages would be present today that were common in Italy thousands of years ago. Instead, some of them, selectively, just died out over time.

A similar situation was revealed in 2005 with the classification of mitochondrial DNA in 24 Neolithic skeletons from Germany, Austria and Hungary. One-fourth belonged to haplogroup G, a rather rare type today. In fact, today's Central Europeans have a 150-times lower frequency (0.2 percent) of this mtDNA lineage. The inference is that sometime between 7,500 years ago and the present day, large-scale population replacement or genetic influx took place in Europe. Today, it is haplogroup H that enjoys dominance. (The study is "Ancient DNA from the First European Farmers in 7500-Year-Old Neolithic Sites," by Wolfgang Haak et al., *Science* 11 310/5750: 1016-18.)

Cases of such discontinuities could be multiplied tenfold or more, especially in the New World. Haplogroup M, a common East Asian lineage, was found in the skeletal remains of two Paleo-Indians about 5,000 years old at the aptly named China Lake in British Columbia, although the message was lost on its discoverers (see R.S. Malhi et al. in *Journal of Archaeological Science* 20:1-7). A study by Pääbo et al. in 1988 proposed the existence of a previously unknown founding lineage on the basis of mitochondrial DNA extracted from a rare specimen of

7,000-year-old human brain matter in Florida. This discovery was almost immediately dismissed as "of no importance." An analysis of the bone remains of 25 pre-Columbian Mayas by Gonzalez-Oliver's group produced one type of mitochondrial DNA that could not even be classified. The Brazilian geneticist F. M. Salzano has remarked that of the 338 ancient cases investigated to date over two-thirds could not be assigned to the conventional six "Amerindian" haplogroups. Researchers found that among the remote Cayapa Indians of Ecuador, one-fifth of genetic variation was "other."

The Etruscan study shows that a whole population can turn over in a few centuries. It doesn't take thousands of years. If this is true, as it seems to be, then the story of the peopling of the Americas has many unwritten chapters. The revised standard version propagated in textbooks and anthropology departments is simplistic and reductive.

September 15, 2009

The Land Bridge to Nowhere

Here we go again. A small number of Siberians crossed the Bering Strait when it was dry land 12,000 years ago and populated the entire New World.

I refer to the article in *PLoS* (Public Library of Science) *Genetics*, titled "Genetic Variation and Population Structure in Native Americans," by S. Wang et al. So far from vindicating a decrepit theory that had nearly drawn its last breath in anthropology circles in the late 1990s, the "new evidence" does nothing but perpetuate geneticists' circular reductionist logic again. Set up a survey the right way, and you will get the results you want.

First, the 678 markers the scientists investigated are those they have determined are associated with Native Americans in the first place. Surprise! They were found in high frequencies in the 29 Native American populations of the study—the very populations that have been studied to death since the 1980s. One much-cited study of alleged Siberian origins of Native Americans used only nine North American Indians, all from the tiny Papago tribe in Arizona. Most of the Indians were from South America. No study will touch an Indian with admixture, although it is likely all Indians have some degree of admixture, at least in North America. Even the reservation-born and -bred Navajos, considered the "purest" have as much as 20 percent by one measure.

In every other population of the world except those in the New World, greater diversity equals source. Thus, geneticists established the "out of Africa" theory because Africa has the most genetic diversity.

But does it? Actually, the Americas have more. Rather than pointing to the source of migrations, the gradual increase in frequency of the 678 markers as we get closer to the Bering Strait could reflect a decrease in diversity as a result of genetic drift the farther removed populations became from a center situated, say, in Central America. The same phenomenon occurs on the fringes of the Atlantic, where one particular genetic type, the male lineage known as the Atlantic Modal Haplotype, reaches frequencies of over 80 percent in Connaught, Ireland. It did not originate in Ireland, though, but rather in Basque lands, which have a high amount of diversity, like Africa.

Geneticists set great store by a concept known as time to coalescence. This is a backtracking calculation of the age of a population according to how many mutations it has accumulated, usually in its mitochondrial DNA. Wind back the mitochondrial clock, and you get to zero hour. Nothing is wrong with this approach, but the geneticists go further and set the time together with theoretical migrations to arrive at a source— say Lake Baikal in central Siberia. They use theoretical migrations to prove theoretical migrations. For Native Americans, supposing even that they had one central origin, it could just as well have been in Panama. But for geneticists, it must be in Siberia or Mongolia because that is the only place it is allowed to be.

The search for one single source is illusory and reductionist anyway. Who would suggest that all Europeans came, say, from Tajikistan, 40,000 years ago?

Rather than keep attempting to reiterate Native Americans' lack of diversity, a colonial attitude with blinkers to begin with, geneticists should study how and why Native American diversity far exceeds that of the Europeans, Africans, Asians and all other world populations.

November 23, 2007

It's Always Darkest before the Dawn

Some reviews of Nicholas Wade's *Before the Dawn: Recovering the Lost History of Our Ancestors…*

By far the best book I have ever read on humanity's deep history.

—E. O. Wilson

Tracing the history and the evolution of the human race from our common ancestor with chimpanzees 5 million years ago in a 300-page book was an ambitious undertaking, but Wade was clearly up to the task. If you are interested in human evolution and the history of mankind, this book is required reading.

—Searching for the Truth

Brings together a selection of early results (from genetics) to tell the story of modern man, Homo sapiens sapiens. We have known a good deal about this topic for decades, from investigations in archaeology, anthropology and comparative linguistics. However, our new, detailed knowledge of the human genome has clarified our understanding dramatically. A good analogy would be the results sent back by the first interplanetary probes, resolving what for centuries had been mere points of light or fuzzy blobs in our telescopes into actual landscapes of craters, volcanoes and dunes, while at the same time demolishing wishful-thinking fantasies (like those about Martian canals).

—National Review Online

More informative than *Nature via Nurture*, more readable than *The Blank Slate*, and proves (contra "The Emperor's New Clothes") that popularizations of population genetics don't have to be deceitful and revolting.

—Bookosphere

Wade's coverage includes the ancient near-extinction of humanity (DNA analysis shows at one point there were only 5,000 humans left after a mass die-off), human migration and intra-human relations between Neanderthals and Homo sapiens. He also talks about the idea of race and the violent war-making nature of human society.

—World History Blog

While "race" is often a dirty word in science, one of the book's best chapters shows how racial differences can be marked genetically and why this is important, not least for the treatment of diseases.

—*Publishers Weekly*

And finally, a big slam from Wade's former employer *Nature* magazine!

Wade's explanations commit various well-known errors, such as equating correlation with causation and extrapolating from individual traits to group characteristics. ... Most of the scenarios he reports have not been rigorously tested, nor is it clear how they could be. The book has many internal inconsistencies, and one can easily find contrary evidence or readily construct alternative "just so" stories that invoke the same genetic scenario and the same kind of reasoning. ... Wade claims example after example of "genes for" traits.

Read the original review in *Nature* 441/7095 (15 June 2006): Kenneth M. Weiss and Anne V. Buchanan review *Before the Dawn: Recovering the Lost History of Our Ancestors* by Nicholas Wade (p. 813).

January 6, 2008

When DNA Second-Guesses History

In a new article in the *European Journal of Human Genetics* (17/5:693), the enigmatic Etruscans of antiquity are again the subject of a DNA investigation. This time, the study, called "The Etruscan Timeline: A Recent Anatolian Connection," uses mitochondrial DNA to probe the ultimate origins of the people, who appeared on the stage of history in about the eighth century B.C.E. We know this time frame is fairly accurate because the Romans started their calendar in 753 B.C.E. with the founding of Rome and dated all records A.U.C. (Ab Urbe Condita,

"From the Founding of Rome"). Roman historians beginning with Ennius and Livy also recount how early Rome was conquered by the Etruscans and made subject to Etruscan kings for the first few centuries of its existence.

That is why it is strange that the present article estimates "an [sic] historical time frame for the arrival of Anatolian lineages to Tuscany ranging from 1.1=/-0.1 to 2.3+/-0.4 kya B.P." Based, then, on the retrospective coalescence of DNA, this calculates the Etruscans' migration from an original homeland in Anatolia (modern Turkey) to as late as 1200 C.E. and as early as 390 B.C.E. What is going on? The Etruscans were clearly seated in Italy 450 years before 390 B.C.E., and by 1200 C.E., they were long since gone as an entity. In fact, by the time of the emperor Claudius, who wrote a lost history of them around 1 C.E., the Etruscans were already considered historical oddities and their language dead.

Are geneticists trying to rewrite history? I think it is a case of a fundamental fallacy in their work. Calculation of a time to coalescence is obviously limited by the validity and reliability of the sample, but it is also very often illusory.

To take the example of Native Americans, just because geneticists arrive at a time to coalescence of 10,000 years before present, doesn't mean the *place* of coalescence has to be in Mongolia/Siberia, where they derive all Native Americans. It could just as well be in the Americas. DNA doesn't necessarily tell us anything about geography. But it is often pressed into service to prop up a theory about human migrations. Let us remember, though, that such constructs are just constructs, so DNA cannot be evidence, only confirmation of someone's historical or racial construct.

If one wishes to speak about evidence in a strict sense, however, it is interesting that the researchers (Francesca Brisighelli et al.) found, by mtDNA sequencing, a "novel autochthomous Tuscan brand of haplogroup U7." This can mean that the same U7 turning up elsewhere may be a sign of Etruscan movements.

November 2, 2009

Armchair Archeology

The explosion in commercial archeology has brought a flood of information. The problem now is figuring out how to find and use this unpublished literature, reports Matt Ford in the current issue of *Nature* magazine.

"I became aware that what I was teaching would be out of date without looking at the grey literature (unpublished reports)," says one professor at the University of Reading in England.

A policy shift in 1990 required all construction projects to document archeological remains in Britain and generated an avalanche of findings that cannot be absorbed by the official academic field. The result is that our picture of the past is very much outdated. Academia is not likely ever to get caught up. Nor are academicians ever likely to warm to new theories of population genetics like diffusionism and transoceanic contact and colonization, since few of those theories ever received a hearing in the halls of academe in the first place.

Read the full story in *Nature*, "Archaeology: Hidden Treasure," vol. 464, pp. 826-27.

April 7, 2010

Archeology Gets Its Feet Wet

American Schools of Oriental Research Annual Meeting: Tracking the Med's Stone Age Sailors
Andrew Lawler

Genetic studies are beginning to fill in the missing pieces in the history and prehistory of seafaring.

"By carefully sorting genetic data from living people, a researcher at this recent meeting covered in *Science* reported that around 6000 B.C.E., early seafarers indeed spread their seed—both agricultural and genetic— from their homeland in the Near East as far west across the Mediterranean as Marseilles, but no farther."

No farther? Could that be because they have looked no farther? Gunnar Thompson's new study, *Ancient Egyptian Maize*, whose first print runs have already been exhausted apparently, builds a well-documented and persuasive case that "Indian corn blossomed with equal vigor along the shores of the Nile River and Gulf of Mexico at the very dawn of history."

You may say that Thompson is not a "real" researcher, but we would counter that 400 corncobs on ancient tombs and papyrus scrolls of Egypt and corncobs depicted with copper weapons on ancient ships are real enough to be remarked upon by anyone with eyes in their head and a brain to think.

"Male regents in the Middle East and India sent mariners overseas in search of the world's purest copper deposits. These were located in the Mediterranean Sea on the Island of Cyprus, in the Persian Gulf on Megan Island, and on Isle Royale in the Upper Great Lakes Region of North America. This worldwide exploration took place in approximately 6000 B.C.E. ... all the way to the shores of North and South America" (p.2).

December 9, 2010

Gene Surfing in Canada

Gene surfing is a process in population expansion whereby certain variations become prominent and dominant in a short time, appearing to skip the slow, steady, uniform accumulation of variegation and diversification. According to a study of the population structure and genealogies of Saguenay Lac-Saint-Jean in Quebec, this type of drastic change accompanied the immigrant wave front that spread over the area in the 17th century. "Deep Human Genealogies Reveal a Selective Advantage to Be on an Expanding Wave Front" in *Science* magazine describes the resulting demographics.

Abstract: Since their origin, human populations have colonized the whole planet, but the demographic processes governing range expansions are mostly unknown. We analyzed the genealogy of more than 1 million individuals resulting from a range expansion in Quebec between 1686 and 1960 and reconstructed the spatial dynamics of the expansion. We find that a majority of the present Saguenay Lac-Saint-Jean population can be traced back to ancestors having lived directly on or close to the wave front. Ancestors located on the front contributed significantly more to the current gene pool than those from the range core, likely due to a 20 percent larger effective fertility of women on the wave front. This fitness component is heritable on the wave front and not in the core, implying that this life-history trait evolves during range expansions.

So gene surfing in an expanding colonization phase can produce a genetic revolution whose effects will be felt for hundreds or thousands of years downstream in history. We wonder if the same wave front demographics might explain some of the following population phenomena:

- Large scale triumph of Norman male lineages following the conquest of England in 1066.
- Selective expansion of Middle Eastern genes in Tennessee (including Cherokee families, Jewish male and female lines and Melungeons)
- Relatedness among Jews and "Jewish diseases"
- Diversity-within-uniformity of Polynesians
- Population replacement of Old European (U, N) by Middle Eastern genes (T, J) in Europe as a result of the Neolithic Agricultural Revolution

Many students of history are puzzled why old populations have the allele frequencies and heterozygosity clines they have. Genetic drift is only part of the answer. Gene surfing and selection in deep history are the rest of it.

November 27, 2011

Obama's Melungeon Ancestry

San Diego State University professor D. Emily Hicks has traced some common ancestry with President Barack Obama. According to the professor of Chicana/o Studies, she and the president share Melungeon roots. Obama is a descendant of Mary Collins of Orange County, Virginia, as well as of Nathaniel Bunch of Louisa County and John Bunch of New Kent—well-known "feeder" counties for what became the Melungeon settlement described in Brent Kennedy's book, *The Melungeons, The Resurrection of a Proud People.*

Obama's Bunch line was found to carry E1b1a haplogroup, a sub-Saharan African male lineage. He is also supposed to have Cherokee ancestry in his mother's colonial genealogies.

Read the whole story, "President Obama and California Professor Share Melungeon Ancestry," at www.prlog.org.

September 21, 2011

Evolution and Ancestry: DNA Mutation Rates

As often happens in the annals of science, two research teams independently reached the same groundbreaking results, and publication to the scientific world occurred simultaneously. The breakthrough in the present case concerned the mutation rate of DNA and has profound implications for human evolution as well as for DNA Consultants' new offerings in autosomal DNA ancestry analysis, specifically our Rare Genes from History Panel. The following two studies are already much cited by geneticists, though they have garnered little attention in the press. They appeared in online versions on the same day, Aug. 23, 2012.

Study or Source	Type of DNA	Sample or Method	Rate per Generation	Time Depth in Years
Sun 2012	autosomal microsatellites	2,477 mutations in Icelanders	.001-.0001	25,000 to 250,000
Kong 2012	single nucleotide polymorphisms	4,933 mutations in Icelandic trios	63.2 or .000000012	Very great
Butler 2009	Core CoDIS STRs (microsatellites)	compiled from studies	.0028-.0001	9,000 to 25,000
Zhivotovsky 2004	Y chromosome STRs	Y haplogroup comparisons	.00069	36,000
Heyer 1997	Y chromosome tetranucleotides	42 males in forensic database	.002	12,500
FamilyTreeDNA 2004	Y chromosome STRs	Estimated from customer base	.004	6,250
Brinkmann 1998	STRs (CoDIS markers)	10,844 Father-son comparisons	0-.007	3,500 to Very Great
Parsons 1997	mitochondrial DNA	134 mtDNA lineages	.000029	862,000
DNA Consultants	Rare Genes from History	average estimate across loci	.001325	19,000

DNA Mutation Rates.

Article sources: James X. Sun et al., "A Direct Characterization of Human Mutation Based on Microsatellites," *Nature Genetics* 44/10 (October 2012): 1161-65. A. Kong et al., "Rate of *de novo* Mutations and the importance of Father's Age to Disease Risk," *Nature* 488 (2012):471-75.

From this, it can be seen that mutation rates vary from a low with SNPs to the high rate of Y chromosome STRs (as much as 0.4 percent per generation). DNA keeps surprising us by proving to be more stable than we would tend to expect, dutifully transcribing its original values from generation to generation without many mistakes or changes. Only Y chromosome seems to be highly changeable, depending on the father's age (Kong 2012). Autosomal STRs mutate at a rate between SNPs and the Y chromosome, between every 19,000 or 25,000 and 250,000 years. For our new autosomal ancestry markers, that is confirmation that the alleles we are examining on a statistical basis are pretty much unchanged for the past 20,000 years. That's about twice the length of what we call world history, hence a meaningful enough time frame for valid inferences about population patterns and ancestry of individuals.

Definition: *mutation* A change in a DNA sequence, either spontaneous within a generation or inherited, sometimes from a very distant ancestor. Mutations usually do not affect our health or cause any differences in our appearance. In other words, they are not genes proper and do not "code" for new proteins. Even though they are non-coding genes, though, they are useful in tracing lineages. From *A Glossary of Common DNA Terms.*

October 23, 2012

Megapopulations

The dictionary defines megapopulation as a very large one, from the Greek suffix mega, the same element as in megabyte. In statistics, a population is the whole field from which you choose a sample or representative segment. Thus, to test American Hispanic/Latinos you might draw a sample of 400 people from a predefined population of everyone with a Hispanic surname in a telephone book. How reliable and valid your sample is depends on methodology. By combining populations you can study a metapopulation (all related populations, for instance North *and* South American Latin or Iberian populations) or megapopulation (all populations with Iberian ancestry in the world). Going from the small to the large, we have, then, individual, sample population, metapopulation, megapopulation, universe.

In a census, the sample and population coincide; everyone is counted.

In population statistics, this hierarchy might look like this: John Doe, Arizona Hispanic study (n=104, that is 104 persons in the sample), U.S. Hispanics, North American Hispanic, Iberian American, Iberian or Part-Iberians in the World.

At DNA Consultants, megapopulations are the broadest ethnic category calculated and reported to you. We look at metapopulations, too, but only as a control measure. Our database coverage is described as follows.

Megapopulation Names and Number of Populations
    African 17
    African-American 28
    American Indian 24
    Australoid 3
    Austronesian 6
    Central Asian 39
    Central European 13
    East Asian 39
    East European 8
    European American 24
    Iberian 32
    Iberian American 61
    Jewish 3
    Mediterranean European 20
    Melungeon 1
    Middle Eastern 36
    North Asian 3
    Northern European 15
    Romani 4
    South Asian 35
    Southeast Asian 12

Beyond Megapopulations: Percentage of Total Populations

These categories correspond roughly to what people used to think of as "race," a now-discredited notion. They are continent-specific, with African and Caucasian extended to African and European populations in North and South America.

African	45	11%
Amerind	24	6%
Austral	9	2%
Asian	67	17%
Caucasian	255	64%

Another Calculation

We created these totals to see what kind of white versus non-white coverage the database has.

White	255	64%
Non-white	145	36%
Total:	400 Populations	

And that's all you need to know about megapopulations!

April 9, 2012

Rare Genes from History

DNA typing has gone from successes in the criminal justice system and paternity testing to new heights in mapping genetic diseases and tracing human history. John Butler in the conclusion to his textbook *Fundamentals of Forensic DNA Typing* raised an important question about these trends: How might genetic genealogy information intersect with forensic DNA testing in the future?

"At DNA Consultants that new chapter in DNA testing arrived several years ago," said Donald Yates, chief research officer and founder. "As we approach our tenth anniversary, examining human population diversity continues to be the whole thrust of our research, and it just gets more and more exciting."

The company's DNA database atDNA 4.0 captures and puts to use every single published academic study on forensic STR markers, the standard CoDIS markers used in DNA profiles for paternity and personal identification. In 2009, the company introduced the first broad-scale ethnicity markers and created the DNA Fingerprint Plus test.

But its innovations didn't stop there. In October 2012, the company announced the launch of its Rare Genes from History Panel.

Why CoDIS Markers?

"Theoretically," noted Butler in 2009, "all of the alleles (variations) that exist today for a particular STR locus have resulted from only a few 'founder' individuals by slowly changing over tens of thousands of years."

How true! Hospital studies have determined that the most stable loci (marker addresses on your chromosomes) have values that mutate at a rate of only 0.01 percent. That means the chance of the value at that location changing from parent to progeny is once every 10,000 generations.

The autosomal clock of human history ticks at an even slower quantum rate than mitochondrial DNA. Like "mitochondrial Eve," its patterns were set down in Africa over 100,000 years ago when anatomically modern humans first appeared on the stage of time.

Though the face value of the cards in the deck of human diversity never changed—and all alleles can be traced back to an African origin—as humans left Africa and eventually spread throughout the world, alleles were shuffled and reshuffled.

Humanity went through bottlenecks and expansions that emphasized certain alleles over others. Genetic pooling, drift and selection of mates produced regional and country-specific contours much like a geographic map.

By the 20th century, when scientists began to assemble the first genetic snapshots of people, it was found that nearly all populations were mixed, some more than others. The geneticist Luigi-Luca Cavalli-Sforza at Stanford University proved that there is almost always more diversity within a population than between populations.

If there is no such thing as a "pure" population—a control or standard—how are we to make sense of any single individual's ancestral lines? Statistical analysis provides the answer. And rare genes are easier to trace in the genetic record than common ones. Their distinctive signature stands out.

It All Began with the Melungeons

About the same time as DNA Consultants' scientists were cracking the mystery of the Melungeons, a tri-racial isolate in the Appalachians, they became aware of certain very rare alleles carried by this unusual population in relatively large doses. The Starnes family, for instance, in Harriman, Tennessee, was observed to have a certain rare score repeated on one location in the profiles of members through three generations. The staff dubbed it "the Starnes gene."

Soon, company research had characterized 26 rare autosomal ancestry markers—tiny, distinctive threads of inheritance that reflected an origin in Africa and expansion and travels through world history. Genes in this new generation of discoveries were named after some distinctive feature associated with the pattern they created in human genetic history—for instance, the Kilimanjaro Gene after its source in Central East Africa. The Thuya, Akhenaten and King Tut genes were named for the royal family of Egypt whose mummies were investigated by Zahi Hawass' team in 2010. The Starnes Gene became the Helen Gene. Because of its apparent center in Troy in ancient Asia Minor and predilection for settling in island populations, it was named for "the face that launched a thousand ships," in the famous phrase by Christopher Marlowe.

All 26 of DNA Consultants' new markers are rare. Not everyone is going to have one. But that's what makes them interesting, according to Dr. Yates. Coming from all sections of human diversity—African, Indian, Asian and Native American—they are like tiny gold filaments in a huge, outspread multicolored tapestry, explains Phyllis Starnes, assistant principal investigator and wife of the namesake of the first discovery. But does that mean that her husband has a connection to Helen of Troy? The markers don't work on such a literal level, but it does imply that Billy Starnes shares a part of his ancestral heritage with an ancient Greek/Turkish population prominent on the page of history.

Over the past two decades, geneticists have worked out the macrohistory and chronology of human migrations in amazing detail and agreement. The Rare Genes from History Panel is another reminder—in the words of an American Indian ceremonial greeting—that "We Are All Related."

These rare but robust signals of deep history can act as powerful ancestral probes into the tangled past of the human race as well as unique touchstones for the surprising stories of individuals. For more information about the science of autosomal DNA ancestry testing, visit dnaconsultants.com or check out its Twitter or Facebook page.

September 30, 2012

Witches in the Attic

Three of my ninth great-grand-aunts were accused of witchcraft. Not exactly what I was looking for, but it is colorful. These were the famous Towne sisters depicted in "The Crucible." All three were accused of witchcraft, and two were hanged. Only one escaped. Of course, neither Sarah Towne Bridges Cloyce, nor her two sisters, Rebecca Towne Nurse and Mary Towne Easty, who were executed, were witches. They just happened to live in a town caught up in a lunacy as bizarre as "The Walking Dead." Except this was real. Rebecca Nurse was hanged for witchcraft July 19, 1692, and Mary Easty was hanged Sept. 22, 1692, as cited by Frances Hill in *The Salem Witch Trials Reader xv.* Sarah alone escaped the collective madness.

Various theories have been discussed as the source of this madness. According to Mary Jane Alexander in her *Los Angeles Times* article, "Family History Proves Bewitching," the truth is probably more complex than hysterical teenagers or hallucinations brought on by ergot-tainted, moldy bread. That image is National Lampoon movie madness—albeit with Puritan clothes. She blames family rivalries and acrimonious land disputes. And some of these theories may have led in varying degrees to the Salem witch trials. But the Indian wars and the perception of Indians as heathens has largely been an ignored, albeit important, factor.

For instance, what of the Indian "witch" cake? One of the famous preachers of the day, Samuel Parris, preached a sermon laying a large part of the blame on a church member, Mary Sibley, who asked an

Indian named John to make a "witch cake," Hill writes. Puritans believed Indians willingly participated in devil worship, according to Elaine G. Breslaw in her *Tituba: Reluctant Witch of Salem: Devilish Indians and Puritan Fantasies.*

Breslaw makes the case that despite many incorrect references to Tituba as an African, there is ample evidence that she was referred to as Indian along with her husband at the time of the Salem witch trials. Like her husband, she was an Indian slave from Barbados who lived in the Parris household. When questioned, she made up wild tales and confessed to witchcraft no one could provide evidence for. For what motive? Survival. I am sure there was little to no love lost between an Indian slave and her white captors who called her a witch.

And the Puritans believed her story because it fit what they thought of American Indians. Though she was not executed, it spun a web of fear that went out of control. That fear was based on a fear of American Indians and their beliefs. Hill says, "That credibility was reinforced ironically by the continued association of Native Americans with witchcraft and a refusal to fully integrate Indians into Puritan society." Indians were considered heathens.

According to a 2004 article, "Salem Witch Trials," in the *Sun Journal,* the Indian wars were a contributing factor as Puritans believed God was punishing them because of the wars—attacks from the Wabonaki and other tribes along the coast: "… It was around this time that the girls started to have fits, believed to be the act of the devil. Since the devil could not do anything without God's permission, the Salem villagers thought that God was punishing them for not being Christian enough. To show that they were truly Christen (sic), they would kill all of the devil's sidekicks to show that when attacked by the devil, they would remain true to God and prove their faith. So, you see, without the Indian war, the Salem witch trials could not have happened."

It is an interesting theory. Of course, that does not mean Tituba and John are completely blameless. Tituba and John's confessions and accusations contributed to the Salem witch trials. But that does not make them witches. They stirred up a lot of trouble in fear of their own fates and accused people left and right. John accused Sarah in court. Sarah (Towne Bridges Cloyce) famously called him out when he testified she was a witch to the court by saying, "Oh, you are a grievous liar," according to Alexander in her *Los Angeles Times* article.

How did it turn around so? It is odd. The court called Tituba a witch after the incident of the "witch cake." But they were not prepared for her powers of storytelling, and she began to point fingers at others. Because the court believed so much in the truth of her testimony that she was a witch, they did not question what else she said or who else she accused. Then it all fell apart like in Yeats' poem, "The Second Coming": "Things fall apart/ The centre cannot hold."

At the core of their belief in her testimony was that an American Indian is a heathen. Until recently, this belief had not completely gone away—if it has now. The song, "Witch Doctor" was released in 1958. "Witch doctor" was a slur for a Native American healer who healed with plants and herbs—properly known among his people as a "medicine man (or woman)."

This slur has been in use for some time though it actually meant the reverse—an aboriginal healer working to protect against witchcraft. The Oxford English Dictionary states that the first use of the term "witch doctor" to refer to African shamans (medicine men) was in 1836 in a book by Robert Montgomery Martin (1803–1868). But evidently the idea that they were evil has been around for a lot longer.

So, how did Sarah escape? On an Easter Sunday, she slammed the door to the church when the preacher, Samuel Parris, insinuated her sisters were witches. Accused shortly thereafter, she was imprisoned and shackled, Hill says. However, her brother, my ancestor Jacob, a lawyer, and other friends spirited her away, and "she surviv(ed) the winter in nearby caves," according to the article "Witch Trials-Victims" on the Miner genealogy website. Later, she and her husband, Peter Cloyce, built their home in a nearby town until she was declared innocent. Eventually, she returned to receive three gold crowns from the town for the persecution of herself and the execution of her sisters.

Sarah's story is depicted, fictionally, as Goody Cloyse, a witch, in Nathaniel Hawthorne's short story, "Young Goodman Brown." She is also portrayed in the movie *Three Sovereigns for Sarah* with Vanessa Redgrave receiving the coins from Salem.

In conclusion, I am in good company being related to three accused witches. Three presidents—Taft, Ford, and Arthur—are "also descended from one of Salem's executed witches or their siblings [as well as] Clara

Barton, Walt Disney, and Joan Kennedy," Alexander says in her article. But no wonder this has been hidden until recent times. It was not until 1957 that "the General Court of Massachusetts resolve(d) that 'no disgrace or cause for distress' be borne by descendants of witch-trial victims." This was about the time I was listening to "Witch Doctor" on the radio.

April 11, 2013

New Euro Test

DNA Consultants' new Euro DNA Fingerprint Test is an analysis of your European DNA in terms of statistical living population matches. Other methods before it generally fail to break down European heritage into meaningful national components.

"For a long time there simply weren't enough data in the literature to support such a product," said Principal Investigator Donald Yates. "European scientists have a different marker standard from the U.S. and that made comparisons difficult. Fortunately, after several years of work by our statistics team, we now have 71 European populations in our database — complete coverage of countries of Europe, including so-called foreign populations like Gypsies."

Like the company's other autosomal line of tests, Euro DNA Fingerprint Test matches your unique DNA profile with populations reported in academic studies. For this, it uses the computer program atDNA.

Other methods of assessing your ancestry are limited by the approach known as "haplotyping." A male or female can take our tests because the markers we look at are genome-wide, and none is located on a sex chromosome. The standard used in this test is the same set of identifying profile adopted by forensic institutes and law enforcement in the European Union, that is European Network of Forensic Science Institutes, or ENFSI.

"Even though there are some tests out there that purport to give you percentages of world or European ancestry," said Yates, "there have been and continue to be fundamental flaws in the methodology."

Most genetics studies, he noted, are still fixated on male or female lineages, those that can be determined by haplotyping.

"If you go back 10 generations," he says, "you have a thousand lines, not just two—no Y chromosome or mitochondrial sex-linked test can tell you anything about your father's mother or mother's father." Autosomal testing has the advantage over sex-linked testing of providing a more complete picture.

May 31, 2013

Cherokee New Year

As Jews conclude a week of Rosh Hashanah and mark Yom Kippur beginning at sunset today, Cherokees worldwide are gearing up to celebrate their own cycle of fall festivals, including a New Year or Great Moon Gathering, which, like the Jewish High Holy Days, appears to begin earlier this year than in the past.

The Alabama Echota Cherokees, for instance, are holding their 16th Annual Festival Sept. 28-29. It will take place at the Oakville Indian Mounds & Park in Danville, Ala.

The Cherokee Nation of Oklahoma got the season off to a gala start with its 61st Annual Cherokee National Holiday on Labor Day weekend. This group of Cherokee descendants traces its roots in Oklahoma back to the days of Indian removal in the 1830s and is recognized by the U.S. Federal Government as a sovereign nation under Indian treaties from the 19th century. It boasts 300,000 card-carrying members today and is frequently in the news for rancorous lawsuits against other Cherokee groups. Their main "rival" is the United Keetoowah Band, also federally recognized and also headquartered in Tahlequah, Oklahoma. A third federally recognized group is the Eastern Band of Cherokee Indians in western North Carolina. This Indian nation's website claims a history of 11,000 years and original territory of 140,000 square miles covering eight present-day southern states, centered on Tennessee, although it has been granted official status only since 1924.

The Echota Cherokee Tribe of Alabama is smaller and quieter than the Big Three. It is one of nine Indian groups or tribes along with Choctaw, Shawnee and two other Cherokee recognized by the Alabama Commission of Indian Affairs. The Echota still live close to the land and maintain tribal property in their communal name to this day, with several thousand families scattered statewide around Falkville, Alabama. When they have a homecoming, it is an emotional event.

We are happy to see that the Echota have an active women's society, and that their current principal chief is female, Charlotte Hallmark. We wish them and their members a joyous New Year. Jews have been celebrating their peoplehood on this date for 5,774 years, but Cherokees can point to many more thousands of years of existence.

September 13, 2013

3 DNA AND APPEARANCE

Six Generations of Cherokee Blood
Donald N. Yates

I wrote this article in 1996, before the advent of genetics in genealogy research. Since then I have determined through DNA analysis that my Native American genetic makeup is probably less than half the figure I arrived at purely with reconstructed genealogy and oral history. Obviously, a lot of those "fullbloods" were not full bloods, but mixed.

The interracial mixing in my family lines goes back further than I originally gauged. I estimate now that you have to go back to the generation of Cherokee born about 1750 to find the first 100 percent Native in my family tree. That is the generation of 6th great-grandparents, approximately the generation of Black Fox and Attakullakulla and Nancy Ward. Evidently, many of my Jewish ancestors exaggerated their admixture with the Indians. Still, a lot of the research hypothesis testing below can be of interest and, I hope, inspire others in their family searches.

As an interesting aside, I also found a tiny degree of East Asian genes. Perhaps Gavin Menzies is correct about the Chinese discovering America ...!

After about seven years of intensive genealogy research, I have established my known degree of Indian blood. I wanted to share the results with the Amerind list, since I know a lot of you are in the same boat. Some of you even have the identical lines and surnames and brick walls. Probably few Indians today have had the opportunities I have had to investigate their family history. I feel anything further I may find will be a case of diminishing returns. I invite your discussion.

How far I have come since that day in 1993 when an older second cousin of mine told me a surprising story about my fullblood Cherokee great-grandmother Elisabeth Yates and first piqued my interest! That was my father's line. Lawden Yates died in 1978, without so much as telling a soul of his Indian blood. As you can see below, I have since determined that he was a minimum of a quarter blood, mixed Cherokee and Creek.

My mother was supposed to have all the Indian blood. She put it all on her maternal grandmother Shankles. I subsequently found out that there was just as much in Mother's paternal line, the Coopers. Does any of this sound familiar?

Wherever I looked, I found Cherokee. Generation after generation, Cherokee or part-Cherokee married Cherokee. Was this an accident? I do not think so.

My investigations suggested that I was the sixth generation since the last of the fullbloods born in the pre-Removal days before 1820. There is a saying that Cherokee elders and clan mothers plan to the seventh generation. Looks like they did—and are still doing—a good job!

In a crude way of reckoning, four out of 32 of my ggg-grandparents appear to have been fullblood Cherokee, making me one-eighth. I found only one ancestor proven to be non-Indian, Louis Graben, from Germany.

Besides Cherokee, which was the overwhelming element, there was a slight mixing with Creek, Cheraw (possible), Saponi (possible), Choctaw and Algonquian tribes such as Powhatan. Putting it all together, I estimate that I am about one-fourth American Indian by blood.

An obvious question is if there was a pattern of Cherokee or part-Cherokee marrying another Cherokee, how to account for it—particularly in cases where spouses might have concealed or even been unaware of their Indian blood? All lived in a predominantly white society in Georgia, Alabama and Tennessee; the chances of randomly marrying another part-Cherokee in the area were slim. It is true that in each generation, most siblings in any given family (Coopers excepted) seemed to marry non-Indians, so that Indian identification gradually disappeared in those lines.

Another factor that contributed to the dying out of Indian identity was moving away from the ancestral lands. The Coopers on the other hand appear to have consciously conserved their bloodline, as I have investigated all my great-grand-aunts and uncles' families. Only Coopers and Sizemores attempted to legally claim their Indian heritage. They filed numerous applications, for instance, in the Guion Miller process of 1906. All were rejected by the government.

A generalization I could make is that when the blood degree drops below one-fourth, there is a tendency no longer to identify as Indian, marry another Indian or raise your children with knowledge of the culture. One reason for the magic figure of one-fourth is that most of us do not interact with family members older than grandparents: typically, mixed bloods have one Indian grandparent and are thus one-fourth Indian. Grandpa or Grandma may or may not have "talked."

Of course, locale and Federal Indian policy have a lot to do with it. During generation one (pre-1820) when the official U.S., state and local government policy was extinction, the only Indians who survived off reservations kept a very low profile. Today, in generation six (Self-Determination), Indians are legally and openly asserting their own identity and culture. As long as the Indian wars lasted (until 1889), it was dangerous to be an Indian.

My Coopers survived because they lived in a "gray" backwoods area on Sand Mountain straddling county lines and state lines. Still, the government managed to repossess their land, in 1892.

Looks played a part, too. Some of the Blevins children were explicitly named Red, White and Black. My great-grandfather Henry Yates was half to three-quarters Indian and didn't look it or admit it. His brothers Josephus and George looked Cherokee to a T and their descendants preserved many of the family memories.

It was a custom for the carrier of certain traditions and stories to select one child or grandchild to pass them to. In some generations, none was qualified, and the tradition died. Stories were considered part of the possessions of a clan—living things that were treated with respect and honor. The Cooper tradition that we are descended from chiefs was passed from Isaac Cooper to his son Isaac, to Isaac's son William, to grandson Peter Cooper. Peter's wife, Lindy Sizemore, was the carrier of the Sizemore secrets. Much in Indian history hangs by a slender thread.

A final thought is that, perhaps, there really is something to "having Indian blood." I have a rare blood type (A negative), as does my wife. It is unlikely that a woman with A negative can have numerous children with a man of positive type (unless modern medicine intervenes with a drug like Rhogam). My wife and I do not have that problem. In a blood type study of indigenous people, A was the predominant type, though it represented less than 2 percent of the overall population. Certain groups like the reservation Cheyenne (Algonquian stock) and Amazonian Indians had 95 percent type A.

May 15, 2000

Idylls of the Caveman
Hominid Harems: Big Males, Small Females and Bodybuilding

Ann Gibbons
Science magazine

A study on page 1443 of the November 30, 2007, issue of *Science* finds that the males of an extinct species of hominid in South Africa took longer to grow up than females and got much larger, suggesting that top males of this australopithecine species invested energy in bodybuilding in order to possess a harem of females, much like silverback gorillas do today.

November 30, 2007

DNA and Emotions
Second Generation DNA Tests Reveal More Than Just Identity

A report on the front page of the April 20, 2008, *Washington Post* by Rick Weiss suggests that "second generation" forensic DNA tests, just like second generation ancestry tests, can do much more than just identify a person. They can yield information on someone's health and even their emotions.

The article's headline is "DNA Tests Offer Deeper Examination of the Accused."

April 20, 2008

Beachcomber Route Confirmed

Since Stephen Oppenheimer's *The Real Eve* suggested that the main out-of-Africa migration of humans proceeded across the mouth of the Gulf of Suez and around the coasts of Arabia, India and Southeast Asia (the "beachcomber route"), controversy has raged about the origin of Asians, whether they split off from the first out-of-Africa groups, sometimes called macro-haplogroup M, in the north central Asian highlands or the Middle East or elsewhere.

A massive project spearheaded by the Chinese has put that question to rest. The 40-institution HUGO Pan-Asian SNP Consortium "strongly concludes the southern route made a more important contribution to East and Southeast Asian populations than the northern route," says Li Jin, a population geneticist at Fudan University in Shanghai, China. Jin was one of the lead authors of a study reported in *Science*, vol. 326, no. 5959, p. 1470, "SNP Study Supports Southern Migration Route To Asia," by Dennis Normile.

DNA Consultants has always followed Oppenheimer's model of the settlement of Asia, but other companies, including the National Geographic Genographic Project with over 200,000 customers purchasing their product, inform their customers differently. Most human migration maps displayed by DNA companies and the news media show Asians splitting off from Europeans and Native Americans in the northern latitudes of Central Asia and do not depict a southern "beachcomber" route at all.

In July of this year, DNA Consultants discovered ethnic markers it released in its 18 Marker Ethnic Panel that prove a southern divide and origin for Asian populations as in the new study.

According to the *Science* report, "Anthropologists, ethnographers, and linguists have long struggled to understand the patchwork-quilt diversity of Asia. Indonesia alone claims some 300 ethnic groups; the Philippines has 180 native languages and dialects. Where did they all come from?"

The previously dominant theory of two major waves of migration from the Middle East must now yield to just one initial migration along the coastal route with populations moving north into East Asia from India and Southeast Asia.

The new study is vindication for the Chinese genetics community, which has often been dismissed and rejected by European and American geneticists. Vincent Macaulay, co-author with Martin Richards of the seminal paper followed by most DNA testing companies, "Tracing European founder lineages in the Near Eastern mtDNA pool," in the *American Journal of Human Genetics*, 67, 1251-1276, when asked about the new findings admitted that the southern coastal route now "seems very strong," as quoted in *Science*.

The study used samples from more than 1,900 individuals representing 73 populations and involved 93 researchers at 40 institutions in 11 countries and regions in Asia. It was "conceived by Asians in Asia and executed, funded, and completed by an Asian consortium," said Edison Liu, executive director of the Genome Institute of Singapore. Researchers screened each individual for more than 50,000 SNPs.

December 10, 2009

Down on the Don

The successful extraction of ancient DNA has been a rare accomplishment in genetic circles until recently. In the journal *Current Biology*, a German-Russian team details how it was possible to avoid the common pitfalls of contamination with modern human DNA in the instance of a 30,000 year-old hunter-gatherer's grave in Russia.

Svante Pääbo, from the Max Planck Institute for Evolutionary Anthropology in Leipzig, Germany, and colleagues used the latest DNA sequencing techniques to study genetic information from human remains unearthed in 1954 at Kostenki, Russia. According to a report by the BBC, the hunter-gatherer's mitochondrial DNA type was U2. Haplogroup U is seen as a predecessor dominant type among Europeans before the arrival of agriculture and Middle Eastern culture about 5,000 to 7,000 years ago. It is hoped the new expertise will help unlock the secrets of other examples of ancient DNA.

The article, "A Complete mtDNA Genome of an Early Modern Human from Kostenki, Russia," is by Johannes Krause, Adrian W. Briggs, Martin Kircher, Tomislav Maricic, Nicolas Zwyns, Anatoli Derevianko and Svante Pääbo.

The article says the recovery of DNA sequences from early modern humans (EMHs) could shed light on their interactions with archaic groups such as Neanderthals and their relationships to current human populations. However, such experiments are highly problematic because present-day human DNA frequently contaminates bones.

For example, in a recent study of mitochondrial (mt) DNA from Neolithic European skeletons, sequence variants were only taken as authentic if they were absent or rare in the present population, whereas others had to be discounted as possible contamination. This limits analysis to EMH individuals carrying rare sequences and thus yields a biased view of the ancient gene pool.

Other approaches of identifying contaminating DNA, such as genotyping all individuals who have come into contact with a sample, restrict analyses to specimens where this is possible and do not exclude all possible sources of contamination.

By studying mtDNA in Neanderthal remains, where contamination and endogenous DNA can be distinguished by sequence, the researchers show that fragmentation patterns and nucleotide misincorporations can be used to gauge authenticity of ancient DNA sequences. They used these features to determine a complete mtDNA sequence from a 30,000-year-old EMH from the Kostenki 14 site.

January 6, 2010

It's All Around You
DNA Found to Extend in a Radius of 50 Miles

According to an Anishinabe elder, DNA has a well-known nonphysical or spiritual dimension. George Starr-Bresette provided three write-ups on some experiments measuring the effect of DNA in the environment.

Electric DNA and Our Living Universe

In Experiment 1 by Dr. Vladimir Poponin, a container was emptied of air, so the only thing left was electromagnetic field. Poponin measured the energy distribution inside the container and found it was completely random. Then, DNA was placed inside the container and the field distribution was remeasured. This time, the energy was organized in an ordered way aligned with the DNA. In other words, the "physical" DNA is connected to the "non-physical" energy field. After that, the DNA was removed from the container, and the order was measured again. The field remained ordered, with the arrangement created by the DNA.

For Experiment 2 by the U.S. military, leukocytes (white blood cells) were collected for DNA from donors and placed into chambers so they could measure electrical changes. In this experiment, the donor was placed in one room and subjected video clips, which generated different emotions in the donor. The DNA was in a different room in the same building. Both the donor and the subject's DNA were monitored. As the donor exhibited emotional peaks or valleys (measured by electrical responses), the DNA exhibited the identical responses. The military wanted to see how far away they could separate the donor from his DNA and still get this effect. They stopped testing after they separated the DNA and the donor by 50 miles and still had the same result. The DNA and the donor had the same identical responses. It means that life everywhere communicates through the electromagnetic cosmic field.

With Experiment 3 by the Institute of Heart Math, human placental DNA (the most pristine form of DNA) was placed in a container from which they could measure changes. Twenty-eight vials of DNA were given (one each) to 28 trained researchers. Each researcher had been trained how to generate and "feel" feelings, and they each had strong emotions.

Starr-Bresette's only comment was, "This still just amazes me ... but some knew it all along!" DNA Consultants has known it since 2004, when the company's webmaster mentioned the Army study to office personnel.

September 23, 2010

In the Land of Thunder

Everyone probably has wondered at some time what makes Aboriginal Australians different from other people, where they came from and how old their ethnic type is. Well, wonder no more. In a groundbreaking article in *Science* magazine, first lead author Morten Rasmussen and 58 co-authors present their study, "An Aboriginal Australian Genome Reveals Separate Human Dispersals into Asia" (*Science* 334, 7 Oct. 2011, 94-98). Rasmussen is with the Feinstein Institute for Medical Research, Manhasset, New York.

In the abstract, they write:

We present an Aboriginal Australian genomic sequence obtained from a 100-year-old lock of hair donated by an Aboriginal man from southern Western Australia in the early 20th century. We detect no evidence of European admixture and estimate contamination levels to be below 0.5 percent. We show that Aboriginal Australians are descendants of an early human dispersal into eastern Asia, possibly 62,000 to 75,000 years ago. This dispersal is separate from the one that gave rise to modern Asians 25,000 to 38,000 years ago. We also find evidence of gene flow between populations of the two dispersal waves prior to the divergence of Native Americans from modern Asian ancestors. Our findings support the hypothesis that present-day Aboriginal Australians descend from the earliest humans to occupy Australia, likely representing one of the oldest continuous populations outside Africa.

This study of Aboriginals will be cited as a landmark case in genetics because the authors took especial care to disarm any criticism concerning possible admixture and contamination, achieved a stupendous rate of success in sequencing DNA sites and used smart comparators to verify their model of what makes Aboriginals different, including Neanderthals, Denisovans, Andamanese, Filipinos, Indians, Papua New Guineans and Melanesians.

First sentence explains the genetic history of Aboriginal Australians is contentious but highly important for understanding the evolution of modern humans. Some mysteries pointed out about Aboriginal Australian DNA by the authors include:

—The Aboriginal population contains a lot of diversity, including specimens of most of the world's haplogroups, male and female.

—Related populations suggested are hunter-gatherers from Nepal and the Philippines, Great Andamanese and Onge from the Andaman Islands, Highland Papua New Guineans and certain peoples from India.

—It was previously unclear whether Aboriginals resulted from a single dispersal out of Africa or multiple-dispersal model.

—The role of hybridization with other archaic peoples was also not clear.

The authors confirm that "before European contact occurred, Aboriginal Australian and PNG Highlands ancestors had been genetically isolated from other populations (except possibly each other) since at least 15,000 to 30,000 years B.P." Also, "our results favor the multiple-dispersal model in which the ancestors of Aboriginal Australian and related populations split from the Eurasian population before Asian and Europeans split from each other" (97). "We find that the European and Asian populations split from each other only 25,000 to 38,000 years B.P., in agreement with previous estimates."

The new study finds that Aboriginals have an amount of admixture with Neanderthals and Denisovans comparable to Europeans and Asians, although they have more Denisovan DNA than other people. "This admixture may have occurred in Melanesia or, alternative, in Eurasia during the early migration wave" (97).

In sum, the Aboriginals are part of the same first-out-of-Africa branch of the human tree as Europeans and Asians, their ancestors splitting 62,000 to 75,000 years ago from Africans, and leaving relic related populations in the Highlands of Papua New Guinea and the Philippines. A second expansion wave through India, Indo-China and Southeast Asia replaced the original stock, while the Aboriginals became stranded and isolated in Australia about 50,000 years ago.

"This means that Aboriginal Australians likely have one of the oldest continuous population histories outside sub-Saharan Africa today" (98).

Properly positioning Australian Aboriginals in the expansion of humans out of Africa opens the way to connecting the dots for all the other prehistoric peoples. The migration map presented by Rasmussen et al. should be carefully studied for clues about the origins of Asians and Native Americans, to begin with.

October 8, 2011

Angel Du Pree commented on Nov. 15, 2013: Back in the seventies I took an anthropology course at the University of Florida. One of the professors had postulated that the Australian aborigines may have had ancestry with the Ainu people off the coast of Japan. Though the Ainu are light skinned he pointed out the fact that both tend to have a similar head shape, notably their foreheads. He showed us a map of ocean currents and one of them led directly from the Ainu down to Australia. Has this been totally debunked? Have there been any genetic comparisons done?

Nat Turner commented on Feb. 15, 2014: A great, intelligent and mystical people, greatly not understood and abused by latecomers, the links between Australia and the surrounds have long been clear, even before the "proof" by DNA. What has not been discussed widely or taken into consideration is the so-called Aboriginals along with the very early inhabitants of India, were both actually preceded by "pygmies" and in both cases these "pygmies" were still extant, at least till recently. They, like the pygmies of Africa, represent the first of humanity, and they remained even in Europe way into historical times. The Greeks pictured them with their protuberant bellies as their gods on pottery, the Vatican documents them as far away as Greenland under the name Skraelings, the Japanese Ainu, entering Japan, sight them as "earth spiders & pit dwellers" under the name Koro-pok-guru. When Pharaoh Seti I wanted to see the "Sacred Dance", ie. heavenly movements, he sent southward to the land of Punt, for a pygmy to perform. The descendants of the first men, even those who were known as the "Gods" are to be found still in pockets across Africa, Asia and elsewhere. The Australian Aboriginals, The Dravidians, the Africans represent the first of man. They must be especially cherished as they purvey the secrets of man's inception.

The Face That Launched a Thousand Ships

DNA Consultants announced in 2012 a new generation of autosomal markers that can help in identifying ancestral lines. In its fall newsletter to customers, the company gave a preview of Rare Genes from History, rare alleles that its research has tied to specific populations in world history. The novel marker discoveries are powered by the autosomal DNA database atDNA, currently in version 4.0A, with 400 populations and more than 115,000 subjects.

The first rare value characterized, the Helen Gene, occurs at a frequency of only 1 in 10,000 people worldwide but is relatively common in Cypriots and other Mediterranean populations. It is 15 times more common in Jews than Asians and is not even reported in many African, Chinese and Indonesian populations. The rare marker is believed to have originated in Africa (along with all other alleles in the human genome) but to have experienced a boost from carriers in ancient Anatolia and the Middle East, spreading to other population centers. The Helen Gene is named for the "face that launched a thousand ships," because its approximate center of diffusion includes ancient Troy, site of one of the world's oldest civilizations and longest wars, as sung by Homer and verified by the excavations of Heinrich Schliemann. Notable appearances are in island populations such as Cyprus, the Balearic Islands and the British Isles.

Models of history tested in the database have linked the Helen Gene with Greeks (often referred to as Hellenes), Italians, Turks, Sephardic Jews, certain North Africans, Celtic migrations, the Rom (Gypsy) people and modern-day Melungeons, though nowhere does its occurrence climb higher than 1 in 500 people, or 0.2 percent.

"These are all extremely rare genes," said Donald Yates, principal investigator. "But because they are so rare their trail on the map of history is very, very clear."

The lineup of new autosomal markers in the first phase of development is as follows:

1. The Helen Gene
2. The Scythian Gene
3. The Kilimanjaro Gene
4. The Thuya Gene
5. The Akhenaten Gene
6. The King Tut Gene
7. The Egyptian Gene
8. The Cochise Gene
9. The First Peoples Gene
10. The Lake Baikal Gene

For descriptions, DNA Consultants' website www.dnaconsultants.com has a section titled Rare Genes in History.

Five of the new products are African (three come from studies of the Amarna mummies in Egypt), two are Eurasian and three are Native

American.

The classic DNA study by the Supreme Council of Antiquities in Cairo, Egypt, is: Hawass Z, Gad YZ, Ismail S, et al. "Ancestry and Pathology in King Tutankhamun's Family." *JAMA*. 2010;303(7):638-647. The feat has also been showcased on Discovery Channel.

August 26, 2012

Out of Asia

"All the lights in the House of the High Priests of American Anthropology are out; all the doors and windows are shut and securely fastened (they do not sleep with their windows open for fear that a new idea might fly in); we have rung the bell of Reason, we have banged on the door with Logic, we have thrown the gravel of Evidence against their windows; but the only sign of life in the house is an occasional snore of Dogma. We are very much afraid that no one is going to come down and let us into the warm, musty halls where the venerable old ideas are nailed to the walls."

These biting words were penned by Harold Sterling Gladwin in *Men out of Asia*, the famous archeologist's most popular, nontechnical work. Published in 1947, Gladwin's book presented a maverick view of the peopling of the Americas identifying five migrations of diverse races including Negritoes and Austronesians. Heretically, he placed the first migration as early as 25,000 years ago and argued that the earliest colonists in the New World were Australoid.

The reaction of his colleagues in the anthropological establishment was stony silence shading off into hrumps and pshaws of injured pride, for Gladwin illustrated "Men out of Asia" with campy cartoons by Campbell Grant making fun of the sacred keepers of knowledge at the Peabody Museum at Harvard, Carnegie Foundation and Smithsonian Institution. In one, the dean of Southwest and Maya archeology Alfred V. Kidder is depicted as Dr. Phuddy Duddy sitting in academic robes atop a whistle sounding the alarm of illogical chronology. In another, a bespectacled Gladwin and his tweedy friend Professor Earnest Hooton of Harvard are shown in the academic doghouse "by request."

The Establishment is still uncomfortable about Gladwin, who died in

1983 after a distinguished career of more than 60 years. Although willing to praise his meticulous fieldwork on the Hohokam at Snaketown and exacting methodologies developed at the research center he founded at Gila Pueblo outside Globe, Arizona, academicians do not know quite what to say about his conclusions and hypotheses, which grew more adamant toward the end.

The destroying angel of unorthodox theories, Stephen Williams of the Peabody Museum, can only think that Gladwin succumbed to his "whimsies" and grew soft-headed in his old age. "I have always regarded 'Men Out of Asia'," Williams scoffs in *Fantastic Archeology* (p. 229), "as a sort of 'hyper-diffusionist' spoof."

We wonder whether the laddies in Harvard Yard and Castle on the Mall do not protest too much.

In a similar high-toned snipe, Williams dismisses the author of *Pale Ink* with its discussions of ancient records of Chinese explorations in America as "a sweet old lady (p. 185)." Louis Leakey, who believed on the basis of the Calico Early Man excavations he organized in California that humans occupied North America as long ago as 400,000 years, comes in for like treatment. Williams implies that Leakey was senile and hurried in his judgment (p. 303).

If a controversial interpretation of the archeological record cannot be debunked as lacking "soil truth" or the hard-won verities of the dig and thus being nothing but "armchair archeology," then one must resort to that time-honored device of the Ivy League sneer. For instance, Jeffrey Goodman's controversial *American Genesis* "mimics a scientific book very well" (p. 303), but it cannot possibly be taken seriously because, well, you know, no person in his right mind subscribes to a North American or Asian origin for humans! In addition to coming from a homely, unpatrician Western background, Goodman was also part Native American, which helped make him a pariah.

But to the happily hidebound Dr. Phuddy Duddys, we say in the words of Blake:

> "Mock on, Mock on, Voltaire, Rousseau;
> Mock on, Mock on, 'tis all in vain.
> You throw the sand against the wind,
> And the wind blows it back again."

The winds of change may have the last laugh.

September 7, 2012

Our Horizontal Ethnic Identities

Andrew Solomon in *Far from the Tree* makes a distinction between vertical and horizontal inheritance or identity. Vertical inheritance is determined by the DNA you receive from your parents. Horizontal inheritance kicks in as we identify laterally with others who are not necessarily related to us. Horizontal identities thus supplement vertical ones imposed on us or expected of us by our parents.

Solomon's startling proposition is that diversity is what unites us all. He writes about families coping with deafness, dwarfism, Down syndrome, autism, schizophrenia, multiple severe disabilities, with children who are prodigies, who are conceived in rape, who become criminals, who are transgender. Many born into such situations forge bonds of common culture with peers that take them farther from the family tree into surrogate families.

We have witnessed this phenomenon with ancestry tests. One sibling will be more oriented toward Native American or Romani or Jewish than another even though both have the same DNA inheritance from their parents. Similarly, one sibling will readily accept an unusual ancestry (such as Melungeon) while another adamantly denies it. The sibling who embraces the offbeat identity often experiences the same sort of "coming out" anxiety as a gay or gifted person. The horizontal inheritance derived from friends and support groups becomes more important than blood ties.

Amazon.com Review of *Far from the Tree*, by Andrew Solomon

Amazon Best Books of the Month, November 2012: Anyone who's ever said (or heard or thought) the adage "chip off the old block" might burrow into Andrew Solomon's tome about the ways in which children are different from their parents—and what such differences do to our conventional ideas about family. Ruminative, personal, and reportorial all at once, Solomon—who won a National Book Award for his treatise on depression, *The Noonday Demon*—begins by describing his own experience as the gay son of heterosexual parents, then goes on to investigate the worlds of deaf children of hearing parents, dwarves born into "normal" families, and so on. His observations and conclusions are complex and not easily summarized, with one exception: The chapter on children of law-abiding parents who become criminals. Solomon rightly points out that this is a very different situation indeed: "to be or produce a schizophrenic ... is generally deemed a misfortune," he writes. "To ... produce a criminal is often deemed a failure." Still, parents must cope with or not, accept or not, the deeds or behaviors or syndromes of their offspring. How they do or do not do that makes for fascinating and disturbing reading.—Sara Nelson

November 9, 2012

The Quest for King Arthur

Tintagel or Trevena (in Cornish: *Tre war Venydh* meaning "village on a mountain") is a civil parish and village situated on the Atlantic coast of Cornwall, England, United Kingdom. The population of the parish is 1,820 people, and the area of the parish is 4,281 acres (17.32 km^2), as cited in Wikipedia. The village and nearby Tintagel Castle are associated with the legends surrounding King Arthur and the knights of the Round Table. The village has, in recent times, become attractive to tourists and day-trippers from many parts of the world and is one of the most-visited places in Britain.

We have previously discussed North African genetics in Britain—for instance, in a blog post on "When Wales Was Jewish." The present article focuses on Tintagel, the fabled home of King Arthur and Mark of Cornwall. It is inspired by the mention of Gormund, the Irish "King of the Africans" in Welsh bardic literature, who was, we submit, a Vandal of the fifth or sixth century.

In British myth and historical tradition, not only Ireland but also Cornwall is the stronghold of "Africans." Mark, the king of Cornwall in Arthurian legend and jealous husband of Isoud or Isolt of Ireland, is portrayed in the Tristan romances as dark-complexioned, rich and of fiery southern temperament. Mark or Marcus is a favored name among Jews, particularly English Jews, in memory, perhaps, of the soldier in Roman Britain who was proclaimed emperor by the army there sometime in 406, in the last death rattle of imperial rule. His sister is Elizabeth (another Jewish name), and his royal residence is fixed in Tintagel on the north coast of the Cornish peninsula facing Ireland. This site's chief fame in medieval literature was as the castle of King Mark in the immensely popular cycle devoted to Cornwall.

A series of excavations began in Tintagel in 1933, uncovering a forgotten chapter in southwest Britain's prehistory. According to O.J. Padel, "the area of Tintagel headland teems with fragments of pottery of a type manufactured in the Mediterranean area (mainly in North Africa and Asia Minor); these fragments are dated between the mid fifth century and the late sixth)."

This researcher at the department of Welsh history at University College of Wales, Aberstwyth, goes on to say: "The importance of Tintagel as a find-site for this pottery cannot be overemphasized. Since being identified there, it has been found to occur at other sites within Dark-Age western Britain and Ireland, including other sites in Cornwall and Devon, Cadbury-south-west Ireland, and as far north as the Scottish Highlands. ... Being imported from so far away, this pottery represents expensive, luxury, goods."

Arthur's Name Arabic?

The origin of the name Arthur has been endlessly debated. It is almost certainly not Celtic, neither from a P nor from a Q dialect, and cannot be traced further back than post-Roman times. The center of gravity for its appearance is the sixth century. In 1998, archeological excavations at sixth-century Tintagel brought to light a find subsequently dubbed the Arthur Stone, mentioning the name Artognou, claimed to be cognate. Although the reading is questionable perhaps this inscription and milieu are on the right track.

Arthur's name has become something of a grail quest for modern researchers. Other theories derive the name from Artorius (Roman or Messapic), Arnthur (Etruscan), Arcturus (the "bear star") or Arto-uiros in Brittonic ("bear man").

Perhaps the Gordian knot of the difficulty can be cut if we consider that many of the names in early Welsh history have Arabic and North African roots. Camlann, for instance, the site of Arthur's final deadly battle with the usurper Mordred, has resisted all efforts to etymologize or locate it. This unidentified place in England has a name that is supposed to mean Crooked Glenn. It may be a corruption of the common Arabic place-name Khamilah, "area of dense trees, low or depressed area with good pasturage." Camelot, the fabled capital city of the Round Table, appears to be little more than the plural of the same term.

Arthur's father is Uthr Pendragon, the epithet following his name meaning Chief, or Head, of the Warriors, or Dragons. Now Arthur's son is Amr, a pre-Islamic tribal name that is meaningless in any Brythonic language. Ar- is a common prefix in Arabic and North African naming conventions, meaning "the." Ar-Rumi, for example, the name of an early Arab poet means "the Greek." Ar-Rahman is "the Most Gracious," Ar-Rabi, "the Master," and Ar-Rashid "the Right-Minded." Many of these are traditional names of God's servants in pre-Islamic religion. If we take Arthur's name as Semitic or Arabic or Kufic Arabic it may be a corruption of his father's name: Ar-Uthr. As to what Uthr might have meant originally, however, we will not venture an opinion here.

April 29, 2013

African Walkabout

It's well-known that scientists believe we all came out of Africa some 200,000 years ago. But some of these first ancestors of ours also returned before Europeans even got there. The migration path went both ways. This is a resounding discovery. Erika Check Hayden in her recent *Nature* article, "African Genes Tracked Back" says this "reverse" or two-step migration meant that these ancestors carried "… genes from the rest of the world [which] were carried back to southern Africa, long before European colonizers arrived." The findings are from a flurry of research enabled by better tools to survey African genomes, Hayden says.

But what does this finding mean? It means that we have not well understood until now the rich diversity of African genetics until superior DNA ancestry testing was used. Hayden quotes Luca Pagani, a geneticist at the Wellcome Trust Saenger Institute near Cambridge, U.K., who says, "Until now, we have been applying tools designed specifically for non-African people to African people." Hayden also quotes Carina Schlebusch, a geneticist at Uppsala University in Sweden, as saying,

"It's a really exciting time for African genetics."

It also explains a mystery. It means that some African groups that were previously thought to be genetically isolated actually "… carry 1-5% of non-African DNA" according to population geneticists at Harvard Medical School in Boston, Mass., who examined the individual genetic variations of some 1,000 individuals (Hayden). And their DNA testing results? This admixture explains why some Africans carry non-African genes. In fact, some carry a lot of them. For instance, the male Y-DNA haplotype R1b1 which is the most common haplotype among Western Europeans is also found among some Africans. Miguel Gonzalez et al. in their 2012 article, "The Genetic Landscape of Equatorial Guinea," in the *European Journal of Human Genetics* says that the human Y-DNA chromosomes R1b1 though "very common in Europe are usually a rare occurrence in Africa," but there have been some "… recently published studies that have reported high frequencies of this haplogroup" in parts of Africa. One wondered why until now.

Hayden isn't the first to propose the idea of an ancient trip from Africa and then back again. There have been genetic clues before this, she says. Gonzalez extrapolates from his R1b1 data "that this represents a 'back-to-Africa' migration during prehistoric times." And Hammer et al. in the article, "Out of Africa and Back In," in the *Oxford Journal of Evolution* postulate that there was more than one African ancestral migration path.

Now that we have determined from these DNA testing results that this migratory path is more multifaceted than previously thought, how are we to think about it? What else can we extrapolate from this? Are we to really think people walked from Africa to other lands and then back again?

September 6, 2013

Raymond commented on Nov. 16, 2013: "It is interesting how much discussion occurs regarding "back migration" and a lack of discussion regarding the "Arab Slave Trade" which brought millions of Europeans to N. Africa as slaves and concubines. (Collusion?) Also, there were many Greek, Roman, Circassian, and other Mamlukes (Arabic for slave) in N. Africa that contributed to the genetic makeup. Their descendants are still found in Berber tribes such as the Kabyle, Rif, etc. Some of these same groups even ruled Egypt for some time, erecting statues of themselves that are mistaken for ancient Egyptian artwork.

Most are familiar with the Zanj (African slaves) but are unfamiliar with the rest of those same documents mentioning the "European" women that were slaves in the harems. This provides a better explanation of European mtDNA in N. Africa than "back migration" from thousands of years ago. Unfortunately, we are only considering works from people during a time of immense racial prejudice as valid references leading to useless debates, conjecture, and falsification of history.

DNA doesn't lie. People's rendering of history only confuses the DNA results. For example, how is it that Native American DNA is found in Africa and Europe? Could it be that Native Americans were taken as slaves to those places, or taken there centuries later in military campaigns? Are Native Americans descended from East Indians who migrated over a "land bridge" (NA's deny this) or did some East Indians intermarry with them when they were brought over as slaves in the 1600s (documented) or came with the Hessian Army (self admissions by the same in the 1800s; documented) and married into some of the tribes? DNA only proves the relation, not the "how" they are related.

Mona Lisa's Smile

Who posed for the Mona Lisa? Mona Lisa's enigmatic smile still stares from the past at onlookers in the Louvre. Her smile has cast a spell on us for centuries, but we are still trying to find the mysterious model.

Now, we are looking to DNA and digging up the dead for help.

According to Carol Kurvilla, in her *New York Times* article, "Researchers Dig for Traces of Mona Lisa's DNA," many art experts believe that Lisa Gherardini Del Giocondo, a neighbor of Leonardo Da Vinci, was the real Mona Lisa. Researchers have opened up the family tomb in search of a DNA match with Gherardini's son, Piero. They are trying to match Piero's DNA to one of the three women dug up at the Saint Orsolo convent and then create a digital image from the skull of the woman. If they can find a match, they can find the real Mona Lisa.

According to Alan Boyle, in his NBC article, " 'Mona Lisa' Skeleton and Her Kins Remains Are Due for DNA Testing," some experts believe her husband commissioned the painting between 1503 and 1506 to celebrate the birth of their second child. After her husband's death, Lisa joined the convent where Italian researchers are hunting through bones for clues.

Boyle says it is a two-part process. First, they are doing carbon 14 tests on three skeletons to determine if they match the time she lived. Then, they are confirming the DNA from the remains of the silk merchant .

However, a bioarcheologist at the University of West Florida says the bones are in poor condition and doubts if anything of value will be found.

If that is the case, will DNA ever uncover the mystery of Mona Lisa's smile?

September 13, 2013

Update: *As of Sept. 24, 2015, DNA is still sought for the real Mona Lisa model, according to a recent CBS news article, "Science Can't Prove Bones in Florence are Possible Mona Lisa Model."*

4 DNA AND SOCIETY

DNA and Music

What do they have to do with each other? Both share the so-called "1/f correlation" structure. Musicians have scored DNA sequences, and you can now hear what the secret code of life sounds like on a Moog synthesizer. A good place to start is the DNA and Protein Music site, which has samples, links, literature and CD suggestions.

April 26, 2005

Don't Ask If You Don't Want to Be Told

Hidden African Ancestors: Hidden secrets of your ancestors
Peter de Knijff
European Journal of Human Genetics 51 (2007) 509-510

The finding of a rare African male line in Yorkshire, England, demonstrates that a Y-chromosome-only reconstruction of geographic origins can be seriously misleading. It also illustrates how a hitherto unknown secret pops up during a rather innocent pedigree reconstruction by means of Y-chromosome testing. As such, it once again shows the importance of a general concept often ignored by participants of pedigree-based Y-testing: If you do not want to know, do not have yourself tested.

August 30, 2007

Cousin Marriages Produce More Children

An Association Between the Kinship and Fertility of Human Couples
Agnar Helgason et al.
Science, vol. 319. no. 5864 (2008) 813 - 816

"Previous studies have reported that related human couples tend to produce more children than unrelated couples but have been unable to determine whether this difference is biological or stems from socioeconomic variables. Our results, drawn from all known couples of the Icelandic population born between 1800 and 1965, show a significant positive association between kinship and fertility, with the greatest reproductive success observed for couples related at the level of third and fourth cousins. Owing to the relative socioeconomic homogeneity of Icelanders, and the observation of highly significant differences in the fertility of couples separated by very fine intervals of kinship, we conclude that this association is likely to have a biological basis."

February 8, 2008

King Tut's Inquest

In 2009-2010, an analysis of 11 royal mummies from around 1300 B.C.E. was carried out by an Egyptian team under the country's chief archeologist, Zahi Hawass. A television special was produced, titled "Unwrapping King Tut." Hawass and his colleagues published "Ancestry and Pathology in King Tutankhamun's Family," in *JAMA*, vol. 303, no. 7. (Feb. 17, 2010).

In a fun review article earlier this year, British journal *New Scientist*'s Jo Marchant summarized much of the resulting controversy in her "Royal rumpus over King Tutankhamun's Ancestry."

We'd be interested in seeing Tut and the other putative family members' DNA fingerprint scores at the bottom of the mystery but are not aware that Hawass and his team actually published the bona fides of their investigations. From a cursory look, it is evident to us that Amenhotep and his descendants, including Akhenaten, Tutankhamun and his unidentified mother (Tiye?), all bore our marker Asian III.

Unsurprisingly, none of the royal mummies seems to have carried a Jewish marker.

It is unclear from the limited data revealed to the world by Hawass whether any had Sub-Saharan African markers.

The new Tut tiff swirls around the question of the pharaohs' African and Western European ethnicity. Without being able to shed light on that, our 18 Marker Ethnic predictor at least suggests they had Asian. Of course, this is not to say they were Asian primarily, since all peoples, ancient and modern, are mixed and may exhibit a variety of ethnic markers in their autosomal DNA.

June 13, 2011

Romanov Remains Verified

Courtney Weaver
New York Times, 07/17/2008, p.12

After extensive DNA testing, Russian officials gave final confirmation that remains found last summer in Yekaterinburg belonged to two children of Czar Nicholas II, his only son, Alexei and his daughter Maria. The remains were located about 70 yards from where the rest of the family's remains were exhumed in 1991. Laboratories in Russia, the United States and Austria conducted the testing.

July 17, 2008

Those Pesky Hunter-Gatherers Again

We have seen there was discontinuity in the genetic record between medieval and contemporary Tuscans. Contradictions keep popping up whenever geneticists seek to show continuity in human populations. The human facts are not obedient to scientific models. The latest example is an article, "Genetic Discontinuity between Local Hunter-Gathers and Central Europe's First Farmers," appearing in *Science* 326/5949:137-40 in October 2009. The authors are B. Bramanti, M. G. Thomas, W. Haak, M. Unterlaender, P. Jores, K. Tambets, I. Antanaitis-Jacobs, M. N. Haidle, R. Jankauskas, C.-J. Kind, F. Lueth, T. Terberger, J. Hiller, S. Matsumura, P. Forster and J. Burger.

Abstract. After the domestication of animals and crops in the Near East some 11,000 years ago, farming had reached much of central Europe by 7,500 years before the present. The extent to which these early European farmers were immigrants or descendants of resident hunter-gatherers who had adopted farming has been widely debated. We compared new mitochondrial DNA (mtDNA) sequences from late European hunter-gatherer skeletons with those from early farmers and from modern Europeans. We find large genetic differences between all three groups that cannot be explained by population continuity alone. Most (82 percent) of the ancient hunter-gatherers share mtDNA types that are relatively rare in central Europeans today. Together, these analyses provide persuasive evidence that the first farmers were not the descendants of local hunter-gatherers but immigrated into central Europe at the onset of the Neolithic.

It is not often that the august personages who collaborate in the field of population genetics admit to surprise, but this study exhibits a few wavering moments of—I don't want to say "humility," but perhaps— slight uncertainty. The main intransigence concerns the large presence of mitochondrial haplogroup U in the skeletons analyzed from Central Europe, 13,400 to 2300 BCE. Whereas these types of U are relatively uncommon in Europe today, they were the dominant population then. Germany and surrounding regions were still very much in the Stone Age. Conversely, the Neolithic types H, T and J, which were supposed to be sweeping across the hinterlands and introducing agriculture from the Middle East (along with a characteristic pottery called Linearbandceramik, German for "linear band ceramics," or LBK) were evidently thin on the ground and held their distance. Nowhere did the twain meet, for "we found no U5 or U4 types in that early farmer sample. Conversely, no N1a or H types were observed in our hunter-gatherer sample, confirming the genetic distinctiveness of these two ancient population samples." In other words, even the entrenched types of populations living as neighbors, U and N1a, were not mixing with each other.

Clinging stubbornly to the "classic model of European ancestry components (contrasting hunter-gatherers with early Neolithic farming pioneers)," the authors explain away the facts in simplistic fashion. With breathtaking generalizations, they assume, and then prove, that the "U types in our hunter-gatherer samples (and their not mixing with the other haplogroups) ... extend beyond the local scale." Do they forget that a study they wrote in the same journal four years ago found a predominance of N1a in skeletons in the same region and time? See Wolfgang Haak, Peter Forster, Barbara Bramanti, Shuichi Matsumura, Guido Brandt, Marc Tänzer, Richard Villems, Colin Renfrew, Detlef Gronenborn, Kurt Werner Alt, and Joachim Burger, "Ancient DNA from the First European Farmers in 7500-Year-Old Neolithic Sites," *Science* 11 November 2005: Vol. 310, no. 5750, pp. 1016—1018.

As T.S. Eliot said, "My end is in my beginning." Thus, the venture ends where it began, with the assumption that there is but one compelling story to be told in Europe for thousands of years after about 6,000 B.C.E., and that is the triumphal march of agriculturalists across the genetic landscape. With false humility, the authors conclude, "The extent to which modern Europeans are descended from incoming farmers, their hunter-gatherer forerunners, or later incoming groups remains unsolved." But circular reasoning is circular reasoning even if it does not beget a strong conclusion.

All such studies presume a starlike and gradual diffusion of people. Hence, they expect to see broad patterns of continuity in time and space. Unfortunately, human history is fraught with disjoint as well as discontinuous phenomena. The ant farm models of population genetics cannot begin to comprehend the complexity of the past or do it justice.

October 1, 2009

Gypsy Migrations

The Gypsies, or Roma, or Romani (so called because of their concentration in Romania) are a far-flung distinctive population with a lot of diversity. In our database, we have samples of four Gypsy populations, plus samples for Romania, Macedonia and Hungary that you can match if you have even a small degree of Gypsy/Romani.

Gypsy DNA can sometimes be conflated or confused with Jewish DNA because both populations originated in the Middle East and often lived in the same Central European areas in modern times, but true Gypsy matches usually come with Indian, especially north Indian matches, because that's where the Gypsies lived around the 900s before they backtracked into Iran and Turkey and eventually crossed the Bosporus into Europe.

The Gypsy language, Romani, shows a strong Romanian influence but its basic vocabulary and grammar point to a north Indian origin.

The Gypsy religion, on the other hand, is not Indian or Hindu but closest to Jewish, Persian and Zoroastrian forms of monotheism.

"It is not known when or why the Gypsies left India but they were living in Iran by the 10th century A.D. The Iranian poet Firdausi (c. 930-1020) wrote of the Gypsies in his epic history of the Iranians, the *Shah Nama* (Book of Kings), that they were originally a tribe of musicians who had been sent to the ruler of Iran by an Indian king. Once they had eaten the ruler out of house and home, the Gypsies took to the roads. By the 11th century Gypsies were living in the Byzantine empire and soon afterwards were spreading through the Balkans. When the Ottoman Turks began to overrun the Balkans in the 14th century, groups of Gypsies dispersed across western Europe, reaching Bohemia in 1399, Bavaria in 1418, Paris in 1421, Rome in 1423 and Spain in 1425. In the early 16th century Gypsies spread to Britain, Scandinavia, Poland and Russia, but the Balkans remained the main Gypsy centre," wrote John Haywood, *The Great Migrations from the Earliest Humans to the Age of Globalization* (London: *Quercus*), p. 142.

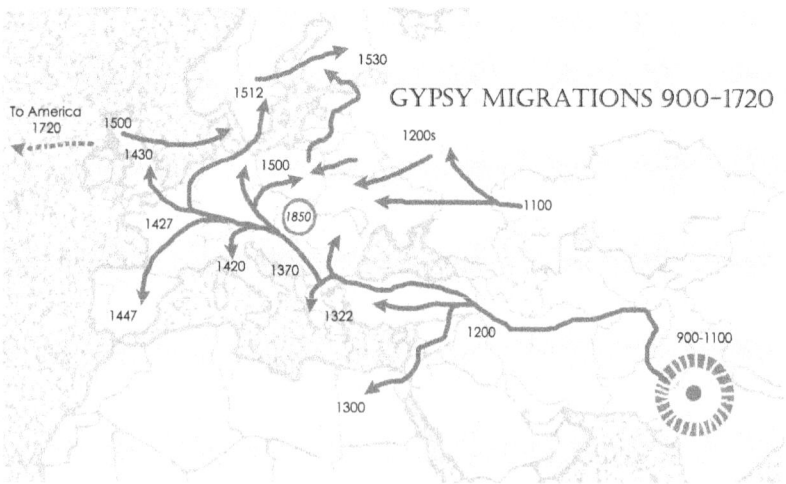

GYPSY MIGRATIONS 900-1720

Shari commented on Oct. 16, 2011: According to my mother's Fingerprint Plus DNA test, both of her parents had Jewish I and Jewish III DNA. One parent had Tatar/Khazar DNA (Jewish IV). India was Mom's Top World Match. Mom's mother was genetically Roma-Gypsy. To date there is no genealogical evidence that Mom's father was either Roma-Gypsy or Jewish. I'm wondering if the combination of Jewish I and Jewish III along with Indian (from India) ancestry is the typical DNA pattern found for persons of Gypsy-Roma ancestry. Perhaps Jewish I and III could also indicate only Jewish ancestry, a possibility for Mom's father's ancestry. Another possibility would be that her father had unconfirmed Gypsy-Roma ancestry. One or the other parent having Jewish IV DNA may provide a clue. I enjoyed reading "Gypsy Migrations."

I've also found the following Internet article to be interesting. Dr. Hancock suggests that Romani had "military" beginnings on the basis of his linguistic and historical research: "An examination of the earliest words in the Romani language suggests a number of things: firstly that there is little in the original, 'first layer' Indian vocabulary that reflects a nomadic or itinerant population, but rather it points to a settled one; and secondly that while there are not many original words for e.g. artisan or agricultural skills, there are quite a few military terms ... " From: "On Romani Origins and Identity," Ian Hancock The Romani Archives and Documentation Center, The University of Texas at Austin, http://www.radoc.net/.

Donald Locke commented on Oct. 18, 2011: "Gypsy DNA can sometimes be conflated or confused with Jewish DNA because both populations originated in the Middle East." I would disagree with this opinion that the Romany originated in the Middle East when we clearly originated in South Asia. India, Sri Lanka, Nepal, parts of Pakistan. I am of the English Romanichal vista "clan" and the Romanichal vista Y DNA results clearly show a high average of our male population carrying Y Haplo Group H1a.

More importantly, I am the researcher who discovered the relationship between marker 425 = 0, null to the Romany H1a male lineages. To date, of all the Romany H1a male lineages identified so far, of all those tested to the 67 marker level, 100 percent were found carrying this same null value marker mutation in common regardless our surnames, and regardless which Romany vista "clan" we hail from. Romany of England, Scotland, Hungary, Bulgaria have found Y Haplo H1a with the 425 = 0 marker mutation, which clearly links the Romanichal vista to the Roma vista's of Europe. mt Haplo Group M5a1, which is also being claimed as South Asian in origin, has also recently been discovered amongst the English Romanichal. I am the administrator of the Y Haplo Group H and Romany DNA projects with FTDNA. To date not a single Asian Y Haplo H1a male has been found carrying the 425 = 0 marker mutation. This mutation, so far, is only found among the European Romany male population. And as far as I am concerned, H1a with the 425 = 0 marker mutation = Romany origins. Donald Locke

Steven commented on May 11, 2012: My name is Steven, and I have found out that my real father was Roma/Gypsy. My mom was Jewish from Morocco. There are a group of people in eastern Turkey called Kurds, and the name Sindh is a common surname with them. I believe they traveled to India, backtracked to Turkey and then went to Germany/Austria. This group became the Sinti Rom of the Rhinelands. That, however, is the Sinti; the other Rom I'm not sure.

Theo commented on July 31, 2013: Hello. While your article is interesting and should be accurate from a scientific point of view, I would like to make some amendments to your cultural references. Back home, gypsies are called Romi, or Rom ethnics, and that distinction makes no linguistic sense in Romanian. This leads me to believe they inherited the name from an older distinction. As a native Romanian, to me the gypsy language makes absolutely no sense. I can't understand a thing until they actually switch to a different language.

Mel commented on May 17, 2014: A few corrections ... The word "Romani" has nothing to do with Romania, as stated in the article above. The word is the feminine adjective form of "Rom," which means man in the Romani language. Our religion tends to be Christian, not "Jewish, Persian and Zoroastrian forms of monotheism."

October 5, 2011

The Sea-Changes of American Genealogy

"For most Americans, the story of their nation's origins seems safe, reliable and comforting."

So begins Elizabeth Caldwell Hirschman's Preface to her new collaboration with Donald N. Yates. Their book *Jews and Muslims in British Colonial America* appeared in February 2012 from McFarland & Co.

Subtitled "A Genealogical History," it presents an iconoclastic new narrative of American history beginning with an examination of the shipping companies organized in England to bring settlers to North America and concluding with a chapter on Georgia, the last "and most Jewish" colony.

Perhaps nothing better epitomizes the sea-change that many genealogies underwent in the American experience than the case of Sir Francis Drake. The original family name was Sephardic Jewish, and the family crest featured six-pointed stars. As the story of America evolved, these were converted to more innocuous, less-revealing eight-pointed stars in engraved book plates and other illustrations.

As well as firing a definite shot across the bow to traditional American genealogies, the study also gives us some delicious smoking guns. In the Massachusetts chapter, for instance, the genealogy of many Pilgrims and Puritans is revealed to contain Jewish or Muslim ancestry. In the Virginia chapter, it is pointed out that Patrick Henry's mother was Jewish, and she was married to a Muslim merchant from Aberdeen, Scotland.

Over 5,000 names are evaluated, and capsule etymologies given for them.

There are illustrations, references, notes and an extensive index. *Jews and Muslims in British Colonial America* is the second in a series of historical studies that began with the same authors' *When Scotland Was Jewish*, published in (2007). The series will conclude with *The Early Jews and Muslims of England and Wales*.

Jews and Muslims can be ordered for $45 from McFarland and Co., Amazon, DNA Consultants Online Bookstore or any bookseller. It is also appearing in an electronic edition that will be available soon.

April 4, 2012

Melungeon Tumult

It used to be that Melungeons (no, I won't give you a link to this hot potato) were seen as a quaint historical ethnic group in the Appalachians. They were followed with a certain quirky interest by a handful of academic researchers and genteel listers on the Internet.

Then the *Journal of Genetic Genetic Genealogy* of the International Society for Genetic Genealogy published an article titled "Melungeons, A Multiethnic Population," by Roberta J. Estes, Jack H. Goins, Penny Ferguson and Janet Lewis Crain. The article got picked up by Fox News and The Associated Press and was trumpeted on one of the genealogy lists like this:

Extra ! Extra ! Read all about it........
http://tinyurl.com/cdb8ama
http://www.foxnews.com/us/2012/05/24/dna-study-seeks-origin-appalachia-melungeons/
http://www.npr.org/templates/story/story.php?storyId=153634685
http://tinyurl.com/c5ebfuv

What ensued was like a box of flying squirrels being dumped in the living room of a trailer during a tornado. Here are some random responses from Melungeon researchers (not to call them Melungeons, per se). We don't have a dog in this fight (or squirrel), but couldn't resist reporting.

In my opinion this "peer reviewed" report contains more *misinformation* than most of the books and magazines that have been published in the last 20 years.

This doesn't hold any more water than Brent Kennedy's theory.

Just not possible to to make an R1a or R1b baby out of an E-3 man and a white woman.

These little oversights are the things that can fuel class action lawsuits it would seem to me.

They wouldn't have wrote [sic] what they did in the article if the paper hadn't been published. I have been saying for years Jack was padding that project with the E haplogroups for his own purposes and this proves it.

And again your "'disagreemen"' is with the "media" in this case the AP and all the Newspaper's that published the article, which there appears to be different version.

Since article was based on the "Peer Reviewed" article by JOGG can you tell us how many of the peers were certified genealogist?

Well, I monitor this list, and I ... have not seen any charactistics [sic] among any of the Melungeons on the ridge, to indicate that they were mixed with Africans. No kinky hair, thick lips, large hips, dark eyes, etc. Mine had long straight hair, slim features, long slim fingers, and a bright turquoise shade of eye coloring. I don't see this at all. Look at the pictures of the Bell family that were published. I also did not see any mention of the haplogroup in the articles. How are we, the public going to see the breakdown of who is who?

My opinion is, I'm not so sure why they have such a personal vendetta to prove the melungeons weren't who they said they were. Maybe it will be solved once and for all now.

They did not bother to mention that they used dna that I had helped to collect for this group or they would never have had any Gibson data. Also they were specifically told on more than one occasion not to use data that I had recruited. Also, secretly, they cut me out of the sharing on this group.
Also, at least one of this group vowed never to agree to have their name attached to it. I guess it isn't any trouble not to keep your word. I think it is past time people stopped trying to make the melungeons to be something they were not and regurgitating old data. When I learned that Roberta and these were doing this article she quickly removed it from her site and represents herself on the site as a "scientist".

You can be sure this is not the last of this. They should be expecting it. I can't believe so many families have been assigned ancestors without any evidence at all. I would call it capricious but it's more akin to arrogance.

You write about Jordan Gibson, yet you omitted the one DNA source from his line. Why? Was it because his haplotype was I1 and not African? Between this article and Wayne Winkler's recent radio interview, you have joined the ranks of Plecker (racist public health official Walter Plecker).

Shame on you.

The genealogies are strung together with no documentation. For instance they write the sons of Gilbert Gibson; Jordan, Gideon and George went to Sandy Bluffs when the records show Gilbert's sons on tax list and court records in Louisa County. The Gideon and Jordan Gibson at Sandy Bluffs were clearly different people, born at different times.

I know of no documentation showing the Collins family from the Pee Dee River related to Vardy Collins or the Collins families on Newmans Ridge. There are also no records that I know of that show Jacob Wolf and James Mooney of Baxter County, Arkansas, were ever in Hawkins/Hancock County. These are just a few of the inaccuracies and misinformation from this report, I am getting together the documents to back this up and will be posting it shortly.

I would like to see a serious discussion of this report with questions answered rather than a back and forth on who should be allowed to do research.

Update: The squirrelly article was apparently the kiss of death for *JOGG*, whose webpage lists no articles published after that issue.

May 25, 2012

Elizabeth Hirschman, Modern Pioneer

We interviewed Rutgers University marketing professor Elizabeth Caldwell Hirschman, author of several books and articles incorporating DNA in her research, to hear her personal story in our continuing series about the people behind the scenes in the field of DNA testing.

When did you first get interested in DNA?

ECH: I got interested in DNA testing around 2000 when I discovered I was Melungeon after reading Brent Kennedy's 1994 book. Brent suggested several different ancestries that possibly contributed to the Melungeon population, and I wanted to find out which of these were correct and which ones I had. I already suspected Jewish ancestry because of the naming patterns in my family over the past 300 years, as well as some of their habits -- e.g., not eating pork, getting married in a home instead of a church, cleaning house on Friday afternoon, no eggs with blood spots, washing all meat, etc. We also had some genetic anomalies -- shovel teeth (sinodonty), palatal tori and large rear cranial extensions, as well as polydactylism.

Tell us more.

ECH: Over the course of the past decade I have been found to have Native American, Spanish, Ashkenazi Jewish, African, Mediterranean and Gypsy/Northwestern India ancestry. My Dad turned out to have substantial Gypsy and African ancestry. He and I share a large cranial rear extension that I believe likely comes from the African ancestry—the photos I have seen of the !Kung Bushmen look just like our head shapes. My Mom has Native American and/or Sino-Siberian ancestry. She also possessed the Asian teeth and palatal tori found in this group.

You've written several books and articles with Donald Yates; how did that come about?

ECH: We shared ancestry from the Coopers, a prominent pioneer family in Daniel Boone's time. In 2000, I wrote him out of the blue when he was a professor in Georgia and introduced myself and asked if possibly the Coopers were Jewish. We began to correspond by email. I told him I was sure one of the reasons I was working so hard to figure out the Melungeon story was because I had to figure out who I am. "Up until last year," I remember telling him, "I thought I was Scotch-Irish, English, white and Presbyterian." It was a big transition to Sephardic, brown and Jewish. It turned out that we were distant cousins and had numerous links in our Melungeon ancestry.

What was a typical publication?
ECH: One article was called "Suddenly Melungeon! Reconstructing Consumer Identity Across the Color Line." This was published by Routledge in 2007 in a handbook on consumer culture theory edited by Russell Belk.

How did the Jewish findings play out?
ECH: On a personal level, both Don and I, as well as his wife, Teresa, returned to Judaism, he and Teresa in Savannah and I in New Jersey. On a professional level, we started the Melungeon Surname DNA Project, which focused on Scottish clan and Melungeon surnames (i.e., male or Y chromosome lines), and we later included Native American mitochondrial DNA. Initially, many people in the genetic genealogy community were frustrated that the incoming Jewish DNA results were not originating in the Middle East, as they had strongly believed and hoped they would, but were showing a lot of Khazar, Central Asian, Eastern European and Western European/Spanish/French input.

Can you elaborate?
ECH: Critics were not happy that DNA was proving a wider and more inclusive picture of the Jewish people. Where Don and I have performed a service, I believe, is by just following the DNA trail and accepting new findings (e.g., the Gypsy/Roma) when they come in, instead of clinging to an a priori theory/belief/wish, for instance, the claim of a Middle Eastern origin for the majority of Jews.

What tests have you ordered from DNA Consultants?
ECH: I ordered every test as they became available over the years, first the Y chromosome and mitochondrial or male-line and female-line tests and later the autosomal or DNA fingerprint tests that analyze your total ancestry. I helped organize the first autosomal Melungeon study by

contributing samples from my mother and brother and obtaining samples from well-known Melungeons like Brent Kennedy and his brother Richard. Increasingly, our testing took on the aspect of a family group study. For instance, I was able by comparing multiple results from relatives to reconstruct my father's ancestry quite satisfactorily, even though he died many years ago. I took the Rare Genes from History for all available family members. There is a streak of the Thuya Gene and First Peoples Gene in all of us, as well as the Sinti Gene (which is Gypsy), while my brother Dick got our father's Khoisan Gene, which is African. Incidentally, it has the same source as the !Kung people and head shape I mentioned before.

If you had H.G. Wells' time machine where would you go?
ECH: I would love to be able to visit my ancestors and see what they looked like, where they lived, how they lived and learn how they got to Appalachia from such disparate parts of the world. I wish I could talk with them. My project now is to visit all the places they are known to have come from and see what the architecture, climate, food, and people are like. That is about as close to "meeting" them as I will be able to get. So far, I've traveled to Scotland, Ireland, Wales, England, Spain, Tunisia and Morocco on the trail of my Sephardic Jewish ancestors. I am trying to get to the Silk Road to see Central Asia, Turkey and Northwest India in the near future.

Professor Hirschman has published over 200 journal articles and academic papers in marketing, consumer behavior, sociology, psychology and semiotics. She has been named one of the Most Cited Researchers in Economics and Business.

December 7, 2012

King Tut's Family Reunion

Some moments really are golden. I felt as if I had stepped off Dr. Who's Tardis spaceship into the 18th dynasty of Egypt and into King Tut's golden tomb in the Valley of the Kings when I saw the famous, golden mask at "The Treasures of King Tutankhamen" at the British Museum in London. But that was years ago.

The frail and golden teenage boy, King Tut, who owned "more than 100 walking sticks" and carried about him "medicinal (herbs)" has since been reunited with his family according to Katie Moisse, in her *Scientific American* article, "Tutankhamen's Familial DNA Tells Tale of Boy Pharaoh's Disease And Incest."

Of course, that reunion wasn't like the day Zahi Hawass describes in his *National Geographic* article, "King Tut's Family Secrets," when the young king and his sister/wife, "Ankhesenpaaten (a daughter of Akhenaten and Nefertiti) return to Thebes reopening up the temples (to Amun)... restoring their wealth and glory." Despite having to use a walking cane, the young, royal pair must have viewed that day as glorious if it did mean a return to paganism and the worship of Amun for Egypt. No, the more recent family reunion was in a basement of all places. According to Moisse, that basement is where the modern Indiana Jones of ancient secrets, Secretary General Zahi Hawass of Egypt's Supreme Council of Antiquities teamed up with paleogeneticist Carsten Pusch from the University of Tubingen in Germany. Their quest was to to map out King Tut's family tree by DNA testing.

This was conducted in two labs. One in the basement of the Museum of Egyptian Antiquities in Cairo, and the other at Cairo University ("King Tut's Family Secrets"). By then, King Tut along with his golden mask and his royal family had flown back to Egypt. Albeit in parts.

What discoveries did Hawass make of the royal bones according to Moisse? Extracting tiny amounts from the bones and using DNA analysis commonly used in "criminal or paternity investigations" they found his family.

They identified "Tut's grandmother, 'Tiye,'" and his parents. His father was "probably named 'Akhenaten'" which Hawass says "... is the most important discovery since the finding of the tomb of Tutankhamun in 1922." However, they determined that his mother was not, as originally thought, Nefertiti but "one of the daughters of Amenhotep III and Tiye," says Hawass (qtd. by Moisse). They have determined "with a probability of better than 99.99 percent" that Amenhotep III was the father of Akhenaten and, therefore, King Tut's grandfather). Hawass says the "story of Tutankhamun is like a play whose ending is still being written." ("King Tut's Family Secrets").

January 10. 2013

Richard III's New Winter of Discontent

Shakespeare painted the last of the York rulers of England as a murderous maniac who was rightly dispatched to hell by Henry Tudor in 1485. But the story of Richard III's skeleton supposedly dug up in 2012 in a parking lot may top that of the Bard for pulling the wool over our eyes. Or it may be the luckiest archeological find since King Tut.

The last of the York dynasty was buried in Greyfriars, Leicester, but Britons are now talking about re-interring the bones believed to be Richard's in Westminster Abbey with England's other beloved monarchs. In 2012, a writer from Edinburgh, Philippa Langley, was walking over a particular spot in the municipal parking lot when she got goosebumps and "absolutely knew I was walking on his grave." Langley helped fund an archeological excavation and on Feb. 4, 2013, the University of Leicester confirmed that a skeleton found in the excavation was, "beyond reasonable doubt," that of Richard III, based on a combination of evidence from radiocarbon dating, comparison with contemporary reports of his slight frame, and a comparison of his mitochondrial DNA with two matrilineal descendants of Richard III's eldest sister, Anne of York.

Hunches and Hunchbacks

According to a BBC article, "Richard III: The Twisted Bones that Reveal a King," the skeleton had a "striking curvature" that could only be that of the hunchback king. But according to a Daily Beast article, "Unraveling King Richard III's Secrets," Shakespeare wrote a century after the fact and had a pro-Tudor, anti-York political agenda. "Portraits made after his defeat that show Richard with a hump—or at least uneven shoulders—are suspect as Tudor propaganda." There is no historical evidence of Richard III having a "striking curvature" of the spine. Or even "uneven shoulders." There is no evidence beyond Shakespeare of his deformity. In fact, there is historical evidence to the contrary. The article, "Richard's Comeback," quotes the historian, Thomas More, as saying Richard III was of "bodily shape comely enough, onely (sic) of low stature." The Countess of Desmond reported that, at a royal ball, Richard was the "handsomest man in the room . . . except for his brother,

Edward, and was very well made."

Despite historical evidence, most articles that discuss remaining doubts about the case like, "Doubts Remain that the Leicester body is Richard III," miss this point and take it as a historical fact that Richard III had scoliosis as does the skeleton that has been found in the parking lot.

What of historical depictions of Richard III's face? "No portraits made during his lifetime have survived and some later copies show signs of having been altered to make him appear more sinister" ("Richard III: The Twisted Bones"). Nevertheless, a 3D scan of the skull was taken, and a 3D face created and painted. John Ashdown–Hill is quoted as telling the BBC in the article, "Richard III Facial Reconstruction Reveals Slain King More than 500 Years After His Death," that it "largely matched" the "prominent features" in posthumous representations of the king. The artist, Janine Aitken said her part was "interpretive not scientific." At least it is a pleasant face. But is it Richard III's face?

Jumping to Forensic Conclusions

And the skeleton includes 10 battle wounds showing Richard III "met a violent death …"—eight to the head and two to the body—which they believe were inflicted at or around the time of death. Since he died in a battle, did not other soldiers meet untimely wounds in such a manner?

Not a few scientists are waiting for peer-reviewed results. But there is none. Instead of waiting for a boring academic conference, the sponsors at the Richard III Society chose to release the results via a Hollywood-style press conference.

What kind of DNA analysis was used? Mitochondrial DNA. According to Bryony Jones in his CNN article, "Body Found under Parking Lot is King Richard III, Scientists Prove," "The mitochondrial DNA extracted from the bones was matched to Michael Ibsen, a Canadian cabinetmaker and direct descendant of Richard III's sister, Anne of York, and a second distant relative who wishes to remain anonymous." Well, end of story and close that book. Right? Not so fast. Some scientists believe that the testing done was not sufficient. Why?

Mitochondrial DNA has limitations. It does reflect the deepest ancestry (see *The Seven Daughters of Eve* by Bryan Sykes), but is also prone to contamination (under such circumstances—LiveScience writer Stephanie Pappas). Especially when we are discussing skeletons reminiscent of "Night of the Living Dead" interred improperly for centuries in damp soil. Timothy Bestor, professor of Genetics and Development at Columbia University Medical Center, is quoted in the New York Academy of Sciences article, "Skeletal Remains of King Richard III Reportedly Discovered," as saying that the possible quality of the mitochondrial DNA "under the given circumstances" was one of his key reasons for skepticism. After 500 years or more in a wet environment like England's, "the microbes are going to degrade the DNA. It's just food to them," says Bestor. And Pappas quotes Maria Avila, a computational biologist at the Center for GeoGenetics at the Natural History Museum as saying, "The DNA results presented today are too weak, as they stand, to support the claim that (the) DNA (sample) is actually from Richard III … more in depth DNA analysis summed to the archaeological and osteological (bone analysis) results would make a round story (She is requesting Autosomal DNA analysis akin to what was done with the hominin discovery of the Denisovans)." And she wonders about contamination with the type of DNA testing that was done. Avila says that, "before being convinced of ANY atDNA study, it should be explicit that all possible cautions were taken to avoid contamination" and … " also warned that people could share mitochondrial DNA even if they share a family tree (Pappas)." In the article, "Doubts Remain that the Leicester Body is Richard III," Mark Thomas at University College London is quoted as saying that "people can have matching mitochondrial DNA by chance and not be related."

And Bestor writes there are other reasons to be skeptical, even though "Richard Buckley, lead archeologist from the University of Leicester, asserts "this is beyond reasonable doubt' based on genetic and historical forensic evidence." Bestor argues that beyond the high risk of sample contamination, there are three other "particularly complicating factors." Of course, it is often an overlooked fact that "the English aristocracy reproduced within a closed gene pool in order to preserve lineages. This inbreeding results in consanguinity" ("Skeletal Remains of King Richard III"). Bestor is quoted as saying, "You may have the same mitochondrial haplotype, but that does not guarantee a lineal descent from a given individual." (Mitochondrial DNA analysis is not the same as Y haplotype DNA analysis because it focuses on deeper ancestry whereas male haplotype DNA analysis is linked to more recent male lines.) He also points out the possibility of adoption. (The chance of an adoption or any type of nonpaternity event increases as one goes back into the distant past of any family tree).

Another confounding factor is that, in the 17-25 generations separating King Richard's sister from her extant relatives, there is a fair chance that children of deceased parents' may have been adopted by their parents' siblings somewhere along the way. After all, medieval lives were short. Such adoptions may have been kept private and excluded from historical or genealogical records.

Moreover, Bestor points out that the "genetic sequences and statistical data are yet to be released" but adds that the "historical evidence is quite compelling." According to this article, forensic evidence of the bones (1455-1540) matches with the time that Richard III was to have died (1485). But didn't many people die at this same time during the same battle with similar wounds?

February 19, 2013

The Surprising Story of India's Caste System

DNA testing reveals how truly mixed we are. Most of the world's populations are much more diverse than most people realize. Just get an autosomal DNA test and look at your results. People have been traveling and intermarrying for thousands of years. However, most would assume that the different caste systems in India would have unmixed and dissimilar genetic backgrounds since they have been strictly segregated for thousands of years.

Nevertheless, that is not the case (Tia Ghose, "Genetic Study Reveals Origin of India's Caste System" in Live Science). How is that possible?

Despite a near 2,000-year strict ban on intermarrying between groups, everyone in India, even "isolated tribes," have DNA from the same regions says Priya Moorjani, a geneticist at Harvard University, and the only thing that differs across India "is the proportion of ancestry" from a particular group (Ghose).

What that means is that even the lowest caste, the Untouchables, are related to the highest, priestly caste.

What is interesting is that their most ancient stories say, except for the Untouchables, they are all connected. In an ancient text known as the Rigveda, the division of Indian society was based on Brahma's divine manifestation of four groups. Brahma is a monstrous-looking creator god with four hands and four heads which was said to give birth to all Hindus. He cast priests and teachers "from his mouth, rulers and warriors from his arms, merchants and traders from his thighs, and workers and peasants from his feet" (ushistory.org, "The Caste System," "Ancient Civilizations Online Textbook"). It almost seems as if they knew long ago of their genetic connections. As it turns out, even the Untouchables are their brothers. We could learn a lot from this discovery.

May 20, 2014

5 DNA AND DAD

Does DNA Produce More Questions Than Answers?

Genealogy doesn't often make news headlines, but in recent months, it has been a debated topic in newspapers worldwide. Why? Genetics and genealogy have merged into an amazing new research field which allows researchers to prove family connections beyond all doubt. The journal *Nature* recently reported that DNA testing was used to support the assertion that Thomas Jefferson fathered the last child of his slave Sally Hemings.

October 15, 2004

Surname Studies and Y-Chromosome Matches

Y-chromosomal STR Haplotype Analysis Reveals Surname-associated Strata in the East-German Population
Uta-Dorothee Immel et al.
European Journal of Human Genetics 14 (2006) 577 582

Abstract. In human populations, the correct historical interpretation of a genetic structure is often hampered by an almost inherent inability to differentiate between ancient and more recent influences upon extant gene pools. One method to trace recent population movements is the analysis of surnames, which, at least in Central Europe, can be thought of as traits "linked" to the Y chromosome. Illegitimacy, extramarital birth and changes of surnames may have substantially obscured this linkage. In order to assess the actual extent of correlation between surnames and

Y-chromosomal haplotypes in Central Europe, we typed Y-chromosomal short tandem repeat markers in 419 German males from Halle. These individuals were subdivided into three groups according to the origin of their respective surname, namely German (G), Slavic (S) or "Mixed" (M). The distribution of the haplotypes was compared by Analysis of Molecular Variance. While the M group was indistinguishable from group G (PhiST=-0.0008, P>0.5), a highly significant difference (PhiST=0.0277, P<0.001) was observed between the S group and the combined G+M group. This surprisingly strong differentiation is comparable to that of European populations of much larger geographic and linguistic difference. In view of the major migration from Slavic countries into Germany in the 19th century, it appears likely that the observed concurrence of Slavic surnames and Y chromosomes is of a recent rather than an early origin. Our results suggest that surnames may provide a simple means to stratify and, thereby, to render more efficient, Y-chromosomal analyses of Central Europeans that target more ancient events.

In one of the first large-scale studies of its kind, a team of German geneticists concludes that the structure of populations as revealed in Y-STR haplotype clusters corresponds to surname groupings by country of origin. Both means of analyzing a population point to a relatively recent historical time frame—as has long been assumed by DNA Consultants and other testing companies.

January 26, 2006

Who's Your Granddaddy?

Paternal Grandfather Confirmed as Main Influence on Family Nutrition and Longevity

Gunnar Kaati et al.,"Transgenerational Response to Nutrition, Early Life Circumstances and Longevity," *European Journal of Human Genetics* 15:784-90.

A study of 271 individuals with 1,626 grandparents finds that there is a sex-specific influence from grandparents on pre-pubescents' nutrition, which, in turn, determines longevity. The influence for boys comes from their paternal grandfather and for girls from their paternal grandmother during the formative years between 8 and 12 years old, so the father's parents set the tone for childhood social circumstances that lead to poor, average or good nutrition.

The study confirms earlier findings, which ultimately mean that your father's ancestral Y chromosome line (with the female marriage partners attracted to it) is a major influence on your health and longevity. So who were *your* paternal grandparents?

April 25, 2007

A Future without Emperors

Y Chromosome DNA Fuels Dynastic Dilemma
David Cyranoski
Nature 439/7079:892-3

The Japanese imperial succession is in trouble because of the Y chromosome. Will Japan have a queen?

February 22, 2006

Not As Old as You Think
Roots of Human Family Tree Shallower Than Previously Believed

Matt Crenson
Santa Fe New Mexican/Associated Press, July 1, 2006

Brooke Shields is probably a descendant of Lucrezia Borgia. Humphrey Bogart is probably a descendant of the prophet Muhammad. And so it goes. Geneticists used to believe that "everybody on earth descends from somebody who was around as recently as the reign of Tutankhamen," according to an AP article titled "The Brotherhood of Man."

But revised thinking on their part puts the most remote common ancestor (MRCA) in a much more modern time frame, so that nearly everyone alive can claim Charlemagne and Ghengis Khan as an ancestor, if statistics and random mating hold true.

The reason is that many, many lines die out. Those that survive have huge numbers of descendants within a few generations, and all of us have huge numbers of ancestors going back to a small genetic pool about a thousand years ago. So, a 13-marker Y-STR match such as those reported in Male DNA Ancestry products from DNA Consultants relates to a relatively recent time frame, unlike previous theories that a match went back many thousands of years to a common male ancestor.

July 1, 2006

Overestimating DNA

Having always been suspicious of the low rate of mutation in Y-STRs used to establish male lineages, blogger Dienekes Pontikos has posted a major critical review of the subject titled "How Y-STR Variance Accumulates: A Comment on Zhivotovsky, Underhill and Feldman (2006)."

His critique suggests, among other things, that the age of male haplotypes has been overestimated by 90 percent, and that the speculative rate of mutation favored by most population geneticists is three to four times the *observed* rate in germline studies of fathers and sons. Furthermore, most estimates neglect rapid expansions of the population and the short-term success of certain male castes.

Read the article on Dienekes' Anthropology Blog.

July 29, 2006

Basque DNA in Idaho

Geneticists seized the opportunity provided by an international Basque cultural event held in Idaho in 2010 to sample volunteers and study Basque DNA. The result was two studies, including "The Y-STR Genetic Diversity of an Idaho Basque Population," published in *Human Biology*.

It was the first DNA study to document the spread of the Basque male chromosome overseas. The Basque people were renowned seafarers.

"The idea is to better understand health risks for Basque people, including an increased incidence of both Alzheimer's and Parkinson's diseases," said Josu Zubizarreta, a Boise State graduate who conducted research with the lead author, Greg Hampikian.

Mitochondrial DNA, which reflects a deeper history, was also studied.

Basques are credited with the invention of the rudder. They provided the crew and navigators for Magellan. Basque names are common on antique maps. The Bay of Biscayne is named for them, and many harbors, points and landfalls on the Atlantic Coast of North America are thought to come from the Basque language, which is known as an isolate and is unrelated to other European languages.

Citation: Zubizarreta, Josu; Davis, Michael C.; and Hampikian, Greg (2011) "The Y-STR Genetic Diversity of an Idaho Basque Population, with Comparison to European Basques and U.S. Caucasians," *Human Biology*, vol. 83, issue 6, article 2. Available at: http://digitalcommons. wayne.edu/humbiol/vol83/iss6/2.

December 10, 2011

Paternity Tests Available in Drugstores

Catherine Arnst
Business Week, 4/14/2008 Issue 4079, p.15

The drugstore chain Rite Aid began selling paternity kits in the U.S. on March 25, 2008. The kits, which are made by company Identigene, allow users to take DNA samples from children and parents, which are then sent to a facility for analysis. The article reports that the tests cannot be used in court, but that they are at least 99 percent accurate.

April 18, 2008

Right Pew, Wrong Church
Do You Have the DNA of Roman-British-Thracian Soldiers in Your Male Line? Probably Not

A member of the International Society of Genetic Genealogy (ISOGG) wrote an article online five years ago, and now, a substantial number of listers on the discussion board DNA-Genealogy-L believe their male lines may go back to a Balkan legionnaire in Roman Britain. This theory has been accepted uncritically thanks to ISOGG members, who contribute most of the material on Y chromosome DNA to Wikipedia articles.

Read our review of this theory from an appendix on Jewish DNA hot spots in England and Wales in our book *The Early Jews and Muslims of England and Wales.*

Steven Bird in "Haplogroup E3b1a2 as a Possible Indicator of Settlement in Roman Britain by Soldiers of Balkan Origin," is, as the title makes clear, most interested in proving a Roman Balkan origin for the haplotype he investigates, now known as Elblbla, the most common type of the haplogroup Elblb (formerly denominated E3b) in Europe. The structure and subclades of this very ancient North African Caucasian lineage have only recently been resolved and overhauled, and the ink is not quite dry. But the data used by Bird with the sometimes confused or outdated nomenclature of older reports can still provide valuable clues for our purposes, although one must proceed with caution in making too many differentiations in the tangled branches of the E tree. We must bear in mind that the target haplotype Elblbla2 (also called E-V13) represents 85 percent of the parent haplogroup Elblb (also denoted as the E-M78 clade) and keep simple E before us without being distracted.

Bird's study appeared in one of the first publications of the *Journal of Genetic Genealogy*, an online journal of the International Society of Genetic Genealogy (ISOGG), founded in 2005 by DNA project administrators of the commercial DNA testing company Family Tree DNA, based in Houston, Texas, "who share the common vision of the promotion and education of genetic genealogy." It is an ambitious work with a very small goal. It uses arguments not only from genetics and statistics but also archeology, geography, history, anthropology and linguistics, often involving such fine points as the epigraphy of a Spanish soldier's diploma from the British Museum issued in 103 C.E. and the detailed movements of Thracian cohorts II and VII in the Roman army. Where angels fear to thread. Bird's theory about the origins of Elblb in England have been enshrined in popular belief. We do not wish to appear ungrateful, but there are problems.

Bird's first mistake occurs in his review of the literature. He misreads Stephen Oppenheimer and represents the author of *The Origins of the*

British as having British E "originating from the Balkan peninsula (26)." If we open Oppenheimer's book to the page cited (207) we see a map illustrating "Near Eastern (British English for American English 'Middle Eastern') Neolithic male migrations via the Mediterranean of E3b (i.e. E1b1b) and J." The vector standing for the migration of these types launches forth from the Peloponnese in Greece at the cropped lower right corner, obviously intending to suggest origins from that general direction, not "the Balkan peninsula." There is no mention of Balkan DNA in Oppenheimer except as part of the bigger picture. The archeological sites Bird adduces as evidence for E settlements in the Bronze Age are not necessarily associated "directly" or solely or chiefly with "proto-Thracian culture," whatever that term may mean. Nova Zagora in Bulgaria is a Stone Age multisite. Ezero Culture occupied most of Bulgaria and extended far north into the Danube region of Romania. Yunatsite, Dubene-Sarovka, and the other "proto-Thracian culture" examples Bird mentions date to before the Thracians or even the Greeks. They cannot tell us anything about haplogroup E. If anything, all these sites vindicate Oppenheimer's theory of the demic spread of Middle Eastern (read Anatolian) agriculture, which Bird calls "flawed fundamentally" (27). The center for the diffusion of E in the Balkans is not in Bulgaria or Thrace but northwestern Greece, Albania, and Kosovo. The Balkan Peninsula does not have to be the only place from which Bird can manage to derive E and get it to Britain in time to become part of the historical record. It is also strong throughout Greece, Cyprus, the Greek parts of southern Italy, North Africa, and even parts of Spain. In fact, its presence in many of those locations is acknowledged to be "due to a founder effect, i.e. the migration of a small group of settlers carrying mostly this lineage (but also a small amount of other North-East African lineages, notably E-M123 and T." (See http://www.eupedia.com/europe/Haplogroup_E1b1b_Y-DNA.shtml.)

Despite these failings relating to statement of thesis and validity of arguments, Bird's work is based on useful data. Three population surveys with frequencies for E in Britain were available to him, the data sets of Capelli, Weale, and Sykes. Notwithstanding the nomenclature confusion, only the Sykes data set has true shortcomings, as the Oxford Genetic Atlas Project at the time contained only 40 E haplotypes, too small for a valid sample. There are problems comparing them, as Bird realizes, but trends and general conclusions are certainly possible. Before attempting to analyze the haplogroup E variation in Britain, though, we must address the matter of time depth.

We have no quarrel with geneticists' and genetic genealogists' methods of gauging coalescence times. Thus, Bird reiterates that the "time to most recent common ancestor" or TMRCA of Cruciani and others led to the "important finding . . . that E-V13 (read 85 percent of E) and J-M12 (read J) had essentially identical population coalescence times (27)." E and J are companion types that expanded from their Middle Eastern homelands together in the same fashion and probably reinforced each other in multiple phases of gene flow. But who is to say in any specific case of a haplotype that it arrived in Britain 4,000 years ago (TMRCA) or at any subsequent time, including the time when our grandfathers lived. The TMRCA sets a haplotype's time of origin but not its place of origin, except by inference. We hypothesize that from a host of other factors, chiefly present-day clusters, genetic distance between types and high concentration of haplotype diversity. Using TMRCA, Bird argues that a specific form of E "could not have arrived in Britain during the Neolithic era (6.5-5.5 kya) if it had not yet expanded from the southern Balkans (27)." We prefer to believe that it came to the British Isles at several critical times, first in Neolithic times but later with the Phoenicians, Jews, Egyptians, Iberians and related peoples.

Bird cherrypicks the data to support his Roman Balkan or what might be called Diocletian thesis, but data are data; these are amenable not only to bearing out the general storyline we present but also to supporting, within the same historical context, the existence of certain hot spots for Jewish and Middle Eastern DNA in England and Wales. We agree somewhat with Bird the Welsh cluster for E is "underestimated by an arbitrary division by Sykes into two geographic regions ('Wales' and 'Northern England') . . . [creating] an impression of a large number of 'Eshu' haplotypes located throughout Northern England, when in fact the Northern English cluster is linked to Welsh cluster geographically (29)." Only, we would see in that Northern English cluster the remains of the historical Welsh Old North (chapters 1 and 7). We would not necessarily see in the Wales-to-Nottingham cluster the fading footprints of "the Ordovices, the Deceangi, the Cornovii, the Brigantes and the Coritani tribes (30)," about whom little is known in any event, but a belt of pre-existing Mediterranean culture reinforced by Roman occupation and somewhat resistant to Anglo-Saxon and Viking intrusions. Another shrinking pocket of the old British culture is shown in the elevated frequencies for both E and J in Strathclyde and Cumbria, part of the Welsh Old North.

Bird has an informative map of Britain illustrating E1b1b distribution according to the Kringing method (34). In this, we can trace all the major pockets of Mediterranean and Jewish DNA. Leaving aside Scotland, and aside from the Midlands pocket already mentioned, our eye is drawn to North Wales (along with a clear wall of high incidence surrounding it as though beating back the forces of history on all sides), Dorset, London, and East Anglia. It cannot be coincidence that these are the very regions where we have diagnosed the presence of Jews and picked up their trail through the chapters of our book.

As a final note, a 2005 paper by Robert Tarín provides phylogenetic analyses of E1b1b haplotypes that cast serious doubt on Bird's assertions and confirm our reading of the evidence. Tarín used 290 individual Y chromosome results to characterize "a separate cluster of mostly Iberian haplotypes which seem to represent a North African entry into Iberia distinct from the E3b (E1b1b) in Europe that may have arisen from Neolithic or other migratory events." He wrote that "it is unknown whether this finding reflects relatively recent gene flow from the Islamic rule of Spain or an older influx possibly from the Phoenicians"—the same quandary about time frame and coalescence we see above. Utilizing the Y Chromosome Haplotype Reference Database (YHRD), Tarín found levels of the Iberian E haplotype as high as 61 percent in one Tunisian population (Zriba, near ancient Carthage), while Andalusian Arabs and Tunisian Berbers both showed frequencies of about 7 percent. We believe this Iberian haplotype is a small, but important Jewish lineage that expanded from Tunisia to the Iberian Peninsula with the Berbers who aided Arab armies in conquering Spain. Interestingly, it accompanied Spanish Jews to Mexico and other places in the diaspora following the events of 1492. Its distribution in Britain should reveal an implantation originally under the Phoenicians reinforced by periodic migrations of North African and Spanish or French Jews throughout the medieval and early modern periods of British history.

January 15, 2012

Articles Cited

Steven C. Bird, "Haplogroup E3b1a2 as a Possible Indicator of Settlement in Roman Britain by Soldiers of Balkan Origin," *Journal of Genetic Genealogy* 3.2 (2007) 26-46.

Robert L. Tarín, "An Iberian Sub-Cluster Is Revealed in a Phylogenetic Tree Analysis of the Y-chromosome E3b [E1b1b] Haplogroup," published online Nov. 2005 and retrieved Jan. 2012 at http://garyfelix.tripod.com/E3bsubcluster.pdf.

DNA Was Discovered by Watson and Crick in 1953, Right?

Wrong. According to a correction in *Nature*, it was discovered by Swiss physician Friedrich Miescher in 1869, nearly a hundred years before. James Watson, Francis Crick and the less-often mentioned Rosalind Franklin were responsible for determining DNA's structure, a double-helix shape.

December 15, 2010

Seeing Red, Seeing Black

Sorry, Jack, no cigar. Your grandpa's Indians are not what you think. And it is not true "most free African American families that originated in colonial Virginia and Maryland descended from white servant women who had children by slaves or free Africans." Negro males did not go around randomly or selectively "fathering" little man-children on "white servant women" in early America.

It is ironic that these fantasies should even emerge in the widely publicized report, "Melungeon DNA Study Reveals Ancestry, Upsets a 'Whole Lot of People." The authors of the report, Roberta J. Estes, Jack H. Goins, Penny Ferguson and Janet Lewis Crain, have spent the better part of 10 years trying to prove they and others with Melungeon ancestry are just plain folks, that is, white folks.

Maybe they are just that, though. Among the conclusions of the report are that Melungeons aren't Portuguese, aren't Native American, aren't Jewish, aren't Romani/Gypsy, aren't . ... On and on. They just have a teeny-tiny bit of Sub-Saharan African in some lines. Not to worry, though, it is just a little soupçon of non-white. And it goes back to a few heroic "negroes" (Estes et al.'s language, not ours) who left a trace of their Sub-Saharan African Y chromosomes in the fathers and sons and grandpas of three Melungeon families.

From an article I published, lo! way back in 2002 in the *Appalachian Quarterly* (now sadly defunct), "Shalom and Hey, Y'all," comes the true story of these "negroes" fathering "multiethnic" babies on innocent white indentured servant women.

In discussing the will of Indian trader James Adair, I point out that Adair did not apparently approve of his daughter Agnes marrying John Gibson (from the selfsame Melungeon family that is creating all the brouhaha today). (Agnes, by the way, was not an indentured servant; her father had a considerable fortune.)

"Notice the harsh treatment Adair accords his daughter Agnes, leaving her and her husband John Gibson the nominal sum of only one shilling (if he had left her nothing, she could have protested to the probate court that he simply forgot her). John was one of the "mulatto" Gibsons of the Great Pee Dee river valley region. Gideon Gibson stands large on the pages of history for his role in the so-called Regulators Revolt. The Gideon Glass Antiques Store today pays testimony to the "richest man in South Carolina" of his time. When members of the Gibson family first moved to the state in 1731, representatives in the House of Assembly complained "several free colored men with their white wives had immigrated from Virginia." Governor Robert Johnson summoned Gibson and his family and reported:

> "I have had them before me in Council and upon Examination find that they are not Negroes nor Slaves but Free people, That the Father of them here is named Gideon Gibson and his Father was also free, I have been informed by a person who has lived in Virginia that this Gibson has lived there Several Years in good Repute and by his papers that he has produced before me that his transactions there have been very regular. That he has for several years paid Taxes for two tracts of Land and had several Negroes of

his own, That he is a Carpenter by Trade and is come hither for the support of his Family" [Box 2, bundle: S.C. Minutes of House of Burgesses (1730-35), 9, Parish Transcripts, N.Y. Hist. Soc. By Jordan, "White over Black," 172.]

"The Gibsons are discussed as Melungeons in Brent Kennedy and as Sephardic Jews in Hirschman. They derive their origins from the Chavis family, one of the oldest Portuguese-Jewish names in America. If they are Jewish (as I believe they were), it is ironic—and probably funnier than any Fanny Brice skit—that historians trot them forth as shining examples of non-slave African American colonials owning land and marrying white women."

The moral of the story? Melungeons have often been hauled into court to prove they are not black. Now, they are being dragged through the Internet to prove they were not Jewish.

Now about those Indians ... That will have to wait until another time.

May 26, 2012

Article cited: Donald N. Panther-Yates, "Shalom and Hey, Y'all: Jewish-American Indian Chiefs in the Old South," *Appalachian Quarterly* 7/2 (June 2002) 80-89.

King-Sized Sheets

The fairytale is over, but it seems that Elvis' DNA may live on. No, I don't mean *that* way. I refer to the DNA extracted from semen- and bloodstained sheets believed to come from the King's Farewell Tour.

I thought the '50s were an innocent time. A time before groupies, drugs and the sexual revolution. Elvis has often been portrayed as a Southern gentleman coming from a simpler time. Girls swooned over him as they had over Frank Sinatra. Priscilla and Elvis were a princess and prince living in the fairyland of Graceland.

But according to these alleged sheets and an article published on March 21, 2010, in the Sunday Times, "Hound Dog: Elvis Presley and His Girls" by Tasha Collins that is far from the truth. Her article reviews the book *Baby Let's Play House: Elvis and His Women*, by Alanna Nash.

Collins states he was a "womanizer obsessed with teenage girls." It seems there was a lot of wild parties, drugs, and sex. A friend of Elvis, Joe Esposito, says, "We talked, smoked grass, drank, went for late swims, and even had orgies.'" Collins quotes Nash as saying, "'The number of his relationships, whether emotional or sexual, bordered on the pathological.'"

But if we can't have a fairy tale, we can have the truth. The DNA analysis of the sheets that belong to Bobbi Bacha of Blue Moon Investigations believed to be Elvis' will be revealed in a new book to come out next year, *Old World Roots of the Cherokee,* by Principal Investigator and author Donald N. Yates.

The DNA analysis and laboratory extraction were performed by DNA Consultants in 2005. The company was able to obtain information on what is believed to be Elvis Presley's ancestry and ethnicity from the Y chromosome, mitochondrial DNA and autosomal DNA fingerprint. The alleged Elvis sample turned out to have a Cherokee-specific form of mitochondrial haplogroup B on the mother's side and a Scottish Y chromosome on the father's. The autosomal profile confirmed these results with high matches for American Indian populations, Scotland and Spain.

The "Elvis" American Indian type, which seems to go back to his mother's Cherokee ancestor White Dove, has been compared with the mitochondrial results of participants in DNA Consultants' Cherokee DNA Project.

How did Bobbi Bacha get Elvis' sheets? Attending a celebrity auction more than a decade ago, she put in the winning bid for some blood and semen stained sheets. Nearly 20 years old, but carefully preserved, they were reputed to come from the hotel room where Elvis Presley stayed on his Farewell Tour in 1977. She won't tell us how much she paid but says, "I could have bought a comfortable medium-sized home."

Bacha is no stranger to high-profile mysteries, crimes and misdemeanors. Part Cherokee, she is also, verifiably, of Melungeon descent. "As you know," she told us from her swanky glass headquarters building in Houston, "Nevil Wayland is my grandfather, and it was he who first coined the term Melungeon." We didn't know, but we soon got an earful. "We believe his wife was the daughter of Chief Red Bird as his son was the Scribe to Chief Red Bird. Nevil built the first church in

Arkansas after the family told of a great war against the Indians and he took them to Arkansas and built Stoney Creek Church. That's the name of it."

Bacha has also been in the movies,or at least her character has. The plucky Texas private eye is played by actress Sela Ward in "Suburban Madness." This film is based on the real-life story of Clara Harris, convicted February 2003 of killing her cheating orthodontist husband by repeatedly running him over with the family Mercedes. Bacha was an eyewitness.

So what of Bacha's expensive king-sized sheet set? She tried for years to extract DNA, to no avail. The discipline had some growing up to do. Finally, she contacted DNA Consultants. Through the efforts of Sorenson laboratory director Lars Mouritsen in Salt Lake City, we were able to succeed where others had failed. We obtained the first DNA profile, Y chromosome and mitochondrial DNA results for what everyone believed was a 30-year-old sample from the King.

If Elvis isn't still around, maybe his hair still is. If it is his. According to Lorraine Bailey in the Courthouse News Service article, this isn't the first time something like this happened ("That's Not Elvis's Hair, Feds Tell Auction Execs"): "Federal prosecutors have accus[ed] three top executives of Mastro Auctions of selling bogus "bona fide" locks (from 2003 to 2008) of Elvis Presley's hair though they knew they were phony."

We don't know. But just maybe this is the real deal. For the rest of the story, and to see what you think, you'll have to read Donald Yates' new book, *Old World Roots of the Cherokee*, where Elvis' genetics is discussed in the chapter titled "DNA."

September 12, 2011

6 DNA AND ETHNICITY

Melungeon DNA Study by Jones with MHA Response

Kevin Jones was a professor of biology at the University of Virginia's College at Wise who used DNA to study the Melungeons. Here is the press release from June 2002 that shocked many Appalachians and threw the Melungeon mystery into high gear.

KINGSPORT, Tenn.—Some of the veil of mystery surrounding the "mysterious" Melungeons was lifted today when the results of a two-year DNA study were announced. New questions have been raised, however, concerning females potentially from Turkey and northern India who are a part of the Melungeon ancestry.

The Melungeons are a group of people of unknown origin first documented in the mountains of Appalachia in the early 19th century. Many believed they were of mixed racial ancestry, and the Melungeons faced legal and social discrimination. As a result, they tended to live in remote areas, most notably Newman's Ridge in Hancock County, Tennessee. In the 1940s and 1950s, sociologists and anthropologists labeled the Melungeons and other similar groups as "tri-racial isolates."

Over the years, numerous myths, legends, and theories evolved to explain the Melungeons' mysterious origins. These legends often involved sailors and explorers from Spain, Portugal, Carthage, or Phoenicia who were stranded on the American continent and intermarried with Indians. The Melungeons themselves often claimed to be "Portyghee."

Most researchers believed they were a product of intermarriage between English and Scots-Irish settlers, Indians, and free African-Americans, and discounted their claims of Mediterranean origin. The DNA results announced today confirmed that the Melungeons have European, African, and Native American ancestry, as well as genetic similarities with populations in Turkey and northern India.

More surprising, however, is the fact that some of these Turkish- and northern Indian-like sequences have been passed through the Melungeons' maternal lines, indicating that their overseas ancestors included not only male sailors and explorers, but females as well.

The results were announced today at Fourth Union, a Melungeon conference in Kingsport, Tennessee, sponsored by the Melungeon Heritage Association (MHA). MHA is a nonprofit organization dedicated to promoting research and understanding about Melungeons and other multiracial groups in the United States. Dr. Kevin Jones, a biologist at the University of Virginia's College at Wise, conducted the study.

The presence of Turkish and northern Indian haplotypes within the mitochondrial DNA samples taken from modern-day Melungeons indicates that women of European/Asian origin were a part of the original mixture that made up the Melungeon ancestry. Mitochondrial DNA comes from the female side of an individual's ancestry.

Previous researchers had assumed that European males intermarried with Native Americans and African-Americans to produce the Melungeons. Although Native and African genes are definitely a part of the Melungeon genetic mix, women were among the overseas settlers who contributed to the Melungeon gene pool.

Dr. N. Brent Kennedy speculated that the Melungeons were of Mediterranean and Middle Eastern ancestry and published his theories in a book titled *The Melungeons: The Resurrection of a Proud People*, published in 1994 by Mercer University Press.

Dr. Jones, a native of London, England, studied at the University of Reading, and did post-doctoral research at Louisiana State University. He is currently a professor of biology at UVA-Wise, teaching courses including cell biology and genetics.

Dr. Jones undertook this DNA study in 2000 at the suggestion of Dr. Kennedy, then vice-chancellor at the University of Virginia's College at Wise. Kennedy asked Jones to analyze DNA samples taken from members of known Melungeon families. Such a study would utilize technology not available to earlier researchers.

"Brent Kennedy ... explained the controversy that surrounded the origins of the Melungeons (and) realized that I had the DNA expertise to look at that," Jones related in an interview with Wayne Winkler, president of the Melungeon Heritage Association and author of an upcoming book about the Melungeons. "The subjects were largely chosen by Brent Kennedy on the basis of pursuing as many of the known Melungeon lineages that existed in the area and taking advantage of his genealogical expertise. People were then asked to donate samples to the study, and in the majority of cases they kindly did so.

"Single hairs were taken to study the mitochondrial DNA which traces the maternal lines of the subject. In other words, the samples represented DNA, which could be traced to the subject's mother, grandmother, great-grandmother, and so on. We also have a smaller number of samples which are cheek cells for looking at male inheritance," said Jones.

"What we get from those is a DNA sequence which we can think of as being about an 600-long letter code, and we can take that string of 600 letters and compare those to what now is literally thousands of samples from around the world. We're interested both in the number of different sequences that we get from the population and also how they appear to relate to other samples worldwide."

About 100 hair samples were studied for mitochondrial, or maternal, DNA, and about 30 samples of cheek cells were taken to study the Y-chromosome, or male, DNA. While more samples might have been taken, Jones said, "That's the beauty of science: One can always subsequently refine and extend the analyses."

The technology available to Jones allowed him to study only the mitochondrial DNA samples; the Y-chromosome samples were sent to University College in London, England, for study. "The 'Y' is technically far harder to do, and indeed, relies on expertise in some other labs in the world to do it, so we're partly dependent on their cooperation and collaboration."

Such testing is not perfect, of course, and does not tell researchers everything about an individual's inheritance. For example, neither test will give genetic information about a subject's paternal grandmother. However,the study was not particularly concerned with individual genealogies.

"We're looking for patterns that exist in the population as a whole," according to Jones. "Now, obviously, each individual sample contributes to that, but I think that for an individual you can say relatively little. Looking at the patterns that occur throughout the population becomes important. And that means the number of samples that are looked at is also significant, and we've tried to do as many as is reasonably possible."

Jones compared these samples to the thousands available through GenBank, an international genetics database, published scientific literature and the Mitochondrial DNA Concordance, databases containing DNA sequence information.

Looking at the maternal lines of the Melungeons who were tested, Jones found considerable variation in ethnicity among the samples.

"It's comparatively straightforward to link particular sequences to particular ethnic groups and different continental areas of the world," he noted, "and the majority of those Melungeon-derived sequences were European in origin. Within those European samples, though, there is significant diversity, and some seem to reflect areas outside the traditional northern European sphere. The ability to tie a sequence to a particular area is dependent upon the historical occurrence of any given haplotype somewhere, and the places that are easy to track are where we've had populations existing for a long time, and not being affected by a lot of different people coming in. So some, perhaps more-isolated, areas of Europe are easier to track than more cosmopolitan (areas)."

While the Melungeons are predominantly European in their genetic backgrounds, they are indeed tri-racial.

"There appears to be a small percentage of both Native American and African-American sequences in there, too," Jones stated, "although they are certainly both in the minority. They're both in there in about equal levels of representation as well."

The long-held belief that the Melungeons originated in Portugal is neither borne out nor negated by Jones' research.

"To date we've found no sequences that can be definitively traced back to uniquely Portuguese sequences. That doesn't mean that they don't exist. A large number of the European sequences are now widely spread throughout Europe, and if one of those genetic sequences happened to come from Portugal, we would not detect that. We can't dismiss that theory at the moment, but we can't provide additional support for it."

Jones finds a stronger possibility for a Turkish or Middle Eastern ancestry for the Melungeons.

"The relatively unusual European-type sequences seem to reflect, perhaps, areas around northern India. It's very hard to say, back in time, what that would have been classified as, and I think one of the problems here is that we tend to think of 'Turkish' in terms of the dimensions of modern Turkey, not of the original scale of people of Turkish origin who, in essence, were spread throughout the European world. Perhaps the best I can say is that some of those sequences are a little more 'exotic' than Anglo-Irish sequences, and some of those could reflect, perhaps, populations that were associated with or moved through Turkey."

The Portuguese and Spanish explorers and early American settlers may well be the key to discovering how these people wound up in America. The Portuguese, in particular, were involved in wide-ranging trade in the 15th and 16th centuries, and had many interests in places such as northern India and the area occupied by present-day Turkey. Both Spain and Portugal had very cosmopolitan populations, with large numbers of people from many parts of the world living within their borders. And Dr. Kennedy and others have suggested the Spanish and Portuguese fort at Santa Elena (in present-day South Carolina), along with a series of frontier outposts, as a possible source for Melungeon ancestry.

With regard to the male lineages investigated, the Y chromosome data also suggests a multiracial origin, including Sub-Saharan African and European components. Of particular interest are Y haplotypes of established Melungeon male lines that possibly reflect Mediterranean and/or Near Eastern populations. This finding indicates that the overseas ancestors of the Melungeons may have come to these shores as part of a male-female family unit, or formed such family units shortly after arrival. Such family units came to America as part of a Spanish/Portuguese colony at Santa Elena in present-day South Carolina.

Theories about when people with this genetic background first came to America are speculative at this point.

"Dr. Jones' work has answered many questions," said Wayne Winkler, president of MHA, "but those answers have raised many more questions. These questions will keep historians busy for some time to come, and we may never have definite answers. The Melungeons may remain one of the mysteries of history."

Statement by Wayne Winkler, President of the Melungeon Heritage Association

The Associated Press mistakenly released the story of the DNA study a day early. We regret that, after two years of hard work, Dr. Kevin Jones was not permitted to break the news of his study himself. The story by Chris Kahn also contained a few inaccuracies. First, the hair samples taken to study mitochondrial DNA, tracing the maternal lines, were taken from men as well as women, contrary to the statement in the story. Secondly, Brent Kennedy was quoted out of context in saying that Melungeons weren't much different from other Americans. In context, he was saying that ALL Americans are mixed to some degree, although not necessarily as much as are Melungeons. And finally, I think the story missed the important news: that women were a part of the original Turkish, Mediterranean, and northern Indian population that came to America. We've always heard the stories about shipwrecked sailors or explorers being the source of our overseas genes; it's now obvious that these genes came, at least in part, from family units of men and women who were attempting to establish themselves in a new land. As the DNA study shows, they succeeded.

June 21, 2002

Ancient Chickens Come Home to Roost

Biological anthropologist Elizabeth Matisoo-Smith of the University of Auckland in New Zealand has come up with proof, in the form of chicken bones, that shows Polynesians were in America long before Christopher Columbus showed up.
Her dating of chicken bones found in Chile, shows before the Spanish arrived, Polynesians had gone there, had given the American Indians chickens and probably got sweet potato in exchange.

An article about the discovery appears in *Nature* 447/7145:620, "DNA Reveals How the Chicken Crossed the Sea," by Brendan Borrell.

The origins and travels of the Polynesians were the lifelong study of

Harvard biologist Barry Fell, a native New Zealander who was widely shunned by other academicians for his theories of diffusion across the Pacific. As early as the 1970s, geographer George F. Carter offered evidence the Polynesians brought chickens to South America and took away sweet potatoes, among many other exchanges of culture.

In 1992, Oxford geneticist Bryan Sykes disproved the theory put forth by Thor Heyerdahl with the balsam boat called Kon-Tiki that American Indians sailed across the Pacific from East to West to found Polynesia. The chicken bones support what diffusionists have maintained for decades, that cultural influences ran in the opposite direction.

June 6, 2007

Double Helix Maelstrom
DNA Pioneer Rebuked For Racist Remarks
BBC News, October 20, 2007

DNA helix co-discoverer James Watson was suspended from the board of the Cold Spring Harbor Laboratory after making disparaging remarks in the *Independent*, a British newspaper, about the intelligence of Africans.

Watson said he was "inherently gloomy about the prospect of Africa because all our social policies are based on the fact that their intelligence is the same as ours—whereas all the testing says not really." He added that people who have to deal with black employees find it is not true that everyone is equal.

Later, Watson, qualified his remarks.

"The overwhelming desire of society today is to assume that equal powers of reason are a universal heritage of humanity," he said. "It may well be. But simply wanting this to be the case is not enough. This is not science. To question this is not to give in to racism."

With Francis Crick and Maurice Wilkins in 1962, Watson received the Nobel Prize in Physiology and Medicine for their discovery of the double-helix structure and nature of DNA, often referred to as "the code of life."

October 20, 2007

Possible Egyptian, Greek, Phoenician and Hebrew Origins of Cherokee

Abstract. A sample of 52 individuals who purchased mitochondrial DNA testing to determine their female lineage was assembled after the fact from the customer files of DNA Consultants. All claim matrilineal descent from a Native American woman, usually named as Cherokee. The main criterion for inclusion in the study is that test subjects must have obtained results not placing them in the standard Native American haplogroups A, B, C or D. Hence the use of the word "anomalous" in the title of a paper prepared by chief investigator Donald N. Yates, "Anomalous Mitochondrial DNA Lineages in the Cherokee." Most subjects reveal haplotypes that are unmatched anywhere else except among other participants, and there proves to be a high degree of interrelatedness and common ancestral lines. Haplogroup T emerges as the largest lineage, followed by U, X, J and H. Similar proportions of these haplogroups are noted in the populations of Egypt, Israel and other parts of the East Mediterranean.

The Cherokee and Admixture

According to a 2007 report from the U.S. Census Bureau, the Cherokee are the largest tribal group today, with a population of 331,000 or 15 percent of all American Indians. Despite their numbers, though, the Cherokee have had few DNA studies conducted on them.

I know of only three reports on Cherokee mitochondrial DNA. A total of 60 subjects are involved, all from Oklahoma. Possibly the reason the Cherokee are not recruited for more studies, I would suggest, stems from their being perceived as admixed in comparison with other Indians.

Accordingly, they are deemed less worthy of study.

In the past, whenever a geneticist or anthropologist conducting a study of Native Americans has encountered an anomalous haplogroup, that is, a lineage that does not belong to one of the five generally accepted American Indian mitochondrial DNA haplogroups A, B, C, D and X, it has been rejected as an example of admixture and not included in the survey results. This is true of the two examples of H and one of J reported by Cherokee descendants by Schurr (2000:253). Schurr takes these exceptions to prove the rule and regards them as instances of European admixture.

The governing logic of population geneticists seems to go as follows: Lineage A, B, C, D and X are American Indian. Therefore, all American Indians are lineage A, B, C, D and X.

The fallacy in such reasoning is apparent. It could be restated as: "All men are two-legged creatures; therefore since the skeleton we dug up has two legs, it is human." It might be a kangaroo.

"The geneticists always seem to cry 'post-Columbian admixture,' " says Stephen C. Jett, a geographer at the University of California at Davis, "but fail to take into account that there are no plausible post-Columbian sources for the particular genetic mix encountered."

"Anomalous Mitochondrial DNA Lineages in the Cherokee" concentrates on the "kangaroos"—documented or self-identifying Cherokee descendants whose haplotypes do not fit the current orthodoxy in American Indian population genetics. Here are some highlights, organized by haplogroup.

Haplogroup H

Although this quintessentially European haplogroup would seem to be the most likely suspect if admixture were responsible for the anomalous haplogroups, there are but four cases of it.

Haplogroup X

Haplogroup X is a latecomer to the "pantheon" of Native American haplogroups. Its relative absence in Mongolia and Siberia and a recently proven center of diffusion in Lebanon and Israel (Brown et al. 1998, Malhi and Smith 2002; Smith et al. 1999; Reidla 2003; Shlush et al. 2009) pose problems for the standard account of the peopling of the Americas. DNA Consultants Cherokee-descended customers include seven instances of haplogroup X. David E. Lewis (whose Cherokee name is Wayauwetsi) traces his unmatched X haplotype back to Seyinus, a Cherokee woman of the Wolf Clan born on or near the Qualla Boundary in North Carolina in 1862. Two cases represent descendants (unknown to each other, incidentally) of the Cherokee woman called Polly who was the namesake for the Qualla reservation (the sound p lacking in the Cherokee language and being rendered with qu).

Haplogroup J

Two other cases, both J's, are related to Polly, tracing their lines back to Betsy Walker, a Cherokee woman born about 1720 in Soco (One-Town). A descendant was the wife or paramour of Col. Will Thomas, the first chief and founder of the Eastern Band of Cherokee Indians located today on the Qualla Boundary. Views about J are still evolving, but it seems to have originated in present-day Lebanon approximately 10,000 years before present. It is a major Jewish female lineage (Thomas 2002).

Haplogroup U

This haplogroup has never been reported in American Indians to my knowledge. In our sample it covers 13 cases or 25 percent of the total, second in frequency only to haplogroup T. One of the U's is Mary M. Garrabrant-Brower. She belongs to U5a1a* (all U5a1a not matched or assigned) but has no close matches anywhere. Her great-grandmother was Clarissa Green of the Cherokee Wolf Clan, born 1846. Mary's mother Mary M. Lounsbury maintained the Cherokee language and rituals. One of the cases of U2e* is my own. This line evidently arose from a Jewish Indian trader and a Cherokee woman. My fifth-great-grandmother was born about 1790 on the northern Georgia and southwestern North Carolina frontier and had a relationship with a trader named Enoch Jordan. The trader's male line descendants from his white family in North Carolina possess Y chromosomal J, a common Jewish type. Some Jordans, in fact, bear the Cohen Modal Haplotype that has been suggested to be the genetic signature of Old Testament priests (Thomas et al. 1998). Enoch Jordan was born about 1768 in Scotland of forbears from Russia or the Ukraine. My mother, Bessie Cooper, was a double descendant of Cherokee chief Black Fox and was born on Sand Mountain in northeastern Alabama near Black Fox's former seat at Creek Path (and who was Paint Clan). All U2e* cases appear to have in common the fact that there are underlying Melungeon, Cherokee and

Jewish connections.

Haplogroup T

"Tara," as she was named by Bryan Sykes, is believed to have originated in Mesopotamia approximately 10,000 to 12,000 years ago and to have moved northwards through the Caucasus and westwards from Anatolia into Europe. The closer one goes to its origin in the Fertile Crescent the more likely T is to be found in higher frequencies. The haplogroup includes slightly fewer than 10 percent of modern Europeans, but accounts for 28 percent of people in the DNA Consultants study. The great-great-grandmother of Linda Burckhalter was Sully Firebush, the daughter of a Cherokee chief who married Solomon Sutton, the stowaway son of a London merchant, in what would seem to be another variation of the "Jewish trader marries chief's daughter" pattern. Three T1*'s are perfectly matching individuals completely unknown to one another before testing who are clearly descended from the same woman. Two of them claim Melungeon ancestry.

The many interrelationships noted above reinforce the conclusion that this is a faithful cross-section of a population. No such mix could have resulted from post-1492 European gene flow into the Cherokee Nation. So where do our non-European, non-Indian-appearing elements come from? The level of haplogroup T in the Cherokee (26.9 percent) approximates the percentage for Egypt (25 percent), one of the only lands where T attains a major position among the various mitochondrial lineages. In Egypt, T is three times what it is in Europe. Haplogroup U in our sample is about the same as the Middle East in general. Its frequency is similar to that of Turkey and Greece. J has a frequency not unlike Europe (a little less than 10 percent). The only other place on earth where X is found at an elevated level apart from other American Indian groups like the Ojibwe is among the Druze in the Hills of Galilee in northern Israel and Lebanon. The work of Shlush et al. (2009) demonstrates that this region was, in fact, the center of the worldwide diffusion of haplogroup X.

On the Y chromosome side of Shlush et al.'s study, male haplogroup K was found to have a relatively high frequency of 11 percent in the Galilee region (2008:2). K (renamed T in the revised YCC nomenclature) has long been suspected to be the genetic signature of the Phoenicians. A TV show by *National Geographic* appeared about a year ago titled "Who Were the Phoenicians?", in which Spencer Wells of the National Genographic Project, unveiled this theory.

Paint Clan and Phoenicians

Without a doubt it was the Phoenicians, whose name among themselves was Cana'ni or KHNAI 'Canaanites', not Phoenikoi 'red paint people' (Aubet 2001:9-12; cf. *Oxford Classical Dictionary* s.v. "Phoenicians"), who are referenced by James Adair when he observes that "several old American towns are called Kan'ai," and suggests that the Conoy Indians of Pennsylvania and Maryland were Canaanites and their tribal name a corruption of the word Canaan. The Conoy Indians are the same Indians William Penn around 1700 described as resembling Italians, Jews and Greeks. By about 1735 they had dwindled to a "remnant of a nation, or subdivided tribe, of Indians," according to Adair (1930:56, 67, 68).

One of the oldest Cherokee clans is called Red Paint Clan (Ani-wodi). So do the two subclades of X and other haplogroups represent Old World and New World branches diverging from each other as long ago as 30,000 years, or do the Native American "anomalous" haplotypes come more recently (but not as late as Columbus) from the same source in the East Mediterranean? The answer probably depends on how open one is to new evidence and revisionary thinking.

According to Jett, "The splits may have taken place well before transfer, with one only or both being transferred to a new place and then one dying out in the home area (and the other in the new area, if both were transferred)." The distinction, at any rate, is irrelevant to the Cherokee who exhibit these not-so-rare haplogroups, although to those denied authenticity on the basis of anthropologists' hardened ideas about the genetic composition of American Indians, it is welcome vindication either way.

References

1. Adair, James (1930). *Adair's History of the American Indians*, ed. by Samuel Cole Williams, originally published London, 1775. Johnson

City: Watauga.

2. Richards, Martin et al. (2000). "Tracing European Founder Lineages in the Near Eastern mtDNA Pool." *American Journal of Human Genetics* 67:1251-76. Supplementary Data. URL: http://www.stats.gla.ac.uk/~vincent/founder2000/index.html.

3. Schurr, Theodore G. (2000). "Mitochondrial DNA and the Peopling of the New World," *American Scientist* 88:246-53.

4. Shlush, L. I. et al. (2009) "The Druze: A Population Genetic Refugium of the Near East." *PLoS ONE* 3(5): e2105.

August 31, 2009

Does Not Compute

Past Human Migrations in East Asia. Matching Archeology, Linguistics and Genetics, ed. Alicia Sanchez-Mazas et al. London: Routledge Taylor & Francis, 2008

According to the reviewer of this compilation of interdisciplinary studies, Frank Roels, writing in *European Journal of Human Genetics* 18:262f., the three approaches are incommensurate because of differing time frames and rates of change. Their models cannot be harmonized with sufficient reliability to write a comprehensive, persuasive history of human migrations and settlements in East Asia.

February 3, 2010

Customs and Beliefs of the Roma and Sinti

Some of our customers have been surprised to get Gypsy/Romani population matches in their results for the DNA Fingerprint Test. Typically, these are combined with Middle Eastern and Indian matches due to the Gypsies' historical migrations. Other customers were not surprised at all and called to tell us about the fortuneteller great-grandmother or mysterious ancestor who traveled with the circus. Gypsy heritage is not unheard of among Melungeons. So, for those who think they may have Roma/Sinti or Romechal (the term used in the British Isles), we have compiled the following list of customs and beliefs taken from an excellent authority.

Strict monotheism similar to Jews
Keeping the seventh day holy
Lighting candles on the evening of Parashat (Friday)
Blasphemy a sin, as is cursing an elder
Beng (Satan) the enemy of God and of the Roma people
The Evil One called bivuzhó (impure) and bilashó

Code of Law
No social classes, only a division into Roma and Gadje (non-Roma)
A court of justice called Kris (Judiciary Council), composed of clan
    representatives as judges
Both men and women serving on Kris
Issues between Roma to be judged only by the Kris, not by Gadje
All Roma equal before the eyes of the Kris
Belief in blood revenge and compensatory payment for clan of
    victim
Banishment from territory of victim's clan for wrong doing
Forfeiture of protection if banished offender re-enters
Roma not even to acknowledge or greet one who is banished
Accursed or banished called mahrimé (impure)
Roma not to ask interest for loans to other Roma, only from Gadje

Sexuality, Marriage and Childbirth
Nudity taboo, allowed only with a husband and wife
Showing naked legs before an elder disrespectful
Homosexuality an abomination
Not allowed to wear clothes of the opposite sex, even as a joke or
    disguise
Virginity before marriage essential
Tokens of virginity shown to the assembly after wedding
Prostitution strongly condemned
Incest taboo, defined in the same way as Mosaic law (including
    step-siblings and in-laws)
Permissible to marry your cousin
Members of the Kris must be married

Lack of a spouse makes a man or woman incomplete

Groom's family pays dowry to the bride's family

Dowry for a widow amounts to half that for a virgin

A man dishonoring a woman should pay the dowry to her family anyway

Runaway couples considered legitimately married

Marriage endogamic, even within the same clan

Clan recognized by a common ancestor within a few generations

Divorce admitted: Husband sends wife out or she leaves

Remarriage expected after divorce

Levirate law practiced (Deut. 25:5-6)

Childbirth impure, must take place outside the home

Mother giving birth isolated with baby for seven days strictly, followed by 33 days of less strict isolation (cf. Lev. 12:2, 4-5)

New mother cannot show herself in public or attend religious services

Both sexes marrying very young (child marriage)

Funeral and Mourning Rituals

Dead to be buried intact (autopsy or cremation sacrilegious)

Close relatives of the dead impure for seven days

Not to touch a dead body

Family and relatives of deceased forbidden to bathe, comb their hair, cut their nails for three days

On third day after a death, relative must wash thoroughly, and then not again until seventh day

All food in house where a person died is thrown away as defiled

On third day after a death, the house is purified ("the ashes of the burning of the sin") and a virgin sprinkles running water

The same ceremony repeated on the seventh day after a death, with food brought to the mourners from another dwelling place

Mourners stay at home

Sitting on low stools

Covering mirrors

Not using oils or perfumes or cosmetics

Not wearing new clothes

Not listening to loud music

Not taking photographs or watching television

Not painting, cooking, and cannot greet people

Day mourning extended after seventh day, remembrance ceremony until 30th day

Another remembrance ceremony on 30th day, closing the strict mourning period

Beliefs in Afterlife

Death is final, no reincarnation or return

Soul goes to Paradise or Hell

Purity and Impurity

Concept of marimé (similar to kashrut)

Lower body and things associated with it impure

Sleeping regarded as an impure state

Not to greet anyone upon waking until washed

Disrespectful to greet anyone in an impure state

Dogs and cats impure

Horses, donkeys or riding animal impure

Carnivorous animals impure

Avoidance of horseflesh

Shoes, pants, hose, skirts, trousers, etc. impure

The camp pure

Restrooms built outside the home

Clothes for the lower body and menstruating women washed separately

Dishes washed in a different place from clothes

Other Practices

Custom of mangel, asking for favors from Gadje

Painting doorposts of dwelling with animal blood to protect against angel of death

Invoking the Prophet Elijah, particularly when seeing lightning or hearing thunder

Firstborn son considered a special blessing to the family

Wearing of whiskers

Left hand related to the public domain (Gadje), impure

Separate dishes and cups for Gadje
Only eating ritually slaughtered animals
Slander considered very, very serious offense, worth taking to Kris
Lack of belief in divination (contrary to general view of Gypsies)
Practice of Tarot cards and crystal balls for Gadje only
Having a Gypsy name besides a civil name
Names that are Hebrew, Greek, Russian, Spanish, Hungarian,
    Persian, never Indian or Hindu
Beef a favorite food
Interest in bullfighting
Middle Eastern music and dance with zithers, etc. (Flamenco in
    Spain)
Fingernails and toenails filed with an emery board, not a clipper
Going to a church called Filadelfia (Brotherhood)
Claiming to be Egyptian in origin
Making pilgrimages to the burial places of your ancestors

Source: Abraham Sándor, "Comparison of Romany Law with Israelite Law and Indo-Aryan Traditions."

September 12, 2010

Autosomal Testing for Native Americans

Reference: Wang, S. et al. (2007). "Genetic Variation and Population Structure in Native Americans." *PLoS Genetics* 3/11 (with good bibliog.) http://www.plosgenetics.org/article/info:doi/10.1371/journal.pgen.00301 85.

If you think haplogroup testing for Native American DNA is in a backward condition, you should look at autosomal testing. It has been practically nonexistent. Even the major 2007 study by Wang et al. has glaring gaps and methodological quandaries.

DNA Consultants' newest autosomal product is the Native American DNA Fingerprint Plus, based on 21 published studies of Native American population groups as well as informal customer data. Results for many individuals were validated and cross-referenced with older haplotyping methodology.

There were data for 3,583 Native Americans available in development of the product.

These test results came from articles published between 1997 and 2009. They included individuals identifying with tribes or nations as follows:

Apache	Lumbee
Athabaskan	Navajo
Huichol	Salishan
Inupiat	Yupik
Kichwa	

The following geographical areas were represented:

Alaska	Ecuador	Minnesota
Arizona	Florida	North Carolina
Brazil	Guatemala	Oklahoma
British Columbia	Mexico	Ontario
Colombia	Michigan	Saskatchewan

No data specifically labeled as Cherokee—the largest Native group in the U.S., with more than 400,000 representatives—have ever been published. In our company database, people of Cherokee descent often receive matches to North Carolina or Michigan Native Americans. The reason for the latter matchup is obscure. North Carolina as the Cherokee's original homeland makes a lot more sense.

May 10, 2011

SNPs and the Emerging Prehistory of Ethnic Groups

No scientific work, to our knowledge, has ever hazarded a guess on what the mutation rate for autosomal CoDIS-type markers might be. Is it like mitochondrial DNA, which has a molecular clock measured in the thousands or tens of thousands of years, or is it like STRs on the Y chromosome, with its much shorter time frame? The question is important if you are trying to extrapolate the history of the human race from today's autosomal population statistics.

From what we can see, putting on diachronistic lenses, the mutation rates for the DYS values on what are commonly called CoDIS markers or the DNA profile for individuals are very small. The values appear to have been set from the beginning of mankind and to have mutated little in the past 100,000 years.

If this is true—and it cannot be a very big "if" or we would have more diversity between populations than what is known—the oldest markers are Sub-Saharan African and the newest European. Statistical divides bear out this reading of the human genetic record, as shown now in our updated map included with the DNA Fingerprint Plus.

We will try to make some notes on the individual markers in future posts.

June 16, 2011

American Indians and Turkic People Share Deep Ancestry

We've known or suspected it for a long time. American Indians and Turkic peoples of the Altai region of southern Siberia share common ancestors. American scientists Thomas Jefferson and Constantine Rafinesque were the first to demonstrate this genetic similarity, long before the days of DNA. Now an article in *American Journal of Human Genetics* has clenched the argument with mitochondrial and Y chromosomal DNA studies.

The groundbreaking citation is: Matthew C. Dulik et al., "Mitochondrial DNA and Y Chromosome Variation Provides Evidence for a Recent Common Ancestry between Native Americans and Indigenous Altaians," *AJHG* 90/2, 229-246.

From *Old World Roots of the Cherokee*, a book appearing June 15 by Donald N. Yates:

— Thomas Jefferson thought American Indians were Turks and Tartars coming across the Bering Sea from Asia, while his contemporary John Filson believed them to be Phoenicians. (See Boorstin, Daniel J. *The Lost World of Thomas Jefferson*, Chicago: U of Chicago P, 1993.)

— (quoting Rafinesque) "Many other empires having begun to rise in the vicinity of Aztlan, such as those of Bali (Indonesia, perhaps Oppenheimer's Eden in the East?), Scythia [Russian steppes], Thibet, Oghuz (Lake Baikal area), the Iztacan were driven eastwards, north of China; but some fragments of the nation are still found in the Caucasus, &c. such as the Abians or Abassans, Alticezecs (Altai Turks), Cushazibs, Chunsags, Modjors, &c.

— "The six Iztacan nations being still pressed upon by their neighbours the Oghuzians (Uigur Turks), Moguls (Mongols), &c. gradually retreated or sent colonies to Japan, and the islands of the Pacific ocean; having discovered America at the peninsula of Alasca (Alaska, a Chinese word), during their navigations, the bulk of the nation came over and spread from Alasca to Anahuac, establishing many states in the west of America, such as Tula [Toltec], Amaquemeca, Tehuajo (Tewa, Tiwa, Tawa), Nabajoa (Navajo), Teopantla, Huehue, and many others.

— "After crossing the mountains, they discovered and followed the Missouri and Arkanzas rivers, reaching thus the Mississippi and Kentucky (26-27)."

How long will it take American history textbooks to catch up to this new proof? Never. A jingoistic Smithsonian has its own version of things, one ingrained in anthropological dogma as deeply as Manifest Destiny. Interestingly, Turkish and Muslim historians have long embraced this finding as a basic fact of history. They have long claimed American Indians as genetic cousins.

June 6, 2012

Anonymous commented on 11-Jun-2012

The people of Iran already have known for eons that the ancestors of the Navajo came from that general area originally. For simple comparison, the similarities between the design elements of Navajo vs. tribal rugs and weavings from Iran, Turkey, Afghanistan, The Caucasus and other areas cannot be simply a "coincidence"; and therefore cannot be summarily ignored. Now, DNA evidence speaks loudly in favor of what has already been known for milennia.

Brian Costello commented on 21-Jul-2012

The ancestors of the American Indians came from Siberia. However, most of Siberia is Yenesian and Tungus not Turkic. Turkic peoples arrived in Siberia very late. The Yakuts were not Turkified until the 15th century A.D.

Elfiya Marat commented on 06-Dec-2013

Brain Costello my friend you are wrong. The first homeland of all Turkic peoples was in southern Siberia, which would be around Lake Baikal. The Yakuts went up north after 10th century. After 10th century millions of Turkic people migrated out of southern Siberia into parts of Russia, central Asia, parts of Middle East and Europe. Remember the word "RUS" comes from Viking Varangians, not indigenous Slavic tribes in Caucasus region. Russia is in Asia.

Ronald Best commented on 31-Mar-2014

I am so surprised. The Phoenicians were famous ancient sailors. Should it really be a surprise there could be ancient land migrators? They really haven't considered the other possibility that Native Americans were in the new world much earlier and simply migrated back towards the middle east. This is still very closed-minded thinking. Native Americans should be claiming that we migrated as far as the Middle East or Siberia.

Identifying by Ethnicity in 2012

Genetics has transformed many of our notions of race, ethnicity and identity. How do *you* report your ancestry when checking off ethnic options on an official form? How do you identify yourself informally with friends and family? Have you ever "changed" ethnic self-identification because of a DNA test? These and related questions were the topics discussed at a 90-minute colloquium at the 12th Annual International Diversity Conference, held on the campus of the University of British Columbia in Vancouver, Canada, June 12.

The title of the panel discussion was "Perspectives on Ethnic Identity: Epigenetics, Marketing, DNA and Genealogy." It was organized by Donald Yates and moderated by Gregory Baskin. Presenters included: Dr. Anne Marie Fine, Scottsdale, Arizona naturopathic physician, who spoke on the emerging field of epigenetics, the multigenerational factors that "turn on and turn off" your genes; Wendy Roth, University of British Columbia, author of the just-published book *Race Migrations: Latinos and the Cultural Transformation of Race,* who presented the results of ongoing surveys of consumers of DNA testing, with an emphasis on changing notions of ethnic identification; Donald Yates, who presented a paper on overlapping ethnic identity in Bernard Malamud's *The People*, George Tabori's "Weisman and Copperface: A Jewish Western" and three early 20th-century poets writing in Modern Hebrew—Benjamin Nahun Silkiner, Israel Efros and Ephraim E. Lisitzky. Yates' paper was titled "Dying Campfires: Jews, Indians and Descendant Organizations" and included a comparison of Marranos (Sephardic crypto-Jews) with so-called Wannabe Indians (descendants of Indians who want to join a Federally recognized tribe but are barred from applying for membership for various reasons). Both categories of ethnic belonging, Yates showed, are often rejected by official authorities like rabbinical courts and the Bureau of Indian Affairs because adherents are seen to be only selectively practicing the group's customs and traditions.

Of the Marranos, for instance, Benzion Netanyahu wrote, "The Marranos ought to be treated realistically according to what they actually were—not *unwilling*, but *willing* converts, and consequently traitors to the Jewish religion and enemies of the Jewish people." In other words, Conversos chose to practice some Jewish, some Christian customs, or to hide their true beliefs with an insincere profession of Christianity.

In the same way, Cherokee and other Indian descendant organizations were criticized by William Quinn in an article that served as a sort of legal brief on the subject of Wannabe Indians published in 1989 in *American Indian Quarterly.* "Wannabe Indians are scorned by 'real' Indians because they pick and choose what customs they will adopt, because they have a 'distorted notion of the way in which Indians live and behave,' " Yates concluded.

June 14, 2012

Anonymous commented on 04-Jul-2012

I think many people tend to lose touch with reality. Ethnicity is and has never been a strictly biological- or genetic-based identity. Rather, it is based on sociocultural upbringing. What so happens to be is that there are some ethnicities that have formed from racial perception and segregation, thus ethnicity is often correlated with the concept of race in this country.

Anonymous commented on 04-Jul-2012

On the whole Jewish/Indian hidden descendant issue. There are crypto-Jews who would probably still be considered Jewish, they seem to have merely "passed", indicating they still considered themselves as Jewish but chose not to admit it to outsiders and also adopted a somewhat syncretic form of Catholicism, resulting from this trying to blend in. Conversos, however, are not Jewish, as they willingly converted and probably never looked back. The whole Tribal enrollment/citizenship in Indian tribes is more political than anything and really doesn't necessarily correlate very well with ethnicity, As there are identifiably ethnic Cherokees who may not be enrolled versus Cherokee citizens who aren't ethnically Cherokee. Just thought I'd clarify.

New Evidence for Old Lines
Old World Roots of the Cherokee Describes 'Anomalous' Haplotypes

Before the DNA Consultants Blog published "Anomalous Mitochondrial DNA Lineages in the Cherokee" in October of 2009, a mere 60 samples, all Oklahoma Cherokee, spoke for the country's largest and "best-known" Indian nation.

Now, the stories of 52 "anomalous" descendants who flesh out the record of Cherokee DNA are available in *Old World Roots of the Cherokee*, a new book by DNA Consultants' founder Donald Yates, with a foreword by Richard Mack Bettis. The 200-page reference work from McFarland is the result of more than 10 years' research and overturns a number of stereotypes about Cherokee history and genetics.

One of the published stories is that of Mary M. Garrabrant-Brower, part of the 13 U haplogroup subjects who had been told by other DNA testing companies they were plainly *not* of Cherokee descent because they didn't have the right haplogroup. "My great-grandmother was Clarissa Green of the Cherokee Wolf Clan," said Garrabrant-Brower, as reported in Yates' book (p. 54). "Her grandfather was a Cherokee chief and my mother maintained the Cherokee language and rituals, even though we moved to the Northeast."

U5 is usually presented as a Eurasian haplogroup. It is among the oldest mtDNA haplogroups found in Homo sapiens in Europe including Cheddar Man, the oldest remains of anatomically modern humans in Britain, with an estimated age of 30,000-50,000 years. Stephen Jett, a geographer who edits *Pre-Columbiana*, maintains if any U haplogroups were to be verified in the New World it would probably be U5.

A second U5 Cherokee in the study is a Scottsdale, Ariz., doctor, who matches only one other person in the world according to the databases available for searching, a descendant of Marie Eastman, born 1901 in Indian Territory. "Because of the precision of the match, he and the descendant of Marie Eastman who was tested are almost certainly cousins in a genealogical as well as genetic sense," reported Yates in *Old World Roots of the Cherokee*. "His own descent is documented from Jane Rose, a member of the Eastern Cherokee Band. Her family is listed on the Baker Rolls, the final arbiter of enrollment established by the U.S. government (p. 54)."

Teresa Panther-Yates, a third U5, of the U5b subclade, has a type that is unique in the world, only found in her own case. "She traces her maternal line back to Isabel Culver, the second wife of Levin Ellis in Hancock County, Georgia. Isabel died about 1838 at the time of the Trail of Tears. There is a tradition in Teresa's family that this line was Cherokee (p. 54)."

An Epochal Shift

As with U, so with T, J and other haplotypes. Whereas the Cherokee descendants who carried these types had been "rejected" by orthodox science and DNA companies who tow the line on such matters, they were vindicated in the new study, often matching other Cherokee and even other participants (previously unbeknownst to them).

Of Yates' findings about the anomalous DNA data, Wake Forest professor Cyclone Covey writes in an introductory note:

> Donald Yates burst onto the scene clarifying Melungeons and went on the explain many more peoples via pioneering researches in DNA. A classically grounded scientist who is

conversant with Hebrew, he detected Cherokee as Greek in what can only be called an epochal shift of thinking, posing the historical challenge of getting Alexandrine and Mediterranean *koine* to Tennessee ahead of Hernando De Soto (p. viii).

Old World Roots of the Cherokee appeared on July 11 and is available from booksellers. Copies signed by the author may be obtained on the DNA Consultants company site.

August 1, 2012

A Red-Hot Tale from the Nation's Capital
By Monica R. Sanowar

I took my first test with Family Tree in 2006. This test showed my mtDNA as L3e2b2 and it went like this:

52% West African
39% European
9% East Asian
0% Native American

I could not believe the East Asian part, and I shrugged it off and thought—that has to be Native American.

So, fast forward—I took another test with Ancestry.com. This was autosomal and showed:

48% West African
44% European
8% Unknown

How can you be unknown?

Neither of these tests really breaks down what country your people may have originated from. So, then I tried 23andMe, their autosomal offering.

49% West African
48.3% European - Central - Northern - Non-specific
and the leftovers were 0.7 East Asian & Native (although the NA

box did not turn red)
and 1.4% unspecified

I knew from family history that NA was on both sides of my fence. I also was aware that I had four of the traits Melungeon people have. I have the ridge in the back of my head that you can lay your finger in; I have ridges on the teeth and I can make the clicking sound on the shovel teeth; I have the Asian eye-fold, and the very high arches. Can't get my foot inside of a boot and if I do, I can't get it off. I was amazed that I got my results in less than two weeks!

Finally, I tried DNA Consultants. Its test was the very first that didn't show "unknown" or nonspecific. Everything was accounted for, although I did find a few shocks. No one told me about Sephardic Jews or the Portuguese. At last, a test verified my Native roots with valid matches to tribes or nations and confirmed Native American autosomal markers— from both parents, as I had been told.

I got into Native culture back in 1983 when I started to go to powwows. I finally felt at home. I enjoyed seeing people that looked like me, mixed. My great-great-great- grandmother was listed on the FREE NEGRO LIST where it asked How Freed? And it was written BORN FREE. Then came a description— a light-skinned black, with long straight black hair and a small scar on her hand. I have a picture of her daughter, Alethea Preston Pinn. Alethea's father was a white man named Allen Preston. Alethea had seven children with James E. Colvin, who was white, and all of their children were put on Walter Plecker's list of "mongrels" not allowed to vote or go to school. That was 1943. Not that long ago.

So, I got a second cousin to take the test with 23andMe who comes directly from Sarah Pinn (the alleged light-skinned black woman). My cousin's haplogroup came in A2N—Native American.

I know that some things may show and some not, but DNA Consultants' test knocked the East Asian right off the page. I've learned a lot of different things with DNA testing, but their test is the best one I have seen and is well worth the money.

I love it when these geneticists and genealogists out there decide what you do or do not have in your family tree, especially the Indian part of the tree. As if this just could not have happened . ... I am proud of all of it. I can just about hang up a flag from everywhere.

I can't praise the DNA Fingerprint Plus enough and wish I'd known about it years ago. I really appreciate all of the knowledge and insight Dr. Yates has about genealogy and history that I was totally unaware of. I actually spoke to him on the phone at length, and he truly made my day. I highly recommend DNA Consultants' service to people who are looking for the truth about their genealogy.

And speaking of spicy mixtures, check out my hot-sauces at Sun Pony. They've got secret, all-natural ingredients just like the family!

July 12, 2013

Elizabeth Colvin, a granddaughter of Alethea Preston Pinn. "Contrary to the belief and convictions of many people, long hair really does exist in my family," says Monica Sanowar. "It isn't a made-up fantasy and this was long before hairweaves. My cousin's hair was down to her calves."

Guest blog author Monica Sanowar is the founder of Sun Pony Distributors Inc., makers of a line of all-natural, wholesome condiments and energy supplements found in stores up and down the East Coast. Her first hot-sauce was Yellow Thunder and her Native name is Sundancer. SunPony's D.C. Redbone Hot Sauce is the official hot-sauce of the Anacostia Indians, D.C.'s little-known indigenous people, who were first recorded by Capt. John Smith in 1608. Sanowar lives in Washington, D.C., not far from the Anacostia's village site, now a national historical landmark. On YouTube you can watch grassdancer Rusty Gillette in a video about D.C. Redbone hot-sauce.

July 12, 2013

Native Americans Traced to Eastern Europe and Northwest Asia, Not Siberia or Mongolia, with Fossil Evidence
Autosomal DNA Set to Rewrite 'Peopling of the Americas'

"In my opinion, the genetic story of the Americas has been botched. Not only are samples flawed but geneticists' times to coalescence are forced into the Procrustean bed of outdated theory." —Donald N. Yates in *Old World Roots of the Cherokee*

A scientific bombshell was dropped at the recent conference on "First Americans Archeology" in Santa Fe, New Mexico. Danish geneticist Eske Willerslev announced that his team had completed the fullest sequencing of ancient human DNA to date—two 24,000-year-old Siberian skeletons. The results showed that the people who lived near Lake Baikal at the dawn of human civilizations and who later expanded as the Native Americans or American Indians of the New World came from Europe, not Asia.

Native Americans today have one-third European genes in their makeup.

"The west Eurasian signatures that we very often find in today's Native Americans don't all come from postcolonial admixture," said Willerslev at his talk, "Some of them are ancient." See the Smithsonian's blog, "The Very First Americans May Have Had European Roots."

See our blog post "Native Americans Have Deep Ancestry in Europe: Yes, It's Official" (October 30, 2013).

Until now it was widely believed that America's First Nations were people like modern-day Mongolians and Eastern Siberian tribes who crossed a hypothetical Bering land-bridge. Study after study has stressed the small number of founder types. With the new revelation of European roots and lack of relatedness to East Asians, however, geneticists are going to have to play catchup with those (including Native Americans themselves) who have insisted all along on diversity and complexity in New World populations rather than sameness and simplicity. The Danish-U.S. team's paper is in final editing by the journal *Nature* and expected to appear very shortly.

DNA Consultants offers a test called Rare Genes from History that will tell you if you have the Lake Baikal Gene, First People's Gene, Cochise Gene or Amerind Gene, all of which are centered in their historical distribution on the motherland of Mongolians and Native Americans in South Siberia. There are also the following specifically European genes: Helen, Circassian, Europa and Scythian. With the tables turning, you can now legitimately have both European and American Indian markers and it may not be a case of admixture.

For an easy way to absorb the latest research on Indians and Mediterranean/Jewish people, listen to the narration by Jack Chekijian for Donald N. Yates' *Old World Roots of the Cherokee*. It is available on Amazon, Audible and iTunes.

October 30, 2013

The Indians Came from Europe

American Indians are now known to have deep roots that are roughly one-third European, genetically speaking. There have always been a few hardy souls that did not swallow the bedtime story that all Native Americans came across the Bering Land Bridge and their oldest origins were thus from East Asia. Now there will be more dissenters.

The "peopling of the Americas" used to be a sacred cow. Anyone who veered from it was seen as stepping into "fringe" science. He or she may as well be discussing UFOs. Nevertheless, genetic testing has now proven that fully one third of Native Americans have deep West Asian and European genetic genealogy roots, and the latter does not just stem from the time of the colonists' arrival. There is an amazing story here.

First, how do we know it? We have long since witnessed that Native Americans have "unexpected" European genes when we do DNA testing, but it has been assumed that this could only be because of trysts between Native Americans and colonists in modern times (Michael Balter, *Science,* "Ancient DNA Links Native Americans With Europeans").

Perhaps that is why so many Cherokee princesses fell at the feet of rugged mountain men in our family history. While I am not discounting this scenario as a possibility in your family tree, it looks like there is a larger story here that has been revealed by DNA testing.

The autosomal DNA testing results of the entire genome of a young Siberian boy who died some 24,000 years ago are revolutionary. What do his DNA testing results show? He is, as is expected, most closely tied genetically with Native Americans. In fact, a portion of his genome only connects with this group.

What was not expected is that his Y haplogroup matches populations "almost exclusively" in Europe and West Asia. Moreover, East Asia, thought previously to be the deepest ancestral link for Native Americans, is not found to be a genetic link in the analysis of his genome. It is now believed that this was an admixture later painted onto this now more-complex genetic picture (Balter).

American Indians have European ancestry? Oh, this will cause consternation in several elite quarters.

August 16, 2014

Shocking Revision

The ecstatic waters ...
Through their ancestral patterns dance.
— William Butler Yeats, "News for the Delphic Oracle"

We've been saying it all along, but it looks as though geneticists may be forced by new findings in ancient DNA to admit that early Siberian people and present-day Native Americans both have strong roots in Europe, only secondarily in Asia. The nuclear genetic bomb was dropped by Danish geneticist Eske Willerslev at a conference on "First Americans Archeology," held October 16-19, 2013, at Santa Fe, N.M. The city that gave birth to the original atom bomb hosted a glittering roster of speakers in a venue better known for its turquoise jewelry, fry bread and avante garde art, including big draws Achilli, Adovasio, Dillehay, Gonzalez and Schurr.

The paradigm-shifting conference program will be commemorated with a book *Paleoamerican Odyssey* ($56) to be published by Texas A&M Press later this year.

Leaked reports in the news media focused on Willerslev's paper, "Genetics as a Means for Understanding Early Peopling of the Americas," which concerned the genetic sequencing of two ancient Siberians' bones discovered in the 1920s and now in the Hermitage Museum in St Petersburg. Analysis of a bone in one of the arms of a boy found near the village of Mal'ta close to Lake Baikal yielded the oldest complete genome of a modern human sequenced to date.

Of the 24,000-year-old skeleton that was Exhibit A, Willerslev was quoted in *The Siberian Times*, as saying, "His DNA shows close ties to those of today's Native Americans. Yet he apparently descended not from East Asians, but from people who had lived in Europe or western Asia." He added, "The finding suggests that about a third of the ancestry of today's Native Americans can be traced to 'western Eurasia.'"

The 4-year-old boy, who died 24,000 years ago in a homeland previously assumed to account for all the Indians who crossed a theoretical Bering land-bridge and founded the First Americans, had a male Y-chromosomal haplogroup of R1b, the most common lineage in modern Europe, and a female mitochondrial lineage of U, the dominant prototype in pre-historic Europe. As it happens, I am the same combination, R1b for male and U for female, as are innumerable others in our in-house study on Cherokee DNA, published, lo, some five years ago.

Whereas previous "peopling of the Americas" stuff has clung to and recycled haplogroup studies (sex-lines), the new shock research relies on autosomal DNA, total genomic contributions from all ancestral lines, not just male-only, not just female-only descent. The title of a blog from Eurogenes rightly emphasizes this: "Surprising aDNA (autosomal) results from Paleolithic Siberia (including Y DNA R)."

When we introduced the 18-Marker Ethnic Panel as an enhancement for our main autosomal product, DNA Fingerprint Plus, lo, again, these five years now and counting, we presented a map of prehistoric human migrations showing without any equivocation that "Native Americans," even as Cavalli-Sforza demonstrated two decades ago, were closer in genetic distance to Europeans than Asians. In fact, we claimed, on the basis of autosomal DNA, that having Native American I or Native American II was a result discrete and separate from East Asian, since Native Americans obtained frequencies of its occurrence as high as 80 percent, and Asians were on the polar opposite of the scale, at the bottom for carrying it. Other methods frequently confused Native American and East Asian to the point of invalidity, particularly those products claiming to arrive at racial or ethnic percentages.

The moral is that autosomal DNA trumps Y chromosome and mitochondrial evidence, and only ancient autosomal DNA can truly explain modern DNA. Even one of the most antipathetic students of American Indian DNA, Theodore G. Schurr, seems to be rethinking the rigid definitions that have built careers and won tenure for geneticists and anthropologists for decades. For the fanatics who have been toeing the party line on haplogroup Q, as set down by Schurr's company, Family Tree DNA, and its followers, we note the following statement of recantation or at least qualification, taken from the Santa Fe program.

Tracing Human Movements across Siberia and into the Americas: New Insights from Mitochondrial DNA and Y-Chromosome Data

In this paper, I present genetic data from native Siberian and indigenous populations of North America that help to address questions about the process and timing of the peopling of the Americas. These new genetic data indicate that Eskimoan- and Athapaskan-speaking populations are genetically distinct from one another, as well as each to Amerindian groups, and that the formation of these circumarctic populations was the result of two population expansions that occurred after the initial expansion of settlement of the Americas. Our high-resolution analysis of Y chromosome haplogroup Q has also reshaped the organization of this lineage, making connections between New World and Old World populations more apparent and demonstrating that southern Altaians and Native Americans share a recent common ancestor. The data also make clear that Y-chromosomal diversity among the first Native Americans was greater than previously recognized.

Overall, these results greatly enhance our understanding of the peopling of Siberia and the Americas from both mtDNA and Y-chromosome perspectives.

"Genetic genealogy" has become a fashionable buzzword, but to my knowledge few research studies or blogs and hardly any commercial tests authentically combine the two concepts. According to genealogy, I myself am about one-quarter Choctaw-Cherokee and three-quarters European. But genetics says my mitochondrial line (U2e) is Eurasian, even though I have traced it to a Cherokee woman who married the Indian trader Enoch Jordan about 1790 in north Georgia. Estimates from other "genetic genealogy" companies for my Native American ancestry, and I've taken them all, range from 0 percent (23andMe) to 8 percent (Family Tree DNA, AncestryByDNA).

DNA Consultants, the company I founded in 2003, does not give percentages of ancestry by policy, but half my top matches in our autosomal analysis are Native American, and North Asian/Siberian is No. 1 in my megapopulation result, followed by Central Asian and Native American (and only distantly by Northern European). On an autosomal approach, if not haplogroup basis, my genes are Native American, which is how I self-identify. If I were to be pulled over for being a brown person in the state of Arizona, where I currently reside, and Sheriff Joe ran my DNA profile numbers through the system he would find that I am 15 times more likely to be North Asian than Northern European, and twice as likely to be American Indian than East Asian, European American or Iberian American (Hispanic).

Read the whole analysis of my personal genetics, with actual reports from various companies, in the Cherokee Results pages on the DNA Consultants website. You may also find an extended study showing what autosomal DNA can do at: "Reconstructing Your Ancestry and Parentage" (blog post, March 14, 2012).

If and when geneticists get serious about identifying the European sources of the American Indian gene pool, and hopefully they will round up not just one suspect (Denmark?), I would like for those who get paid and promoted to study us to please consider the following points:

— First New Cherokee Data Published in More Than Ten Years

(announcement, August 1, 2012)—in-house study described numerous instances of U, findings published in Donald Yates' *Old World Roots of the Cherokee*.

— Stephen C. Jett, who taught geography at The Ohio State University 1963-1964 and then at the University of California, Davis, serving thrice as geography chair and becoming emeritus in 2000, current editor of *Pre-Columbiana*, has frequently pointed out that just because Native American haplogroups match Siberian haplogroups it doesn't mean the population of either Native America or Siberia was the same in remote history as today. He considered this a big fallacy of Big Science.

— Constantine Rafinesque, whose *History of the American Indians* was the first and most comprehensive treatment of the subject, believed all the early settlers of the Americas came "through the Atlantic," and only beginning about 1000 B.C.E. did the Iztacans and Oguzians (Central Asian Turkic peoples) arrive. See our blog: "American Indian and Turkic People Share Deep Ancestry" (June 6, 2012).

— Canadian environmentalist Farley Mowat, the author of thirty-seven books, has constantly challenged the conventional knowledge that Vikings were the first Europeans to reach North America. In *The Farfarers* he describes the Alban people of Old Europe as visitors and colonists from the time when walrus hunters discovered the sea routes to the West before the Bronze Age. America's original name of the White (or Beautiful) Land is mentioned by Rafinesque and in Hindu, Greek, Egyptian, Mesopotamian, Arabic, Algonquian Indian, Irish, Norse and Chinese accounts. See "An Interview with Farley Mowat" on YouTube.

— Cyclone Covey of Wake Forest University, among other historians, has noted that Clovis Culture appears fully formed without any antecedents in America, with the most perfect examples of Clovis points traced in a cline of occurence in archeological sites to the Atlantic Coast.

— The earliest Americans clearly practiced the same Mother Goddess religion elaborately documented in the east Mediterranean and Old Europe by Marija Gimbutas. Their ideas of matriarchy or gylany (in Riana Eisler's coinage) did not come from Asia. See "Archeologist of the Goddess" (webpage) and "Syncretism, Not Animism" (PPT), a presentation given at the Sandy, Utah, conference, March 29, 2011.

— When customers of DNA Consultants with various degrees of Native American admixture have their European population matches analyzed, a frequent top result is Finland or Finno-Ugric or Uralic. This "false match" could be explained by shared ancestry between the present-day Finns (where U is a common haplogroup) and ancestors of Native Americans coming from Europe.

— John L. Sorenson and Carl L. Johannessen in *World Trade and Biological Exchanges before 1492* (2009) document several plants that originated in the Eastern Hemisphere (not Asia) and traveled early by human hand to the Americas. For instance, Cannabis sativa (marijuana) moved from Western Asia or Europe to Peru by 100 C.E., and mugwort (Artemisia vulgaris) was brought to Mexico from across the Atlantic Ocean by 1500 B.C.E. Both grow in profusion in Europe and temperate parts of Central Asia. Goosefoot (an important Ohio Valley Moundbuilder staple), cotton, coconuts, bananas, turmeric and North American myrtle likely took the same route. In the opposite direction, Mexican agave spread to the Mediterranean world by 300 B.C.E.

Archeologists described the recent news from Santa Fe as jaw-dropping. We expect that when the definitive report on the Siberian boy's 24,000-year-old genome appears in the journal *Nature*, where it is at press, their hair may fall out! At any rate, the European origins of Indians is going to be a game changer not only in genetics, but anthropology, archeology, government and, perhaps most significantly, in the self-awareness of millions of Americans who count Native Americans among their ancestors.

October 30, 2013

Who Are the Cherokee?

My Cherokee grandfather was a jack-of-all trades my grandmother told me. I thought he was about as tall as God.

The Cherokee are the New York Jews of the Native Americans. They are proud, intelligent, rich and worldly. Think Cher (Turkish/Cherokee). Some sent their daughters to Paris to attend universities. According to the Cherokee National Historical Society, after the Civil War, the Cherokee government "operated its own legislature, court system, jails, newspapers, an orphanage, and schools including seminaries of higher learning for both men and women." Living in cabins and often intermarrying with whites, they lived in a westernized fashion—modern for the time.

The story of their origins depends on who you ask. It is common knowledge that the Cherokee have a mixed ancestry because they intermarried with whites. And they have been mixed for some time.

However, the theory has long been that the First Americans all decided to cross the Bering Land Bridge walking out of Siberia, hunting Ice Age elephants, and trekking across the tundra to people the new world some 13,000 years ago or more. This theory still holds sway over most scientists. It was taken for granted in a recent *Nature* article, "Reconstructing Native American Population History" by Harvard's David Reich.

Despite this, not everyone thinks this is the story. Melissa Ludwig, in her article, "Bering Land Bridge Theory Disputed," quotes Michael Collins, an archaeologist with the Texas Archaeological Research Laboratory at the University of Texas at Austin, as saying there is mounting evidence otherwise. Ludwig writes that, "for more than 20 years, Collins and other scientists have been digging up artifacts from Chile to Texas that convince them the first Americans didn't walk here at all, but came by boat, and arrived much earlier than thought."

"The Bering Land Bridge theory has held sway for 70 years or so," Collins says, "But a few of us for the last 25 years have come to seriously doubt that theory." If this is correct, the Cherokee as well as other American Indians must have come by a different route to the Americas. Did they?

Much of what we now take for granted in science has had to fight for recognition against what was once considered the established science of the day in Galileo fashion. Some historians and others have suggested that the origins of the Cherokee (and other Native Americans) might not be explained by just the Bering Land Bridge theory. If this is true, their ancestry could indeed be very different.

What do the Cherokee say about their own origins? Vine De Loria, in the book, *Red Earth, White Lies,* asserts that the Cherokee—if not from America—must have come from across the ocean as their creation stories suggests.

And science is beginning to suggest they did just that. Though DNA is an evolving science, through a combination of autosomal DNA studies and historical analysis some have suggested possible Mediterranean, European, and Mid-Eastern connections. (See Dr. Yates' book *Old World Roots of the Cherokee.)*

Two leading anthropologists, Michael Collins and Bruce Bradley, in the book, *Across Atlantic Ice,* conclude that some of the earliest inhabitants to America came from the Mediterranean (Spain) and other parts of Europe by boat.

The *Smithsonian* article, "New Book Reveals Ice Age Mariners from Europe were America's First Inhabitants," (March 2012), quotes Bradley as saying, "We now have really solid evidence that people came from Europe to the New World around 20,000 years ago. Our findings represent a paradigm shift in the way we think about America's early history. We are challenging a very deep-seated belief in how the New World was populated. The story is more intriguing and more complicated than we ever have imagined."

There is now clear evidence of other routes as well.

"There are more alternatives than we think in archaeology and we need to have imagination and an open mind when we examine evidence to avoid being stuck in orthodoxy," Dennis Stanford of the Smithsonian's National Museum of Natural History adds. "This book is the result of more than a decade's work, but it is just the beginning of our journey."

The origins of the Cherokee—and other Native American peoples— seems to be a more complex story than is generally assumed.

August 17, 2014

7 DNA AND EVOLUTION

Monkey's Uncle
Nature Alert, 28 June 2006

According to an article titled, "Genetic Evidence for Complex Speciation of Humans and Chimpanzees," by Nick Patterson et al., humans and chimpanzees may have once interbred.

The evolutionary split between humans and chimpanzees is much more recent than was thought, according to a new comparison of the complete genome sequences of the two species. The data show that the split occurred no more than 6.3 million years ago, probably less than 5.4 million years ago.

The speciation process appears to have been rather unusual possibly involving an initial split followed by later hybridization before a final once-and-for-all separation.

June 30, 2006

A'Changing Times
Check Out Heredity's Free Podcast Series

Each month Elli Leadbeater and Steve Le Comber present a free audio show. The podcast features interviews with the people behind the science and a digest of breaking news from *Heredity* editor Richard Nichols at Queen Mary University of London.

Featured in January's episode is H.-J. Bandelt of the University of Hamburg, a leading statistician in the area of population genetics. His commentary on mutation rates, "Clock Debate: When Times are A-Changin': Time Dependency of Molecular Rate Estimates, Tempest in a Teacup," appeared in *Heredity*, vol. 100, in February 2008.

According to Bandelt, many current studies on the molecular clock use outmoded data and confused models to arrive at their conclusions regarding human evolution and ancient migrations. Experts do not always distinguish between hot spots in DNA sequences and an average rate of mutation. They may base their suppositions about changes in coding regions (genes) on variation in non-coding regions such as the control loop in mitochondrial DNA. For these reasons, many of our theories about time to coalescence of genetic types may be essentially flawed. Bandelt mentions the model of Amerind migration as an example.

You can subscribe to the service and receive the latest episode in your email or browse the archives.

January 8, 2008

Not in the Genes

A major new multidisciplinary study in the October issue of *Genetics* concludes that human variation and evolution were sped up by learned behavior and culture, and that molecular variation and gene expression—long thought to be the primary driving forces—work in tandem with environment to shape human beings.

The article is titled "Explaining Human Uniqueness: Genome Interactions with Environment, Behaviour and Culture" and is contributed by Ajit Varki, Daniel H. Geschwind and Evan E. Eichler, all at the Center for Academic Research and Training in Anthropogeny (CARTA).

"Rejecting any 'genes versus environment' dichotomy," they write, "we consider genome interactions with environment, behaviour and culture, finally speculating that aspects of human uniqueness arose because of a primate evolutionary trend towards increasing and irreversible dependence on learned behaviours and culture—perhaps relaxing allowable thresholds for large-scale genomic diversity."

The new approach in evolutionary theory overcomes an oft-voiced criticism that the rate of spontaneous, random mutation is too slow to accommodate the changes in species, especially quantum leaps.

"Anthropogeny" is a newly dubbed field that seeks to explain the origin of humans.

October 2, 2008

Scary Findings

In a report published in the October 30 number of *Science*, Chinese paleoanthropologists claimed that the jawbone and teeth unearthed by them recently in the southern province of Guangxi represent a form of man 100,000 years old. Their interpretation of the fossil challenges the Western theory that claims our ancestors peopled the world in a migration out of Africa late in the last Ice Age, about 50,000 years ago. But there is more. American and European scientists stand to lose even more face if China's insistence is true that this early human is a hybrid with *H. erectus*, a more primitive species also known as Peking Man.

Discoverer Jin Changzhu pointed out that the jawbone curved outward, whereas that of the older species of H. erectus had an inward-sloping chin, and modern human chins generally jut out farther than the Guangxi specimen's. Such an intermediate chin, he said, suggested interbreeding with *H. erectus*.

In the West, paleoanthropologists and geneticists for the most part vehemently deny that any interbreeding between species of man could have taken place before our type emerged as the sole and supreme species favored by evolution. Neanderthals, they claim, were replaced by modern humans in Europe and died out without a trace in the genetic record about 30,000-40,000 years ago.

"The initial publication makes shaky claims based on preconceptions," scoffed Tim White, a paleoanthropologist at the University of California, Berkeley.

The title of the report is "Signs of Early *Homo sapiens* in China?" It was written by Richard Stone and published on page 655 of vol. 326, no. 5953, of *Science*.

The confrontation reminds many of the battle over Peking Man dated to about 500,000 years old. At first, the Chinese maintained that East Asian people were descendants of Peking Man (and not of Africans or out-of-African humans). Later, they modified their view and held that modern Asians represent a hybridization with Peking Man. Possibly all the "races" or continent-specific forms of modern man are the result of anatomically modern humans interbreeding with more primitive hominids in their part of the world.

We wonder why Western scientists are in such a huff about the conclusions of Chinese paleontologists since there is solid proof of admixture between modern humans and archaic human groups like Neanderthals, Homo erectus and Homo floresiensis (the fossils of "hobbits" discovered in Indonesia in 2004). One instance among many of publications demonstrating this possibility is: Jeffrey D Wall, "Detecting Ancient Admixture and Estimating Demographic Parameters in Multiple Human Populations," *Molecular Biology and Evolution*, vol. 26, no. 8 (August 2009), pp. 1823-27.

Perhaps the right hand of genetics doesn't know what the left hand of anthropology is doing.

October 30, 2009

We Are All Part Neanderthal

The bombshell arrived with the May 7, 2010, issue of the journal *Science*. Titled "A Draft Sequence of the Neandertal Genome," it presented the years-long attempt of an international team of scientists to derive DNA from ancient female Neanderthal bones and determine if there was any genetic overlap with humans. The news was so sensational that the journal made the original scientific report and all collateral materials free to everyone, along with a podcast, multimedia presentation "The Neandertal Genome" and slew of links and forums for comments.

The original press release from the Max Planck Institute for Evolutionary Anthropology in Leipzig was embargoed for May 6, 2010, 8 p.m.

Discovered in a quarry in Germany in 1856, 40,000-year-old Neanderthal man became the first recognized early human fossil. The debate immediately began whether Neanderthals were a separate species or sub-species of Homo sapiens. German language orthographic reforms rendered the spelling of the name Neandertal in the 20th century, although most people even today prefer to stick with the *th* of the original word. Neanderthals are named after the Neander Valley (German *thal* or *tal*) in which they first came to light.

More and more of them turned up over the years: in Belgium (1886), a nearly complete skeleton in southern France (1908), Israel (then-Palestine, 1930) and Iraq (1953). The first ambitious genetic work was a partial sequencing of their mitochondrial DNA based on highly degraded specimens: Krings et al., *Cell* 90, 19 (1997). A second mtDNA sequence was achieved in 2000. The complete mtDNA sequence came in 2008: Green et al., *Cell* 134, 416 (2008).

In the meantime, Neanderthals were found to have red hair and fair skin, body paint, customs, societies, rituals and art. They used fire, tools and weapons. They hunted bison, horses and other large animals and made bread of acorn meal. With their short arms and weak shoulder sockets, however, they probably could not throw spears. Before they were conquered by their smaller human cousins, they had colonized an area extending from Spain to Western Siberia and the Middle East. They were acclimated to northern Europe's icy temperatures and flourished especially before and during the last Ice Age. Then, suddenly, about 30,000 years ago, the fossil record goes silent. Their last holdout appears to have been in Spain.

Our picture of Neanderthals is likely to change radically now that we know they were among ancestors of ours, not a dead-end, primitive race. Some writers had already speculated, in fact, that Neanderthals were more advanced in many ways than their rivals, Cro-Magnon Man. Certainly, their religion was highly adumbrated. Some carried Venus figures on necklaces. According to the author of *The Neanderthal's Necklace*, Juan Luis Arsuaga (co-director of the World Heritage Site Sierra de Atapuerca in Spain), at one burial in Russia, a 60-year-old adult had 3,000 beads of drilled mammoth ivory sewn onto his clothes. A boy in the same burial wore a belt decorated with 250 arctic fox canines. There were also shells, armbands, head ornaments, bracelets, pendants, assegais, ceremonial staffs and other artifacts made of bone, antler, ivory and stone (p. 294).

The blockbuster draft of the Neanderthal genome just published noted genes linked to cognitive abilities, geo-spatial skills, language and motor coordination as well as strength, reproductive advantages and (what we knew already) cold adaptation. Much attention is likely to focus on the Neanderthal's signature occipital bun, noticed in isolated or vestigial populations like the Berbers, Saami, Canary Islands, Native Americans, Australian Aborigines and Melungeons. These populations probably preserved greater proportions of Neanderthal admixture than others.

Because the genetic legacy of Neanderthals (so far) has not been detected in the mitochondrial record, it is believed that gene flow came from males mating with human females. No male Neanderthal lines survive—not surprisingly. Only autosomal DNA reveals the Neanderthal contribution to human populations.

May 12, 2010

Neanderthals Had Language

Michael Balter, "Neandertal Jewelry Shows Their Symbolic Smarts," *Science,* vol. 327. no. 5963 (15 January 2010) 255-256 doi: 10.1126/science.327.5963.255

A handful of marine mollusk shells, possibly used as necklaces and paint cups, shows that Neanderthals expressed themselves symbolically, say the authors of a paper published online this week in the *Proceedings of the National Academy of Sciences.* They argue that the findings suggest that social and demographic factors, rather than cognitive differences, best explain why so-called modern behavior was relatively rare among Neanderthals.

January 14, 2010

100,000-Year-Old Neanderthal Artifacts Found in Britain

Ancient flint meat-cutting tools found in sediment along a highway in Dartford, Kent, prove Neanderthals were present in Britain before the beginning of the last ice age (and possibly before the previous two or three cold periods or interglacials), when the British Isles were joined to mainland Europe due to low sea levels. The find pushes back the earliest known evidence for Neanderthals in England by 40,000 years, according to a June 1 report in the *Daily Mail* headlined "Neanderthal Man Was

Alive in Britain at Start of Ice Age.

Previous notions postulated "pre-Neanderthals" (perhaps *H. heidelbergensis*) sparsely occupying Britain in the Pleistocene Period but vacating it when the climate became prohibitively cold about 200,000 years ago. It was thought that true Neanderthals did not reach Britain until about 60,000 years ago, not far in advance of their cousins, *H. sapiens sapiens.*

Neanderthals were known to be in northern France and Belgium about the same time. Their range must now be extended to Britain, and that country's prehistory recast. As in the rest of Europe, it appears Neanderthals were living in Britain hundreds of thousands of years ago since a date like the Dartford find is only a *terminus ante quem*. If Neanderthals were living 100,000 years ago off the present-day junction of the M25 and A2, they probably had not just arrived. Similarly, they probably did not die off or move out of Britain in the next generation. It was not an excursion or isolated incident. The chances are unlikely that the first, and so far only, evidence of Neanderthals in Britain would be from the exact beginning or precise end of their population's stay.

Current thinking about interglacials is changing, well, like the weather, but there have undoubtedly been many eons-long stretches of time during which Britain was temperate and hospitable to humans. Since the Middle Pleistocene about 600,000 years ago, significant advances of continental ice sheets in Europe have occurred at intervals of approximately 40,000 to 100,000 years. These long glacial periods were separated by more temperate and shorter interglacials. The age of Neanderthal humans is conventionally set at 400,000 years before present but may be revised backward as we learn more and encounter more fossil evidence.

June 30, 2010

Are Neanderthals Out of the Doghouse?

We predicted as much: Anthropologists are beginning to have a more positive attitude toward the role of Neanderthals in human prehistory. According to an article in today's *Washington Post*, "Scientists are broadly rethinking the nature, skills and demise of the Neanderthals of Europe and Asia, steadily finding more ways that they were substantially like us and different from the limited, unchanging and ultimately doomed inferiors most commonly described in the past."

The article by Marc Kaufman is titled "Anthropologists Adopt a More Favorable View of Neanderthals," and appeared in the October 4, 2010, edition of the newspaper.

Earlier research this year noted that Europeans have, on average, 1 percent to 4 percent Neanderthal genes. That began the wheels of scientific thinking rolling.

"Our picture of Neanderthals is likely to change radically now that we know they were among ancestors of ours, not a dead-end, primitive race," we wrote in the blog post "Most Humans Part Neanderthal" on May 12. DNA Consultants introduced its Neanderthal Index in June.

An important paper that is helping restore Neanderthals' position in prehistory is "A Niche Construction Perspective on the Middle-Upper Paleolithic Transition in Italy," by Julien Riel-Salvatore (*Journal of Archaeological Method and Theory*). Riel-Salvatore also has a blog on prehistoric toolmaking and related subjects. Among his perceptions is that Neanderthal DNA was probably strong at first but got watered down in the course of time. That is confirmation for our targeting archaic populations to measure your Neanderthal Index.

October 5, 2010

Cave of the Mare's Nest

Jewelry from the Grotte du Renne "reindeer cave" at Arcy-sur-Cure in Central France has long been assigned to Neanderthals, helping rehabilitate them in the picture paleontologists paint of early mankind. But these artifacts have now been questioned thanks to a redating of the lowest levels of the cave, where Neanderthals were presumed active, according to the journal *Science* (vol. 330, no. 6003, p. 439) in October 2010.

In the new study, a team led by dating expert Thomas Higham of the University at Oxford in the United Kingdom reports 31 new radiocarbon dates from the Grotte du Renne using novel methods designed to avoid contamination. The dates, obtained on materials such as bone tools and ornaments made of animal teeth, paint a disturbing picture: While upper layers attributed to modern humans clock in at no older than 35,000 years, artifacts from the Châtelperronian levels range from 21,000 years old, when Neanderthals were long extinct, to 49,000 years old, before the Châtelperronian began. About one-third of the dates were outside the expected range.

What we don't understand is the dichotomy between "modern humans" and Neanderthals to begin with. If the dating was mixed (contaminated) how can paleontologists be sure, since "modern humans" and Neanderthals were mixed themselves according to all proven genetic analyses. These retests seem to be splitting hairs to prove or disprove pet theories that no longer apply.

It is particularly nonsensical to maintain, as the archeologist Randall White of New York University has said, "This key site should be disqualified from the debate over (Neanderthal) symbolism." The debate over Neanderthals' symbolic intellectual and communicative capacities has already been settled to 99 percent of the scientific world's satisfaction. It is a non-issue. Paintings, jewelry and art from over 32 Neanderthal grave sites and camps settled it several years ago, and since the Neanderthal genome was sequenced in draft form earlier this year, it has been demonstrated that they had the same or similar intellectual genes as "modern humans." In fact, their brains were bigger, so if anything they had a greater capacity to conceptualize the world.

We wish archeologists and science writers would work on their own symbolic thinking a bit and let the bones of tired myths and ethnocentric fallacies rest in peace. And pu-lease spell the word wrongly right. It's N-e-a-n-d-e-r-t-h-a-l. Neandertal is purist and is not going to become popular. Trust us.

Next thing you know they're going to be talking about Visigots and Ostrogots.

October 21, 2010

Is Evolution Accelerating?
What Happens When 'Pressure of Natural Selection' Ceases?

Maybe I do not understand "classic Darwinism," but I am puzzled by the claims of numerous articles in leading scientific journals that evolutionary change in human beings is "accelerating." Since when?

One such article was published several years ago by the National

Academy of Sciences, "Recent Acceleration of Human Adaptive Evolution." It claims an acceleration in evolution in the last 1,000 years, on an inferred and theoretical, not observed basis, needless to say. Its projections are based on the 4 million SNP database called HapMap, a collection of mostly European and American gene irregularities. Several books and a slew of articles in journals like *Nature* and *Science* have jumped on the scientific bandwagon. But we have a few embarrassing questions.

If the "classic" model of Darwinism was based on primitive man, New Guinea orchids and Galapagos turtles, wouldn't we need to redefine what selection and survival of the fittest and its other tenets really mean for, say, a Harvard or Stanford white male Ph.D. who sits in front of his computer most of the time or the current population of the United States, few of whom qualify as hunter-gatherers or tribal or starved or threatened by predators?

Isn't there rather a disconnect today between surviving and your genetic composition? Pretty much anyone can marry and have children with pretty much anyone they want to, and few people are dying off without issue from bad genes or fatal decisions.

If genes are supposed to mutate and produce more "fit" and eligible marriage partners because of "pressure" from the environment like drought or low-protein diets or too much or too little sunshine, how does evolution operate on a level playing field where the environment is a shopping mall or suburbia?

World population has increased from 200 million in the year 1 to nearly 7 billion. Perhaps the presumed rise in SNPs is an effect of population growth? Evolution was supposed to operate on the basis of selection, not indiscriminate proliferation.

Evolution, it is claimed, has now sublimated beyond the physical realm. It is internal, or mental, or spiritual, or cultural, or social, or political, or even (God help us) scientific. This was the super-clever dodge favored by Teilhard du Chardin, a Catholic missionary to China who wrote *The Phenomenon of Man*. He called it the noosphere. Whatever else, this latest Big Science Fair project requires a new definition. There are no measures possible beyond our present lifetime. We have no way of judging a Neanderthal's consciousness.

I'm sorry but I just don't get it. I wish someone would please explain it to me so I do not need to feel guilty about not reading all those pat-me-on-the-back and pip-pip-don't-you-know-old-man Victorian recidivist articles that I find, somehow, very self-serving.

I am an avid follower of the Darwin Awards, but the only thing I can see accelerating are modern American and European scientists' delusions of grandeur.

December 14, 2010

Neanderthals and Other Friends
Interbreeding with Neanderthals and Denisovans Conferred Immunity to Diseases, Aided Spread of Humans in Asia, Europe

According to a professor of immunology and microbiology at Stanford University, humans were able to survive, spread and expand their populations once they left Africa because of immunities to disease they acquired from Neanderthals and Denisovans, who had lived in Europe and Asia already for hundreds of thousands of years.

A review of the new research appears in the online science magazine *Discover* under the date of June 20, 2011. The professor's name is Peter Parham.

Crux of the Matter, according to Royal Society Report

- Parham began by taking a close look at a family of genes called human leukocyte antigens (HLAs), which play a central role in our body's immune responses. We are able to react to a wide array of diseases because our HLA genes are highly variable, each containing dozens of alleles (forms of genes).
- Our ancestors in Africa, however, would have had a small number of HLA alleles because they likely traveled in small bands and had little contact with other groups. Moreover, their HLAs would have only protected them against African diseases.
- When Parham compared the HLAs of modern humans with those of Neanderthals and Denisovans, he noticed some overlaps. In particular, he found that HLA-C*0702, an allele common in Europeans and Asians but nonexistent in Africans, was also present in the Neanderthal genome. Similarly, HLA-A*11,

153

which is found in modern Asians but not in Africans, popped up in Denisovan DNA.

- Overall, about 50 percent of HLA Class I alleles in Europeans seemed to come from Neanderthals, 70 percent to 80 percent in East Asians from Denisovans, and 90 percent to 95 percent in Papuans from Denisovans, Parham said at a recent *Royal Society* meeting.

The latest revelation about the true nature of Neanderthals shows how fast current scientific and popular thinking is moving on the subject. Two years ago, it was still debated whether "humans" could interbreed with Neanderthals, or whether Neanderthals were even a human species. Denisovans were only discovered in the last year.

July 1, 2011

Victorian Caveman

Did you know that Charles Darwin's *On the Origin of Species*, published in 1859, does not contain a single mention of the word "evolution"? I am reading it for the first time and was struck not only by the absence of that term in Darwin's first edition (it does begin to creep in after 20 years in later editions) but many other discrepancies between the historical Darwin and modern Darwinism.

For those inclined to believe conspiracy theories—for instance, that it was not Darwin, but Darwinists or even anti-Darwinists who invented the theory of evolution—here are some items to consider:

- Darwin was a mediocre student at Cambridge, where he learned little science or mathematics and preferred theology. "He passed the final examination in January 1831 at an undistinguished position, 10th in the list of candidates who did not seek honours" (Richard Keynes, in the introduction to the first edition of Darwin's *On the Origin of Species*, published by the Folio Society, 2006).
- Darwin's disordered notebooks and papers, in which he supposedly developed the theory of evolution, only to keep it "secret" for 20 years, were not transcribed and published until 2000.
- The modern reconstruction of Darwin's theory of evolution evolved

itself. It began to take iconic form only after the 1960s, when historians of science began to "read between the lines" of Darwin's work. His career was divided into an initial period of 10 years when he was a biologist on the Beagle, then a "secret period" of 20 years until he and Alfred Russel Wallace "simultaneously" broke the theory in 1858, and finally another 20-year period until "evolution" began to appear in his writings by name shortly before his death.

- Darwin's interests were erratic, not to say eccentric. He spent eight years studying the sex life of a Peruvian barnacle. He published four extensive monographs on the subject between 1851 and 1854. During this period he wrote nothing on "transmutation of species," the early term for "evolution." His young son asked a playmate, "Where does your father do his barnacles?" (Keynes, xxi.)

- After barnacles, Darwin turned his attention to fancy show pigeons. He joined the Philoperisteronic Society and added an aviary to his house.

- The scientific establishment at the time did not exactly acclaim Darwin's *On the Origin of Species*. The geologist Adam Sedgwick wrote Darwin a scalding letter. "Many of your wide conclusions are based upon assumptions which can neither be proved nor disproved," Sedgwick said, accusing Darwin of "deserting the true method of induction." (xxiv). The astronomer Sir John Herschel called Darwin's work "the law of hiddeldy-pigglety."

Finally, no matter what you might decide about the "evolution of evolution," both Darwin and Darwinists reject the idea that Neanderthals might have been, in the words of the subtitle of *On the Origin of Species*, anything like a "Favoured Race." They died out, right? This being so, it is interesting to me that a portrait of Charles Darwin exhibits most, if not all, the characteristics of a Neanderthal: sloping forehead, powerful jaws, craggy brow, occipital protuberance and large nose. We don't know quite what to make of it but wish Darwin had willed his skull as well as his thoughts to science.

September 30, 2011

Were Neanderthals the First Artists?

Were Neanderthals capable of creating art? The idea seems shocking to us. After all, we learned in school that these were brutish savages without higher thinking and symbolic thought or expression. The picture of a Neanderthal making hand prints in Spanish caves or making shell necklaces is odd indeed because art is largely "considered evidence of sophisticated symbolic thinking, (and) has traditionally been attributed to modern humans, who reached Europe some 40,000 years ago" according to the recent Wired Science article, "First Painters May Have Been Neanderthal Not Human."

In the article, a stunning panel of hands is displayed in the El Castillo Cave, Spain, dated to 37,300 years old. Perhaps, an ancient Neanderthal kindergarten. In fact, they remind me of my daughter's kindergarten painting of her hands dipped in red and blue finger paint and pressed onto manila paper.

Where did they get artistic expression? And was it genetic? Or was it learned from humans? We are not yet clear whether there were Neanderthal-human babies. The initial publication in the May 7, 2010, article in *Science*, "A Draft Sequence of the Neandertal Genome" suggested that many of us have "1-4% Neanderthal DNA." Clearly, Neanderthal and human hanky-panky must have been going on.

But now some scientists are doubting it, positing we only share a common ancestor.

A recent article in *Discovery* discusses this in "Humans, Neanderthals Did Not Have Babies," as does a *Smithsonian* article, "Hot for Hominids Did Humans Mate with Neanderthals or Not?" The latter takes the middle ground, quoting science writer Ed Yong as saying that it was probably a "rare" occurrence and every population has that "weird" person in the group which is not indicative of the actions of a community. If this theory is correct, perhaps, it wasn't the popular thing to do. Sort of like dating an artistic Bigfoot.

But whether your ancestor went to bed with Neanderthal Jane or not, many now think Neanderthals, not humans, may have been the earliest artists.

A recent *Daily News and Analysis* article, "Neanderthals Learned to Make Jewelry and Tools from Modern Humans," says an international team from the Max Planck Institute in Germany suggests there was a "cultural exchange" between the two species, and there is evidence that Neanderthals "learned how to make jewelry and sophisticated tools" from the early ancestors of humans. The reasoning behind this is based on the fact that artistic relics were found near Neanderthal remains, but the artwork was "clearly" indicative of human hands. So, the conjecture is that it must have been learned. Perhaps that Neanderthal kindergarten had a human art teacher?

Whether this is true or not, Neanderthals were creative. According to Kate Wong, in her *Scientific American* article, "Oldest Cave Paintings May Be Creations of Neanderthals, Not Modern Humans," according to archeological evidence, Neanderthals not only wore feathers but "painted their skin" and "made jewelry from teeth and shells."

So that human art teacher had Neanderthal kindergarteners that resembled some early American Indians?

But was their artistic expression learned? There are those that do not think so. According to Eric Wayner, in a recent *Smithsonian* article, "Do Feathers Reveal Neanderthal Brain Power?", Neanderthals wore feathers as personal adornments that showed them to be "capable of symbolic expression." And Wong says, there are Spanish and French caves thousands of years old with cave paintings long thought to be the artwork of early humans that are now thought to be the work of Neanderthals.

Why? She says because of recently refined techniques of radiocarbon dating that these paintings are "significantly older" than once thought. In fact, some may be older than the date when the first humans arrived in Europe around 41,500 years ago. When there were thought to be only Neanderthals.

"A large red disk" on one of the Spanish caves, El Castillo, is "at a minimum 40,800 years old, making it some 4,000 years older than the Chauvet paintings which were previously thought to be the oldest in the world." This and a "stretch of limestone wall with dozens of hands" in the same cave are both thought to possibly be the handiwork of Neanderthal painters because the "estimates" are considered to be at best a conservative "minimum." According to Ker Than, in the recent *National Geographic Daily News* article, "The new dates raise the possibility that some of the paintings could have been made by Neanderthals who are thought to have lived in Europe some 30,000 or

40,000 years ago."

If Neanderthals were painters where did they get their creative expression if not from humans? Wong says that both "modern humans and Neanderthals might have inherited their capacity for symbolic thinking from their common ancestor." And she says, if that is the case, "the roots of our symbolic culture go back half a million years."

Either way, artistic impression has been in our DNA for a very long time.

December 4, 2012

Neanderthals in America

Yes, Virginia, there is a Neanderthal fossil record in America. And apparently a Neanderthal hybrid record.

No genetics publication has put all the evidence together: the genetics establishment is still in denial about most things Neanderthal. The evidence is scattered and mostly unrecognized, but, in our opinion, conclusive and compulsive. Consider the following article:

Frank L'Engle Williams and Gail E. Krovitz, "Ontogenetic Migration of the Mental Foramen in Neanderthals and Modern Humans," *Journal of Human Evolution* 47/4 (October 2004) 190-219.

The mental foramen (literally "mind's little hole") is an anatomical trait very pronounced in Neanderthals, a small dimple in the lower jaw of the skull beneath the teeth, or mandible. It is found sporadically in humans, where it is classified as archaic. Among the places where it has been identified are Oleniy Island in Northern Russia and the Baltic region, where Cro-Magnon like Europoid and Mongoloid types have been found, along with "large and massive" torus occipitalis or Anatolian bumps (Alexander Mongait, 1959; Marija Gimbutas, 1956). Neanderthal bones were also reported in Bakhehisarai in the Crimea (Alexander Mongait, 1959), in the Joman or Ainu of Japan (Carleton Stevens Coon, 1962) and the "race of giants" continually being unearthed in West Coast, Ohio Valley and New England archeological sites, caves and mounds.

Archaic giant skeletons with mental foramina, occipital bumps, double rows of teeth and other Neanderthal features are reported, in fact, all over the Americas. Fritz Zimmerman has gathered a lot of the evidence in a new book titled *Nephilim Chronicles*, of which a small excerpt was published in *Ancient American* magazine, issue 91, pp. 24-27. Here is one of the newspaper reports he cites:

Evening News (Ada, Oklahoma), November 8, 1912.
PRIMITIVE MEN OF GIGANTIC STATURE.

Eleven skeletons of primitive men, with foreheads sloping directly back from the eyes and two rows of teeth in the front of the upper jaw, have been uncovered at Craigshill at Ellensburg, Washington. They were found about twenty feet below the surface, twenty feet back from the face of the slope, in a cement rock formation over which was a layer of shale. The rock was perfectly dry. The jawbones, which easily break, are so large that they will go around the face of a man today. The other bones are also much larger than those of the ordinary man. The femur is twenty inches long, indicating a man of eighty inches tall (6'8"). The teeth in front are worn almost down to the jawbones, due, it is believed, to eating uncooked foods and crushing substances with the teeth. The sloping skull shows an extreme low order of intelligence.

We note that the female mummy clutching a child known as The Thing on display at a roadside attraction on Interstate 10 north of Tombstone, Arizona, has a double row of teeth. It supposedly was one of three skeletons sold to the operator of the original site for $50 by a Chinese gentleman passing through. The Thing is discussed in several works by David Hatcher Childress. My son and I paid our two bucks and saw it last Christmas on a road trip.

October 18, 2011

Kathryn Halliday commented on 19-Oct-2011
Very interesting article. What caught my eye is the article from Ada, Oklahoma—where I was born and now live in my old age. It is the center, after the removal, of the Chickasaw Nation.

Fritz Zimmerman commented on 01-Feb-2012

There are many cases of "archaic" type skulls that are associated with the Maritime Archaic who migrated to North America (by boat) from 7000 to 2000 BC. They eventually migrated into the Great Lakes region. These are a few of headlines of giant skeletons with Neanderthal like skulls in the Great Lakes: http://gianthumanskeletons.blogspot.com/ 2012/01/giant-human-skeletons-with-archaic.html This link will take you to headlines from the coastal regions, where more of these Neanderthal-looking skulls were uncovered. http://gianthumanskeletons. blogspot.com/2012/01/giant-human-skeletons-headlines.html.

Anonymus commented on 02-May-2012

Yes, there is overwhelming evidence to support this. I'm one of the rare people who have in my possession a skull showing brow ridge, teeth, leg bones, and hundreds of tools.

Chris L Lesley commented on 24-Nov-2013

Anonymous, I sure would like to get a replica of that skull preserved and displayed in the GAWM Museum.—Chris L. Lesley

What Lurks in Our Family Tree?

Some of our earliest grandmothers may well have known or even dated other kinds of humans at the original café of rocks.

Charles Choi, a Huffington Post Live Science contributor, in the article, "Denisovan Genome Sequenced, Reveals Brown-Eyed Girl of Extinct Human Species," says that "…our planet was once home to a variety of other human species." It sounds like something out of *The Lord of the Rings*.

Though we have tossed some out as invalid, there are others according to a recent *Smithsonian* blog, "Four Species You've Never Heard of," like "some hominid fossils in China that are hard to fit into any of the known species." Who knows what mysteries are in those boxes waiting to be discovered and elsewhere?

Recently, with high DNA sequencing, we have discovered a lot from the bones of archaic hominids like the Denisovans and Neanderthals.

Another Huffington Post Science article quotes Svante Pääbo, a

Swedish biologist who specializes in evolutionary genetics, as saying, "... there is ... no difference in what we can learn genetically about a person that lived 50,000 years ago, and from a person today, provided that we have well-enough preserved bones."

And those bones tell stories. We just can't agree yet on the entire story. Pääbo, in the article "Denisovan Genome Sequenced," says that "Denisovans diverged from modern humans in terms of DNA sequences about 800,000 years ago." We can now say some archaic humans are in some way related to us. But how?

Some think we are the Neanderthals. The *Smithsonian* article, "Virus Fossils Reveal Neanderthal's Kin," says that humans and Neanderthals are "so close, in fact, that some researchers argue the two hominids might actually be the members of the same species."

But the consensus appears to be that "Humans and Neanderthals are cousins" ("Virus Fossils"). That article also quotes researchers from *Current Biology* as saying, "Denisovans seem to be more closely related to Neanderthals than we are." At any rate, meet your new cousins. Not quite what one expected. Imagine those relatives over for the holidays.

Whether we were kissing cousins or remote relatives with Denisovans or Neanderthals continues to be debated. A *Smithsonian* article, "Top 10 Hominid Discoveries of 2011," concurs that the finding that "Denisovans bred with several lineages of modern humans—people native to SE Asia, Australia, Melanesia, Polynesia, and elsewhere in Oceania" was one of the most astounding discoveries of 2011. Katherine Harmon, in her article in *Nature*, "New DNA Shows Ancient Humans Interbred with Denisovans," further bolsters this position when she claims the latest evidence from high-sequencing DNA analysis shows that Denisovans and humans interbred.

However, the author of another article in the *Smithsonian* "Denisovan Genome Sequenced" is unconvinced and says that these Denisovan "genetic variants" probably reflect interbreeding with our closest relatives, the Neanderthals, not the Denisovans.

So, the question remains. How are we related? How different are we? Are we all the same happy human family?

We don't have the answer to all of these questions, but it seems that we have a common mystery ancestor. Why? "Researchers discovered that most of the ancient viruses found in Denisovans and Neanderthals are also present in humans, implying that all three inherited the same genetic material from a common ancestor" ("Virus Fossils").

If that is true, that means we have the same original grandmother. A bit shocking. Not quite what one expected or the picture one might want to put over the dining room table. Sort of reminds one of Worf on Star Trek with a slightly feminine touch. According to the *Smithsonian* article, "Top 10 Hominid Discoveries of 2011," "Australopithecus, 1.97 million years old," is our possible direct ancestor.

So how do we differ? For one thing, we have a much greater genetic diversity. We have traveled and been around for quite a while. According to the author of the article, "Humans Have Been Evolving Like Crazy Over the Past Few Thousand Years," a recent *Smithsonian* blog quotes *Nature* as saying, "Over the past 5 to 10 thousand years ... the genetic diversity in the human population has exploded."

In contrast, the Denisovan DNA that was sequenced exhibits a very low genetic diversity, probably showing it was initially small but grew quickly—too quickly for much genetic diversity ("Denisovan Genome Sequenced"). The article says that if the Neanderthal genome shows the same information that this likely means one population from Africa gave rise to both the Denisovan and the Neanderthal populations.

And some researchers conclude that what makes us unique is our brain power—though not all. Pääbo, in the article, "Denisovan Genome Sequenced," says it was our "brain function and development" that helped us increase and dominates as a species and made all of "human history possible."

However, John Hawks, a biological anthropologist at the University of Wisconsin, in the *Nature* article, "New DNA Analysis Shows Ancient Humans Interbred with Denisovans" counters this idea and says it is "too early to begin drawing conclusions about human brain evolution from genetic comparisons with archaic relatives" and "decoding the genetic map of the brain and cognition from a genome is a long way off."

The latest science sounds like a tale from "The Munsters." We have a few not-quite-human cousins to put on our family tree. And it looks like we have the same brown-eyed Jane as a grandmother who ran around the rocks with some real characters.

October 22, 2013

Hominin Trysts

Our most ancient cousin has been found through genetic testing. Personally, I have never wanted to find any more cousins and have done my best to run from the ones I have. Nevertheless, a lot of people have looked for cousins through DNA testing. Now, science has found our most ancient cousin who lived some 400,000 years ago in northern Spain. They found what was left of him in a place that sounds reminiscent of the zombie movie, *World War Z:* Pit of Bones ('Sima de los Huesos'). But the finding has raised more questions than it has given us answers (Ewen Callaway, *Nature,* "Hominin DNA Baffles Experts").

Why? Though they did not find zombies, they still did not find what they expected to find. The team, led by Svante Pääbo, a molecular geneticist, thought they would attribute the ancient bones to early Neanderthals as had long been supposed; however, DNA testing results did not show this to be the case to the scientists' surprise (Callaway).

Instead, they discovered Denisovan DNA, another ancient human-like species that was curious because the "only remnants of Denisovans come from Siberia—a long way from Spain" (Elizabeth Landau, *CNN,* "Oldest Human DNA Found in Spain"). Denisovans were not our direct ancestors but cousin-related (Landau). But how did Denisovan DNA arrive in Spain? No one seems to be asking that looming question.

That is not all. When they examined the femur bone, they did so with mitochondrial DNA testing (which shows only the female line of ancestry). However, since the mitochondrial line only reflects one parent's line of ancestry, it is not the entire story. The team hopes to be able to do nuclear DNA testing, also the basis of autosomal DNA testing, within a year (Callaway). Nuclear DNA provides a clearer picture of both sides of one's ancestry and could also provide a clearer picture to how it is connected to other human or human-like species (Landau). Right now, it is not clear at all.

Pääbo believes the femur bone, which was found in the '90s and just recently tested, may reflect a common ancestor to both Denisovans and Neanderthals. However, Chris Stringer, anthropologist at the Natural History Museum in London, believes the femur may reflect a mystery hominin. Why? Not far from the caves, they found more hominin bones that are even older—some 800,000 years. Stringer believes that this was a founding group to both the Denisovans and the group belonging to the mystery femur that may account for the mystery species Denisovans had some hot dates with at the original Hardrock Café.

Our ancient ancestry just gets stranger.

August 31, 2012

Ancient Bones

It isn't as easy to do DNA testing on ancient bones as "CSI" and other television shows might make it appear. Most people probably have the idea that an Indiana Jones type just snaps everyone into shape in the lab, and the results are whipped up magically. Why would anyone think that? Genetic testing has become so advanced and so routine in our everyday lives. However, genetic testing of ancient bones is a much more difficult process.

What is the difference? Contamination is more likely. When genetic testing is done by a professional, certified ISO lab, chances of contamination are next to impossible as proper procedures are followed. No matter the type of genetic testing one has, very careful steps are part of the process to ensure contamination does not happen for most modern-day medical and genetic testing. If proper standards are followed, that is not a problem. Your DNA is sitting in a container, nice and secure, before it is analyzed. The DNA of an ancient human is sitting in dirt and has been for many thousands of years. This means the DNA has often largely been broken down and degraded in warm soil where bacteria have made a cozy home (Jocelyn Kaiser, *Science*, "Cleaning up Ancient Human DNA").

Of course, there are exceptions, but the conditions are exceptional. The difference is found in cold conditions versus warm ones. There have been cases of "human DNA samples taken from human bone, tooth, hair, or other tissue typically preserved in frozen soil, ice, or a chilly cave" like "Ötzi the 5300-year-old iceman and even Neanderthals."

However, that isn't the norm, and we cannot change the conditions the DNA that we find has undergone because we want the best DNA testing results. How do we fix the problem of genetic testing of bones and other usually an expensive process, but scientists are thinking of ways they can filter away the "non-human DNA." Then the bones can tell us more stories. For instance, using this method a team determined a "500-year-old Peruvian mummy did not have European ancestry as claimed" (Kaiser).

Cleaning ancient bones using this method will mean better DNA testing results despite bad conditions. And it just might mean that we can use whole genome sequencing on samples that we did not think possible before (Meredith L. Carpenter et al, *American Journal of Human Genetics,* "Pulling out the 1%- Whole Genome Capture").

What does this mean? If we are able to get better results from Neanderthals, from Denisovans, from the little ancient hobbit-like people of Indonesia called the Homo floresiensis, we might have a better picture of our origins as well as other hominins and how we connect to those different hominins. We would have a clearer story.

November 12, 2013

8 DNA AND FAMILY

A Moral Imperative
Genealogists Have Responsibility to Communicate with Other Family Members: Genetic Disease Experts

Communicating Genetic Information in Families—A Review of Guidelines and Position Papers
Laura E Forrest et al.
European Journal of Human Genetics 15 (March 2007) 612-618.

According to genetics educators at Royal Children's Hospital in Australia, you have a moral responsibility to let other family members know of any genetic risks or predisposition to disease you have learned about in your genealogy research. The study is the first to assemble guidelines for genetic genealogists in this area.

March 30, 2007

Acadian Anomalies
Anomalous Native American Lineages Now Identified in Micmac

After posting "Anomalous Mitochondrial DNA Lineages in the Cherokee," and after being interviewed on the subject by an Internet radio show host, I was contacted by participants in the Amerindian Ancestry out of Acadia Project who were struck by similarities in results for the two groups.

Established in 2006, the Amerindian Ancestry out of Acadia DNA Project mission is to research and publish the mtDNA and Y chromosome genetic test results of site participants who descend from persons living in Nova Scotia and surrounding environs in the 17th and 18th centuries, focusing specifically upon the early population of l'Acadie. As part of the mission, the Project develops a database of published mtDNA and Y Chromosome test results and encourages the sharing of this information among other similarly focused studies for the purposes of comparison and the advancement of science and research.

According to Project Administrator Marie A. Rundquist, "We descend from both Amerindians (mostly Mi'kmaq) and the early French settlers who arrived in Port Royal in the 1600s, many of them single French men who married Amerindian wives, whose families would become pioneers of the New World. Our family lines have extended well-beyond the original boundaries of what was known to the French as Acadia, but to our AmerIndian ancestors as Mi'kma'ki, as our ancestors settled the outer-reaches of Nova Scotia, including Cape Breton, Newfoundland, New Brunswick, Prince Edward Island and Quebec. Our family lines continue to extend, traversing the entire North American continent and beyond."

She adds, "Many who live in the United States trace their genealogies back to the first Acadian AmerIndian immigrants who arrived in Louisiana after being deported from Nova Scotia by the British in 1755 (in the Grand Deportation)—and belong to a 'Cajun' community known worldwide for its food, flair, fun, and love of all things French."

Several members belong, as it turns out, to rare haplogroups X, U, and other "anomalous types" as compiled by me for DNA Consultants customers and reported in the previous blog post. Some highlights from the study of Cherokee descendants are:

— H, the most common European type today, is virtually absent, demonstrating lack of inflow from recent Europeans
— J present in lines explicitly recognized to be Cherokee
— X the signature of a Canaanite people whose center of diffusion was the Hills of Galilee, hypothetically correlating with Jews and Phoenicians
— U suggesting Eastern Mediterranean, specifically Greek

— K also suggesting Eastern Mediterranean or Middle Eastern, hypothetically correlating with Jews and Phoenicians

— T reflecting Egyptian high frequencies found almost nowhere else

According to Elizabeth Caldwell Hirschman, the Cumberland Gap mtDNA Project with overlapping territory with the Cherokee and Melungeon homelands in the Southern Appalachian Mountains also shows elevated frequencies of T. Project administrator Roberta Estes recently published the results of a large study of Native American Eastern Seaboard mixed populations "in relation to Sir Walter Raleigh's Lost Colony of Roanoke" in the online *Journal of Genetic Genealogy*, 5(2):96-130, 2009.

Estes is a board member of the Melungeon Historical Society and has an introduction with links to the study and its data on the society's blog, titled, unfortunately, "Where Have All the Indians Gone?" The Indians are not themselves aware of having gone anywhere but are still here, Roberta. ...

Harvard University professor Barry Fell in his book *Saga America* first published in 1980 presented historical, epigraphic, archeological and linguistic evidence suggesting links between Greeks and Egyptians and the Algonquian Indians of Nova Scotia, Acadia and surrounding regions around the mouth of the St. Lawrence Seaway, particularly the Abnaki ("White") and Micmac Indians. He noted as early as 1976 in his previous study *America B.C.* that the second century C.E. Greek historian Plutarch recorded "Greeks had settled among the barbarian peoples of the Western Epeiros (continent)." Fell inferred from Plutarch's passage "these Greeks had intermarried with the barbarians, had adopted their language, but had blended their own Greek language with it."

In an appendix, he assembled extensive word-lists comparing Abenaki and Micmac vocabulary in the areas of navigation, fishing, astronomy, meteorology, justice and administration, medicine, anatomy, and economy with virtually identical terms in Ptolemaic Greek. One example is Greek *ap'aktes* Abenaki/Micmac *ab'akt* English "a distant shore." Fell's work was significantly corroborated and expanded by John H. Cooper, "Ancient Greek Cultural and Linguistic Influences in Atlantic North America," *NEARA Journal* 35/2.

Acadia project's website is: http://www.familytreedna.com/public/Aca
dianAmerIndian/default.aspx.

November 30, 2009

Melungeon Riddle Solved by Autosomal DNA

*After many years in development, the results of a DNA ancestry project
enrolling 40 Melungeons were published and made public, marking the
end of an attempt to solve the mystery of a Southern U.S. ethnic group
with autosomal DNA.*

Seeming to lay to rest an old controversy in American history about
Melungeons, the scientific data supporting a genetic mixture of white,
American Indian and Sub-Saharan African were placed online today by
the organizers of DNA Consultants' Melungeon DNA Project.

The data report a sample of 40 Melungeons' DNA fingerprints.
Population analysis of the participants' DNA fingerprints was used in an
article for *Appalachian Journal*. Titled "Toward a Genetic Profile of
Melungeons in Eastern Tennessee," the study was co-authored by
Donald N. Yates, principal investigator of DNA Consultants, and
Elizabeth Caldwell Hirschman, a professor at Rutgers University in New
Brunswick, N.J.

"This is a giant stride forward in understanding the mixed ancestry of
Melungeons," said Donald Yates, co-author. "Never before has
autosomal DNA been used in attacking the problem."

The 40 participants' names were: Anonymous, Mabel Bentley, Judy
Douglas Bloom, Leah Laura Bulgariev, John (Dick) Caldwell, John R.
Caldwell, Sr. (deceased), Virginia Caldwell, William Collins, Mary
Goodman, Floyd Milton Grimwood (deceased), Ann Reagan Haines,
Linda Barnett Hall, Nancy E. Hammes, Elizabeth Caldwell Hirschman,
Pat Goin Jones, Brenda LaForce, Everett LaForce, Jessica Kiely Law,
Bonnie J. Lyda, N. Brent Kennedy, Richard Kennedy, Margaret E.
Kross, MeriDee Orvis Mahan, Karen Mattern, Sebenia Ann Milbacher,
Nicolas J. Millington, Holli Starnes Molnar, Nancy Sparks Morrison,
Teresa Panther-Yates, Billy Starnes, Julia Starnes, Keely Starnes, Phyllis
Starnes, Richard Stewart, Doretha J. Thornton, Kaye M. Viars, Celia
Wyckoff, Wayne Winkler, Betty Yates Adams, Donald N. Yates.

Participants were qualified by their genealogies and included many names familiar to those who follow Melungeon genealogy discussion groups on the Internet, including Brent Kennedy, author of the 1996 book that started the Melungeon Movement, his brother Richard Kennedy; Elizabeth Hirschman, a native of Kingsport, Tenn., along with several members of her family; Wayne Winkler of the Melungeon Heritage Association and author of *Walking Toward the Sunset*; and Nancy Morrison, creator of the online Melungeon Health referral service.

September 16, 2010

Surprises in English and Irish DNA

Over a year ago, there appeared one of the few studies of autosomal DNA in Ireland and Britain. If you have English/Welsh, Irish, northern Irish, Highlands Scottish, Lowlands Scottish or Swedish matches, you will want to read this post. Here is the original article and abstract.

> C. T. O'Dushlaine, D. Morris, V. Moskvina, G. Kirov, Consortium IS, M. Gill, A. Corvin, J. F. Wilson, G. L. Cavalleri
> Population Structure and Genome-wide Patterns of Variation in Ireland and Britain
> *European Journal of Human Genetics* 2010 Nov;18(11):1248-54. Epub 2010 Jun 23.

Abstract. Located off the northwestern coast of the European mainland, Britain and Ireland were among the last regions of Europe to be colonized by modern humans after the last glacial maximum. Further, the geographical location of Britain, and in particular of Ireland, is such that the impact of historical migration has been minimal. Genetic diversity studies applying the Y chromosome and mitochondrial systems have indicated reduced diversity and an increased population structure across Britain and Ireland relative to the European mainland. Such characteristics would have implications for genetic mapping studies of complex disease. We set out to further our understanding of the genetic architecture of the region from the perspective of (i) population structure, (ii) linkage disequilibrium (LD), (iii) homozygosity and (iv) haplotype diversity (HD). Analysis was conducted on 3,654 individuals from

Ireland, Britain (with regional sampling in Scotland), Bulgaria, Portugal, Sweden and the Utah HapMap collection. Our results indicate a subtle but clear genetic structure across Britain and Ireland, although levels of structure were reduced in comparison with average cross-European structure. We observed slightly elevated levels of LD and homozygosity in the Irish population compared with neighbouring European populations. We also report on a cline of HD across Europe with greatest levels in southern populations and lowest levels in Ireland and Scotland. These results are consistent with our understanding of the population history of Europe and promote Ireland and Scotland as relatively homogenous resources for genetic mapping of rare variants.

Though the focus was on genome-wide association studies (GWAS) and linkage disequilibrium, or medical aspects of DNA, this study was groundbreaking in using supercomputing and has enormous implications for the history of the British Isles. It used data from over 3,000 individuals from seven populations:

1. Ireland/Dublin
2. Scotland/Aberdeen
3. Bulgaria
4. Portugal
5. Sweden
6. South/Southeast England
7. Utah

Data came from several sources: the International Schizophrenia Consortium, Wellcome Trust Cast Control Consortium 1958 Birth Control Data set, Utah European ancestry population (CEU) and HapMap project.

The study aimed to describe, statistically, four measures of the Irish and English populations:

1. Population structure
2. Linkage disequilibrium, with consequences for the study of common Irish and English genetic disorders
3. ROH, or runs of homozygosity, essentially a reflection of

inbreeding and the remoteness of a population

4. Haplotype diversity (based on SNPs in atDNA)

The main conclusion was that Irish/English forms a separate and unique population since the Ice Age very different from either Bulgarian (SE Europe) or Portuguese (SW Europe), with great affinities to Sweden or Scandinavian populations (p. 1250). For instance, "the breakdown and patterning of LD (linkage disequilibrium) ... is virtually indistinguishable among the Irish, Scottish, southern English, Swedish..." (p. 1250).

"Diversity across Britain and Ireland is reduced in comparison with mainland European populations, with Scotland and Ireland having lower levels than southern England (p. 1251)."

The study postulates that Irish and English linkage with genetic disease came about as a result of population stasis or unchanging conditions. The agricultural revolution swept in a lot of additions to the gene pool in most of Europe, including Southeast England, but in areas like Ireland, Scotland and Sweden the same population stayed on the land with little increase. There was a negative effect on population size during the Norse migrations of the 10th century and in the Irish Potato Famine of the 19th century. The study mentions a "kinship effect" apparent in Irish and Scottish clan histories (p. 1254).

The surprising suggestion is that there may now be a groundswell of research into "Irish" and "Scottish" and "English" diseases comparable to the attention devoted to Jewish diseases.

A related study is: A. Auton, K. Bryc, A. Boyko, K. Lohmueller, J. Novembre, A. Reynolds, A. Indap, M. H. Wright, J. Degenhardt, R. Gutenkunst, K. S. King, M. R. Nelson and C. D. Bustamante, "Global Distribution of Genomic Diversity Underscores Rich Complex History of Continental Human Populations." It appeared in *Genome Research*, February 2009.

May 22, 2011

More Light on the Melungeons

Phyllis Starnes drew together many threads of Melungeon research when she delivered her presentation on autosomal DNA validation studies at the Fifteenth Melungeon Union, held at Warren Wilson College, Swannanoa, N.C., July 15-16, 2011. Sponsored by the Melungeon

Heritage Association of Kingsport, Tenn., the conference was appropriately titled, "Carolina Connections: Roots and Branches of Mixed Ancestry."

Starnes, who is administrator of DNA Consultants' Melungeon DNA Studies as well as an assistant investigator responsible for authoring reports, began her presentation by telling her own story. In 2002, she read an article about the occurrence of familial Mediterranean fever in Appalachia, where she grew up.

"This article was the catalyst for me to address my own health and ancestry," she told participants.

She had met N. Brent Kennedy, author of the touchstone book *The Melungeons: The Resurrection of a Proud People*, and soon became acquainted with both Elizabeth Hirschman (*Melungeons: The Last Lost Tribe in America)* and Donald Panther-Yates, both speakers at Melungeon Fourth Union in Kingsport. The resources she needed for understanding her peculiar heritage were coming together.

Starnes summarized the Hirschman-Yates study of Melungeon DNA results published last December in *Appalachian Journal* and went on to reveal the results of a validation study of the Melungeon data in which the DNA profiles of the 40 participants were fed back into the database atDNA, expanded to reflect the world's only autosomal DNA Melungeon sample.

Many Melungeon DNA project participants, not surprisingly, had Melungeon as their No. 1 match, including Starnes.

In 1990, physical anthropologist and chemist James Guthrie analyzed blood sampled from 177 Southern Appalachian people identifying as Melungeon tested by Pollitzer and Brown in 1969. Guthrie's analysis was consistent to a remarkable degree with the Hirschman-Yates study.

All studies to date have verified and confirmed repeatedly that Melungeon descendants carry an unusual mix of Jewish, Mediterranean, Turkish, Iberian, Native American and African DNA. They also inherit genetic predispositions toward developing familial Mediterranean fever and other disorders.

This overarching thesis explaining what makes Melungeons different was advanced over 20 years ago by Brent Kennedy. It has now been re-examined, probed, tested and validated by unimpeachable follow-up studies. Little has turned up to change Kennedy's original thinking. It would be wrong to say that Melungeon origins today are controversial or mysterious. There is much we do not know about them, but their genetic and medical profiles are clear.

August 30, 2011

Peter McCormick, Cultural Geographer

Describing himself as "a cultural geographer by training," Peter McCormick contributed an interesting chapter mentioning Melungeons to a recent volume of political science and anthropological essays. Titled *Border Crossings: Transnational Americanist Anthropology*, the collection is edited by Kathleen Sue Fine-Dare and Steven Rubinson and published by the University of Nebraska Press (2009). It may be the first time a practicing academic historian has committed to a considered opinion on the subject since Melungeons first appeared on the radar of Americanists with Price's "tri-racial isolate" definition in the 1950s.

McCormick has a Ph.D. from the University of Oklahoma and is an associate professor of Southwest studies and Native American and indigenous studies at Fort Lewis College in Colorado. His most recent work has been on the autogeography and autohistory of his extended family in the Plains, Southwest, Appalachia, Iberia, South America and Mediterranean.

Here's how he describes Melungeons (p. 286): *The Melungeon population of Appalachia has been the subject of a tremendous amount of interest and controversy lately. A consensus appears to be building that this population, once thought to be small, is rather large and is a result of the mixing of Iberian and Middle Eastern settlers who had been part of Spanish and English trading parties with the indigenous population of the American Southeast. Later migrations into the Piedmont and upper South by refugees of the Inquisition (Sephardic Jews and Moors) in the seventeenth and eighteenth centuries supplemented this population (see Hirschman 2005; Kennedy and Kennedy 1997). Our families were of this mixture.*

Professor McCormick goes on to write about his personal Melungeon genealogy: *Sephardic names include Cuba, Pillo Monnis Callahin, Jorgas, Nassi, Khanadi, Rosa, David, Baez, Santos and Gascon. The families that were at one point crypto Jews include Kieffer, Mayabb,*

Dula D'Aultun, Baigne and Ball. Our Melungeon families are Sizemore, Yates, Brashears, Collins, Lucas, Noel, Bass, Kennedy, Davis, Nash, Mullins, Center and Carrico. The family names on the Miller-Guion and Dawes rolls include Tunnell, Mabe, Waller, Yates and Doolin.

McCormick's testimony and evaluation of the evidence, together with his willingness to name names and self-identify as a Melungeon in academia, are important signs that the Melungeon thesis advanced by Kennedy and further documented by Hirschman and others is winning the day.

We thank McCormick for his part in bringing the true story of Jewish and Middle Eastern ancestry in Appalachia to wider attention.

February 15, 2012

Reconstructing Your Parentage

Every year in the United States about half a million paternity tests are performed proving or disproving whether an alleged father is the true parent of a child. Sometimes there is a court order to do this; at other times, it is sheerly for personal information. The determination of parentage is made based on a simple comparison of a small rock-hard number of genetic markers in the DNA fingerprint of the child and alleged father.

Samples are extracted from a 30-second cheek swab and processed at any of an estimated 2,000 forensic labs across the country. The standard in place since about 1997 has been a set of 30-32 biallelic or double values each person carries on loci spread across their chromosomes, making for a virtually unique identification signature that reflects the equal DNA input of mother and father (and, in fact, all grandparents and all ancestors).

Often termed CoDIS markers (standing for Combined DNA Identification System), these alleles or variations are the magic numbers underlying the popularity of paternity tests as well as the national passion for jailing or exonerating crime suspects. If a value is found in the DNA profile of the child and is not present in the two observed values of the alleged father on the same locus, this constitutes what is known in the paternity business as an exclusion: The alleged father is almost certainly *not* the true father. Conversely, if *all* the alleged father's values can be detected in the child's on each location, one after another, that male is judged to be the child's biological father to a 99.999 percent certainty. Paternity tests are simple math.

A famous paternity test involved proving who was the true father of the baby born to Anna Nicole Smith in 2006. After her death in early 2007, several men came forward claiming to be the father, including a European prince, Anna Nicole's bodyguard and a convict who had been a former boyfriend. Larry Birkhead pressed his case. When the results came in, he was declared by Bahamian court to be the baby's biological father. The child's original birth certificate was amended to show this.

Can paternity testing be used in a reverse process to establish the identity of a father, given only the child's DNA profile? No, but with enough DNA profiles available for comparison the missing member of a family group can be reconstructed by comparing alleles they must share, called obligate. Doing so is a matter of logic and statistics, mostly just either-or, deductive logic.

I became interested in reconstructing a parent's profile after many of DNA Consultants' customers inquired if such a calculation or estimate was even possible. Some were adopted persons who had no recourse to testing their parents, some knew one parent but not the other, and some had no access to parents. They were either uninterested or unavailable. In a special category were females who were only-children with both parents deceased who wanted to know something about their father, but who could not take a Y-chromosome haplotyping test, as they did not carry a copy of their father's male DNA. In this respect, autosomal DNA testing is the great equalizer.

My father, Lawden Henry Yates, died in 1978. My mother, Bessie Cooper Yates, lived to the advent of DNA tests, but I failed to obtain any sample from her before her death in 2006. I had siblings and half-siblings still living, however, so in 2010, I constructed a family group autosomal DNA study with the help of Crystal Wagner at Chromosomal Laboratories/Bode Technology. The results were very satisfying. This paper and blog post will serve as a report to those who are interested.

Step One. I was fortunate to have the participation of three half-sisters by my father, along with his second wife, their mother. Comparing mother and daughters I was able to verify the obligatory alleles each daughter *must* have received from the mother.

Autosomal alleles are fixed in our genealogy, have little or no mutations (unlike YSTRs, which mutate from generation to generation, as do mitochondrial nucleotide positions, though more gradually over time) and derive from both parents equally by recombination at the moment of conception. They are copied and preserved without change in every cell of our bodies. The mother is responsible for half of the equation. By a process of elimination the other number on each row of the lab report *must* represent the father's contributions. This method is completely logical and unequivocal. There can be no other answer to the problem. No studies suggest these pieces of our double helix DNA change significantly in transmission from one generation to the next or mutate over time in genealogies. Their values and patterns are strictly attributable to heredity.

Step Two. The father's alleles are confirmed by a comparison with three children by his first wife, my mother.

Step Three. By the same process we can now reconstruct the mother's DNA profile. In it, we can expect to visualize the final piece of the puzzle, proceeding from the known to the unknown according to the immutable laws of autosomal DNA and genetic inheritance.

We have arrived at my mother Bessie Yates' DNA profile by a multistep process of extrapolating it using three of her children and three children by her husband's second marriage, along with the test results of my half-sisters' mother. Seven tested profiles yielded two reconstructed ones. In the process, we have also recovered my deceased father's DNA profile.

Separating Mother and Father's Contributions

Having overcome these hurdles, I was most interested in the utility of the results. I felt confidant about the method. But what excited me most was to see how my own autosomal ancestry results might be, respectively, apportioned in my parents. For this, I ran a DNA Fingerprint Plus on them both. The findings were very satisfying to me personally, helping solve many questions I had always had about what ancestry I got from my father, what from my mother and what from both.

Let's start with American Indian admixture. My own DNA Fingerprint Test, as well as percentage tests through another company, suggested a relatively large amount, perhaps one-quarter all told by various measures, but family tradition had placed Native American heritage solely on my mother's side. To be sure, my mother gave me a Native American mitochondrial haplotype, indicating a female line going back to a Cherokee woman in Georgia, traced as far back in records as 1790. Extensive genealogy research showed, however, that my father's great-grandmother was also a Cherokee with the surname Thomas from North Carolina. What did the new autosomal DNA profiles say?

On a rough measure, I have received a "double dose" of Native American II, a marker co-relating with 80 percent of 24 tested American Indian populations in the atDNA 4.0 database. (Two siblings and one half-sibling received only single doses.) This seemed to indicate that I had some degree Native American (not possible to say how much) from both parents. True enough apparently, judging from the top world matches for my mother and father. I give here the top ten for comparison.

MOTHER

Rank World Population Matches

1 **Russia - Chukchi (n = 15)**
2 White - Maine (n = 151)
3 **Native American - Athabaskan (n = 101)**
4 Swedish (n = 311)
5 Hispanic - U.S. (n = 199)
6 El Salvadoran (n = 296)
7 **Native American - Choles - Chiapas (n = 109)**
8 Portuguese - Azores (n = 100)
9 Argentinian - Patagonian - Chubut (n = 320)
10 Korean - Western U.S. (n = 63)

FATHER

Rank   **World Population Matches**

1   Melungeon (n = 40)
2   White - Canadian (n = 164)
3   Belgian - Flemish (n = 231)
4   **Native American - Saskatchewan (n =1**
5   India - Indo-Caucasoid - Brahmin (n = 11
6   **Native American - Minnesota (n = 191**
7   India - Indo-Caucasoid - Kayastha (n = 1(
8   Japanese - Central (n =164)
9   Argentinian - Santa Fe (n = 562)
10   Brazilian - Sao Paulo (n = 113)

My mother's Native American population matches were slightly higher and more numerous than my father's, including more peoples like the Chukchi and Mongols, but my father's were not inconsiderable in their own right. Here's how their two megapopulation rankings look:

MOTHER

North Asian	1 in 35 billion
Northern European	1 in 632 billion
Central Asian	1 in 747 billion
American Indian	1 in 827 billion
European American	1 in 856 billion
Iberian American	1 in 1 trillion
Iberian	1 in 1 trillion
Central European	1 in 2 trillion
Melungeon	1 in 2 trillion
Mediterranean European	1 in 2 trillion

FATHER

European American	1 in 20 trillion
Northern European	1 in 185 trillion
Jewish	1 in 204 trillion
Iberian	1 in 274 trillion
Iberian American	1 in 728 trillion
Central European	1 in 919 trillion
Middle Eastern	1 in 924 trillion
American Indian	1 in 1 quadrillion
East European	1 in 2 quadrillion
Mediterranean European	1 in 2 quadrillion

These results confirmed that my father did have some Native American, although evidently not as much. They also suggested that although both bore about the same mixture of European and Native American ancestry (including high matches to Melungeon), my mother had a more pronounced Native American cast, her highest match being to North Asian, one of the supposed Asiatic feeder populations of Native Americans, whereas my father's top match was European American. Based on profile frequencies, my father was five times more likely to be European American than American Indian if subjected to forensic profiling, whereas my mother was 18 times more likely to come out as a Siberian Native than Northern European. Sometimes, it seems, exotic ancestry rises to the top. My overall conclusion was that my mother probably had 3/8 and my father 1/8 Native American heritage, which corresponds to their proved genealogies.

In my own profile, combining those of my parents, here are my megapopulation results:

SELF

North Asian	1 in 3 billion
Central Asian	1 in 12 billion
American Indian	1 in 25 billion
East Asian	1 in 42 billion
European American	1 in 42 billion

Northern European	1 in 44 billion
Iberian American	1 in 50 billion
Central European	1 in 70 billion
Iberian	1 in 75 billion
Melungeon	1 in 103 billion

According to these frequencies, my mother and father's Native American ancestry reinforced each other in me to make my top four matches Native American (or Siberian-Mongol-Turkic), so that I am about twice as likely to be graded into the Native American category by population statistics than the European. Similar conclusions emerged from my siblings' tests, and a diminished presence of Native American indicators was confirmed in my half-siblings, although their mother seemed to evince some Native American as well as my father, the shared parent. All participants in this study had grandparents born in North Alabama.

Further observations are possible. For instance, I was surprised to see a large indication of Jewish ancestry in my father's profile. Genealogy confirms as much, as the family surname is Hebrew (an anagram of Ger Tzedek similar to Katz, Kohen Tzedek). The emigrant Yates figure was reportedly an English Jew in 17th-century Virginia. My mother also showed Jewish ancestry. Both parents matched Melungeons, an Appalachian ethnic type suspected to have Sephardic Jewish forebears. My father's family included uncles named Josephus, Manaen, Irbin, Azariah, Lazarus and Sherith—apparently his Middle Eastern matches were truthful to a partial Muslim background. My mother's mother was named Palestine, and the names Isaac and Jacob were ubiquitous in her family tree. But neither side of the family claimed any Jewish heritage. It was left to autosomal DNA to reveal that hidden inheritance.

Although never performed before to my knowledge, this method of reconstructing autosomal profiles can be useful to others seeking to recover unavailable relatives' genetic fingerprints and to separate parents' contributions to their children's ethnic and ancestral stories. Since it is based on immutable markers in DNA, it rests on more solid ground than Y chromosome alleles or mitochondrial mutations. The challenge in exploiting the method is to have enough subjects in your family group study. In my case, I was fortunate to have a prolific father with six living children. I would like to conclude by thanking all my siblings, half-sisters and my father's widow. Their participation made it possible to present a true first in DNA genealogy.

March 14, 2012

Melungeons Forever

As the sponsor of the only published study to date on the genes of Melungeons, "Toward a Genetic Profile of Melungeons in Southern Appalachia," by Donald N. Yates and Elizabeth C. Hirschman, the owners of this blog naturally have an interest in Melungeons, a controversial American ethnic type.

Imagine our surprise at coming upon "Melungeons, A Multiethnic Population," put online by the International Society for Genetic Genealogy in their *Journal of Genetic Genealogy*. The authors are Roberta J. Estes, Jack H. Goins, Penny Ferguson and Janet Lewis Crain. It appeared ... well, we're not really sure when it appeared.

Roberta J. Estes, the lead author of the new article about Melungeons, was honored with the Prestigious Paul Jehu Barringer, Jr. and Sr. Award of Excellence in grateful recognition of her Dedication and Devotion to Preserving and Perpetuating North Carolina's Rich History. This award was conferred for her academic research paper, "Where Have All the Indians Gone? Native American Eastern Seaboard Dispersal, Genealogy and DNA in Relation to Sir Walter Raleigh's Lost Colony of Roanoke," published by the *Journal of Genetic Genealogy*. It can be read here: http://www.jogg.info/52/index.html.

We are glad to see Melungeons receiving long-overdue attention on the Internet but cannot recommend the new "review."

"Where have all the Indians gone"?! We're all still here, thank you very much.

But consider this excerpt from the "review": "Furthermore, as having Melungeon heritage became desirable and exotic, the range of where these people were reportedly found has expanded to include nearly every state south of New England and east of the Mississippi, and in the words of Dr. Virginia DeMarce, Melungeon history has been erroneously expanded to provide "an exotic ancestry ... that sweeps in virtually every olive, ruddy and brown-tinged ethnicity known or alleged to have appeared anywhere in the pre-Civil War Southeastern United States."

Concerning Melungeon heritage becoming "desirable and exotic," Estes et al., and our readers, may wish to consult the more recent study by Elizabeth Hirschman and Donald N. Yates:

"Suddenly Melungeon! Reconstructing Consumer Identity across the Color Line," *Consumer Culture Theory* (Research in Consumer Behavior, Volume 11), ed. Russell W. Belk and John F. Sherry, Jr. Amsterdam: Elsevier, 2007. Pp. 241-59.

This study interpreted the notion of "slightly mixed" differently from Virginia DeMarce in the 1990s.

But the co-authors and Virginia DeMarce are seriously mistaken if they think any marginalized or disadvantaged members of society go around trying to prove themselves outcasts. They've got the shoe on the other foot. Their language with its condescending mention of color tones is offensive. I, for one, am offended, and any sponsoring or supporting organization, ought to be. In fact, they ought not to allow such views to be published.

And that's my two cents' worth on Melungeons writing about Melungeons who don't believe they or anybody else is Melungeon.

Melungeons—real people in history—suffered enough to have their memory dishonored by lies and falsehoods hundreds of years later. I believe the same about Native American peoples and the descendants of slaves. No one should be able to write the history of other people for them. Let them speak in their own voices.

Anonymous commented on 26-May-2012
I don't know about the researchers' methodology, but I do not agree with the conclusion. Or I think they have it backwards: The Melungeons are Iberian/North African with possibly sub-Saharan as well. For example, one of my 5cM segments at 23andMe is Melungeon (Collins identified as a name). On this same segment is a distant cousin with four Greek grandparents. This is clearly a Sephardic segment. My question for the researchers is: Why are so many Melungeon descendants testing positive with obvious ties to the Iberian Peninsula and Hispanic territories? Please tell me if I can help your studies further. Ellin

Joseph commented on 16-Jul-2012
If you compare the melungeon dna projects to the portugesse dna protects ... you shall see a nearly 75 percent match to then projects ... compare this with the portugesse ancestery the melungeosn stated ... you have a match. http://www.ourfamilyorigins.com/portugal/dna.htm http://www.familytreedna.com/group-join.aspx?Group=portugal

May 23, 2012

Much Ado about Melungeons

Mountains are in labor, and an absurd mouse is born.

In a previous post, we drew attention to "Melungeons, A Multiethnic Population," published by the International Society for Genetic Genealogy in *Journal of Genetic Genealogy*, its authors Roberta J. Estes, Jack H. Goins, Penny Ferguson and Janet Lewis Crain.

The article is a bit forbidding at some 100 pages, and it fairly bristles with self-importance and personal DNA results, but we will review it.

Here are some highlights from the article, misspellings and all, with our comments in *italics*.

Summary: Many sources exist where the Melungeons identify themselves variously as Indians and Portuguese. Only one family, the Goins, are identified orally as having negro heritage. Given the physically dark appearance of the Melungeons, they have unquestionable heritage other than European.

This seems to be an unsurprising conclusion, until you realize that after limiting their sights to Melungeons who called themselves Portuguese, preferably only those in the 37869 ZIP code, and Goins who already identified themselves as having Sub-Saharan African (please, not the n-word in 2012, or at least capitalize it), the authors are going to draw a further veil on proceedings and deepen the mystery. Read on.

Every Melungeon core family is identified in multiple records as being "of color."

We won't comment on the equivocation going on here. Please read on.

DNA evidence identifies several lines conclusively as having African roots, specifically, Bunch, Collins, Goins (3 separate lines), Minor and possibly Nichols. Gibson has one line who has tested and shows

184

haplogroup E1b1a, but they also match another Louisa County affiliated family, Donathan.

Of these families, the Collins family has four different haplogroups within the same family group, a situation not unexpected based on the commentary by Will Allen Dromgoole wherein she states that of the Collins that while "they all were not blood descendants of Old Vardy they had all fallen under his banner and appropriated his name."

The Collins and Gibson founding lines, meaning Vardy Collins and Shephard "Buck" Gibson were said to be Cherokee and stole the names of white men in Virginia. Their DNA indicates that if they were Native, it was not via their paternal line.

Comma splice. Hate to be petty. How do you steal a white man's name? I certainly hope no Melungeons are going to steal mine. This is one of the funniest conclusions I have read so far. But do continue, Gentle Reader.

Dromgoole reportedly stayed with Calloway Collins who stated that his grand-father (sic) was a Cherokee Chief. His Collins grandfather was Benjamin Collins who lived on Newman's Ridge and did not remove in 1835. There are no known Cherokee who lived on Newman's Ridge. The Cherokee Nation was significantly further south prior to removal in 1835, as shown in Figure 12.

After making fun of other people who claim Cherokee chiefs and princesses in their family tree, the authors seem willing to entertain an exception with their own relatives, or friends. We will not quibble with their Cherokee history but would have said "farther" rather than "further." Maybe that is a regionalism.. Don't give up yet.

The Mullins line was reputed to be Irish and is confirmed genetically to be European. However, "Irish Jim", the progenitor is listed as a "free person of color", a very unusual classification for an immigrant from the British Isles. Droomgoole (sic) states that the Mullins will "fight for their Indian blood." No Indian heritage is evident in historical records or DNA.

We would like to remark that Irish, like other undesirables in early America, were often considered non-white and persons of color. Please recommend the book by Nell Irvin Painter titled The History of White People *to the Hancock County Public Library.*

The Denham line was said to be Portuguese and oral history indicates that the line originated "further south" or possibly from a shipwreck, yet the Revolutionary War pension application of David Denham says he was born in Louisa County, Virginia. The Denham line may connect with the Gibsons as early as 1627 in Charles City County. The Denham DNA is European and the Denham descendant who DNA tested has no Spanish or Portuguese matches. Denham is not Portuguese on the paternal Y-line.

Watch that distinction between "further" and "farther." The latter is to be used of distance; the former of degree or depth. I was born pretty far south but not fur.

A significant amount of oral history regarding Portuguese heritage exists, but no historical, genealogical or genetic evidence has been discovered to corroborate the oral history. Some historical information refutes the oral history.

Really? Who have you been talking to?

Claims of Portuguese ancestry are a pattern that stretches beyond the Melungeon families and is found explaining a "dark countenance" across the eastern half of the U.S., providing a European answer to the question of why.

Oh, no. Now we have "dark countenances." Please get your hands on that book I mentioned.

One possible source of the pervasive Portuguese oral history is that the Portuguese were heavily involved prior to 1642 in the early importation of African indentured servants, some of whom would eventually become free and some of whom would become slaves.

So that's it!

On the 1880 census, several Melungeon families claimed Portuguese as their race. An analysis of the families so claiming reveals that none of them were descended from the Denham line. Some, but not all were descended from the Sizemore and Riddle Native families. Of the 22 adults listed initially as Portuguese, more than half, 12 are descended from either the Goins or Minor families with African haplogroups, 11 are descended from the Sizemore family, 4 from the Riddle family, 4 are not descended from any of the above and 3 are unknown.

Tsk, tsk. The word "none" requires a singular verb. You should write, "None of them is..." I am not even going to attempt to straighten out your punctuation or sentence predication. Gentle Reader, please persist. The best is yet to come.

Ironically, the Sizemore family is not identified as Melungeon in Hancock/Hawkins Counties, but is ancestral to many Melungeon families and settled there are well. The Sizemore family is proven genetically to be Native, haplogroup Q1a3a. Furthermore, there are two Native Sizemore lines, although only one is known to be ancestral to the Melungeon families. A European Sizemore line also exists, and the Bolins match the European Sizemore lines, suggesting that these families may have had a common genesis or that these Sizemores may in fact be Bolins. Both families are found in early Virginia along the North Carolina border.

I always wondered about that. Now I know less than I thought I did before.

A link has been found through the Goins family to the Lumbee. The "Smiling" Goins family was not thought to be an original Lumbee family, but subsequent research has shown that even though the group in 1915 was thought to be an "outside" group, the ancestors of this group were found in 1770 with other founding Lumbee families. The Moore and Cumberland County Pocket Creek Goins groups have always claimed kinship with the Lumbee. Other links to the Lumbee have not yet been found. The Lumbee Tribe has been reticent to support DNA testing and common surnames between the Lumbee and the Melungeon Core group have not all been tested.

I don't blame the Lumbee tribe for being reticent to support DNA testing. Most folks I know are pretty reticent about DNA. A better word would have been "reluctant."

The Riddle family who is also ancestral to the Melungeon families is genetically European, haplogroup R1b1b2, but is documented historically to be Indian from a 1767 tax list where they are noted as such. Furthermore, they are found in other "Indian Communities" such as Pocket Creek in Moore County, N.C., tied to the Goins family. In 1820 several Riddle families are found beside a Goins family whose first name is illegible. In 1830 in Moore County, William Riddle is found beside both Levy and Edward Goins, believed to be the Goins family of the Lumbee.

That Riddle family! And now we find out they are living next to "Smiling" Goins.

Edward Goins is later found in Sumter County, S.C., a progenitor of one the Smiling Indian families in Sumter County, S.C., also known as Red Bones. This Goins family moved from Sumter County and settled in Robeson County, N.C. in 1907. The progenitor of this line, Frederick Goen, is found with the Lumbee much earlier, on the 1770 Bladen County tax list. Testimony regarding this family in 1915 states that the father's line is Melungeon.

Are you sure this is the summary?!

The Goins family is found in multiple locations in Virginia, North Carolina, South Carolina and Tennessee, several of which are involved with legal proceedings relative to their race. There are three genetic Melungeon Goins family lines, two E1b1a and one haplogroup A, all three being of sub-Saharan African origin.

Wait a minute. Aren't we just talking about male lines, and only one family at that, and only three cases at that. That doesn't seem like a fair summary.

In Hawkins/Hancock County, Tennessee, Sumter County, S.C., and Spartanburg District (Georgetown County), S.C., these Goins families are referred to as Melungeon. Genetically, they share a common ancestor, probably John Goins found in Hanover County in 1735.

Indeed! So to carry this to its logical conclusion, Jack Goins is descended from John Goins. John Goins was a white man. So is Jack Goins. Did I miss anything?

The Sumter County, S.C., Goins family is found in Bladen in 1770 ... where Louisa County families later settled. *(several paragraphs omitted for brevity's sake)*

Turning to autosomal genetic testing, no Native heritage was found using marker D9S919, although this finding does not disprove Native heritage.

Absence of evidence does not mean evidence of absence. That's what my father always told me.

It is possible in some cases that haplogroup E1b1ba could be found in rare instances in Europe through historical invasions such as the Roman Legions. However, given the Louisa County cluster, it's unlikely that a large cluster of haplogroup E1b1a of European origin would be coincidentally found together in the colonies. It's much more likely that this cluster is a result of people with a common bond living in close proximity and intermarrying. Furthermore, if haplogroup E were to be found in Europe, it's much more likely to be E1b1b, the Berber haplogroup, not E1b1a. No Melungeon families are found with haplogroup E1b1b or subclades.

Thank goodness those Roman legions didn't make it to Tennessee. But it seems like no North Africans did either, which is strange. See our post "Right Pew, Wrong Church."

Marriage partners in colonial Virginia were legally restricted beginning in 1691 with the passage of a law that forbid the English intermarriage with Indians, mulattoes and negroes. Prior to that, interracial marriages and encounters outside of marriage occurred regularly. This restriction, along with increasingly severe penalties in the event that the intermarriage did occur was repeated in various laws in 1705, 1753 and 1792 in Virginia and in 1715 and 1741 in North Carolina, in essence, requiring anyone who was other than white to intermarry within their own group or groups of racially similar individuals, meaning others "of color." Legal marriages between whites and other races would have had to predate 1691, although illegitimacy certainly knew no boundaries. In marriages occurring after 1691 in Virginia, in couples where one individual was "other than white," both partners could be presumed to have at least some recognizable non-European heritage.

This is one of the most hilarious and bigoted parts of this article, so be sure you read it several times to absorb it in all its unintended humor.

Given the proven Native ancestral families to the Melungeons combined with cultural styles that are perhaps suggestive of a maternal culture, Native or African, via illegitimacy, one would expect to find Native or African mitochondrial DNA. However, all mitochondrial DNA to date has been European. This was not expected given the very high levels of consanguity and intermarriage within this group from at least the mid- 1700s through the mid-1900s. However, Heinegg's analysis of mixed race families in early Virginia and his discovery that the predominant pattern of African or mixed men fathering children with white indentured female partners may explain these findings.

Typo: consanguinity. And sorry, but we don't buy your and Heinegg's theory about African men "fathering children with white indentured female partners." Those weren't African men, for one thing. But that is a whole other story, and it happened in Spain, and besides the wench is dead.

No evidence, historical, oral, genealogical or genetic has been found to support a Turkish, Middle Eastern, Jewish or Gypsy heritage.

Paydirt! The end! So what are they? You're not going to cop out and tell me they are just plain old folks. Or are you? Shucks, I guess that would make sense, though. Start out with a bunch of plain old folks, test them, and you can prove they are plain old folks. Your conclusions come from your premises. And your premises come from your conclusions.

I am normally all in favor of any DNA test or genealogy subscription or genealogical resource that can help the family researcher discover their ancestors. But "Melungeons, A Multiethnic Population," published by the International Society for Genetic Genealogy in *Journal of Genetic Genealogy*, by Roberta J. Estes, Jack H. Goins, Penny Ferguson and Janet Lewis Crain is without doubt one of the most pretentious, portentous and poorly conceived articles I have ever read in just about any field, and I will read almost anything. If you want a laugh, though, check it out. You may find out why Smiling Goins is smiling.

May 24, 2012

I Found My Grandmother's People

I grew up thinking I was English on my father's side and French-Cherokee on my mother's. But I also knew there was a mystery in my Southern Gothic family. And no one was talking about it. It took me most of my life to document my Melungeon and Jewish ancestry. I learned a lot on my journey through perseverance, but I learned the most from the DNA Fingerprint Plus. That journey has been a long one.

Our family reunions were bizarre. Grandmother, named Etalka Vetula, and my Aunt Elzina would arrive from an obscure town in Alabama looking just like two witches from Oz. At least that is what I thought when I was 6. They wore clothes from another time and place. Sweeping black Victorian dresses with high collars, long sleeves and black lace-up boots in the heat of summer. It was the 1950s. No jewelry adorned their necks, and no smiles appeared on their faces.

It was especially discomfiting since my mother looked and acted like an early Cher, wearing flamboyant Spanish skirts and always tossing out witty and outrageous remarks. The witches and she did not get along.

Whispers about "Black Dutch" circulated around the dinner table with slivers of pecan pie. Fried chicken was the main dish with tall, silver candles at the center lighted at dusk. We never ate ham. We never went to church.

If it was Christmas, it was spelled Xmas. There was no mention of the Baby Jesus. Ornaments, songs, and holiday cards were generic and dissembling. My mother once shooed away carolers singing "Silent Night." Her favorite holiday tune was "Rockin Around the Xmas Tree," although we did not sing this around my severe aunts. The tree was decorated with redbirds and popcorn. My father would use the comics to wrap presents, not because we were poor. He was a doctor.

My maternal grandmother, Luta Mae, would have had taken me out of the Indian calico floral dress before my aunts arrived, putting on a black velvet one, to my dismay. I was the very picture of Wednesday from the Addams Family.

Of course, my family seemed normal to me. I did not know why most mothers I knew looked more like Doris Day, or why they shunned my beautiful, dark-olive skinned mother at the country club, or why their children would not play with me. This was the South in the '50s before integration. Wasn't I white? I was confused.

When I married Don, we became very interested in our genealogy and would spend many weekends researching in archives and libraries, trying not to go blind looking at microfiche. On one of these weekends, we visited my Aunt Elzina. I have six Elzinas in my family. As amiable as ever, she jeered, "You will never discover the truth about my mother's people."

She did let slip an interesting story about Great-grandmother Redema. When Redema was a young girl they "arrived on a rich man's farm to train his horses." Afterward they managed to buy the farm. Were they Gypsies? Aunt Elzina's arrogant challenge made me the more determined. There were several clues that I began to add up. I remembered that my mother took me aside when I was 12 (the age when a young girl is given a bat mitzvah in the Jewish faith). Did I know I was named after a little Sephardic Jewish girl "running around in the Arizona desert"? I thought it an odd remark, as I did not know any Jews.

We read Brent Kennedy's book and discovered many of our ancestors were Melungeon. Don looked up the name Etalka and determined it was Hungarian and Yiddish. My father gave me two paintings of my great-great-great-grandparents. I said, "Dad, these people are not English or French. They are too dark. What are they?" He said, "You ask too many questions. If you keep digging, you will find people you do not want to be related to."

A second cousin sent me a fragment of a letter she discovered a great-granduncle of mine had written. It was in Demotic Egyptian. After my Aunt Elzina died, I discovered from documents my father shipped on to me in cardboard boxes that before my line of Rameys were in France, they were in Egypt.

Before I got these records, we had just started the business DNA Consultants. Don ran my profile. Egyptian was at the top. I laughed, "This has got to be wrong!" But DNA doesn't lie.

I got expected information as well as some surprises. Melungeon was near the top. My grandfather had always told me he was part Cherokee. I got a Native American marker. But I also got things that threw me. My mother probably loved Spanish skirts for a reason. I have a lot of Iberian results. I did not know I had Turkish ancestry. This is a common match for anyone with Melungeon ancestry. And English? Near the bottom! It turns out I have more Mediterranean and Eastern European ancestry, as well as Jewish. I have two out of three Jewish markers, one a Sephardic Jewish marker, confirming my mother's talk to me, and proving one of the reasons that led to my open return to Judaism. I do have Hungarian matches as well as Romanian/Transylvanian matches (originally Hungarian territory), which confirmed Don's suspicion that Etalka was Hungarian. It turns out to be a diminutive of Adele, a Hebrew name.

So, I found my grandmother's people, despite Aunt Elzina and other naysayers.

September 22, 2013

Roberta commented on 04-Oct-2013
You wrote of your "open return to Judaism". If you don't mind my asking, does this mean that you were accepted by the religion as Jewish because of your DNA and didn't have to convert?

Donald Yates replies: No, not really. The rabbi said DNA didn't matter and he wasn't quizzed about it anyway. The important thing was that my immediate ancestors were practicing Jews, some buried in the Jewish cemetery. And that I wanted to re-embrace Judaism. He said other rabbis might require that I "convert" instead of return. His name is Rabbi Arnold Mark Belzer, now retired.

You Will Never Discover the Truth

"You will never discover the truth." That is what my Aunt Elzina told me as we sat in her Victorian parlor when I asked her about my Grandmother Etalka Vetula. I was determined that I would.

My family was a mystery. And I only had a few pieces to the jigsaw puzzle. For many years, I settled with those pieces instead of real answers. First, I dabbled in genealogy. I did find some answers. After reading about the Melungeons, I thought this might explain some of the mystery around my family. But my family wasn't talking. And my family tree was lined with dead ends at every turn. And I only added more pieces to my jigsaw pile. One great-aunt of mine in Alabama hung up when I asked her what she could tell me about her line. And my father said, "If you keep digging, you might find people you don't want to be related to."

Next, I thought I would turn to science and get more complete and reliable answers by purchasing a DNA test.

After all, DNA holds the answers to our entire genetic code. We are our DNA. John Wilwol, in his recent NPR article, "A 'Thumb' on the Pulse of What Makes Us Human," quotes Sam Kean, author of *The Violinist's Thumb And Other Lost Tales of Love, War, and Genius, As Written by Our Genetic Code*, that DNA is what makes us who we are. Wilwol also quotes Kean in helping us understand what DNA is and how it differentiates from genes: " 'While DNA is a *thing*—a chemical that sticks to your fingers, he writes, genes are more conceptual in nature, ...'' 'like a story with DNA as the language the story is written in.' " So if DNA holds some keys to discovering who I am ... why not get it analyzed? And see if I can better understand the story? But how?

At first, I considered getting one of the haplotype sex-linked tests. These are known as maternal or paternal tests and examine the parts of our DNA known as Y- DNA and MtDNA. This type of test analyzes either your biological father's direct male line—his father's father's father, and so on, all the way back—or your biological mother's direct line—her mother's mother's mother, all the way back by testing the variations or mutations in her mitochondrial DNA. A male can take either the male or female test, but a female can only take the female test. She must conscript a male relative, such as a brother, father, or uncle, if she wants a male haplotype test. These tests are good for assisting with genealogy and finding male and female cousins. But it wasn't what I was looking for. I wanted more information than just one line. Was that possible?

Finally, I discovered autosomal DNA, and I was amazed! Wow! Autosomal DNA is the new wave of science. The second generation of DNA testing. I could have never found all the answers without using autosomal DNA. Why? Autosomal DNA tests reflect the entire spectrum of ALL your ancestral lines. Not just one line at a time.

How is that possible? All parts of our genetic code are readable and meaningful. Your marker locations (loci) are spread across your entire genome, not confined to your male (Y chromosome) or female (mitochondrial) DNA. That means there is no junk DNA after all, according to Rose Elvereth's recent Smithsonian blog post, "Junk DNA Isn't Junk and That Really Isn't News." It also means DNA is an important key to your discovering you.

September 22, 2013

Digging Up Jesse James

Any outlaw skeletons playing a Tennessee jig in your ancestry closet? We are always looking for famous people—in our ancestry or in the news. But there are other ways to get famous. So, you might find both. And you might find a DNA mystery. Many hope to find royalty. But some of us have some unusual branches to our family tree. Like Jesse James. Jesse James was a mystery in life and remains a mystery in death. We can't decide whether we should honor him or curse him, and we can't decide where he is buried (*San Antonio Express-News* columnist Roy Bragg). We can only agree that he was a man of mystery.

And I know someone, a Jon Watkins, who is related to Jesse James.

A Southerner who resembles James somewhat, Watkins says as a kid he played and crawled in Jesse James' getaway tunnels under a house James lived in in Tightwad, Missouri. This was long after the shoot-out in front of the white, two-story house where his grandfather told him Jesse and his gang surrounded four marshals and hanged them in a tree in the front yard. He says his grandfather cut the tree down, stacked it into pieces, and burned it—refusing to use it for firewood because it was "blood money." Jesse James was his great-uncle on his maternal (mother's) side. "My grandmother might have even met some of the family," he says in a faraway voice, seeming to waver for a moment, as we all do, from thinking of Jesse James as something of a thug to someone like Robin Hood.

Why do we feel so conflicted about Jesse James? In his article, "The Afterlife of Jesse James," Bragg quotes Martha Grace Duncan, a law professor at Emory University, who wrote the book *Romantic Outlaws, Beloved Prisons: The Unconscious Meanings of Crime and Punishment* (which analyzes why some outlaws are treated with reverence in folklore) as saying "he fits the mold of the noble bandit."

Throughout history, according to Bragg, some have viewed James as a "murderous, bank-robbing thug. Others, however, speak of him as (a) hero who never surrendered after the Civil War. To them, he fought the institutions—banks and railroads—that ran roughshod over the common man." Some of us are torn. But Bragg quotes Duncan as saying his "die-hard fans" overlook the fact that "James was no Robin Hood" and kept the loot for himself and his gang and gunned down innocent people who got in his way. But there are those of us who overlook that.

Why? Duncan explains, "For some Southerners … there was a situation of moral inversion. The legal government, after the war, was in the wrong in their minds." Moreover, Duncan says, even though "we may not want to be revolutionaries or rebels or criminals," we are "fascinated with those who do step over the line and succeed in hiding out (from the law). A part of us is attracted to this." In fact, many of us are so attracted to it, he has become part of our culture and seems to be part of all of our families.

Beyond the myth and mystique of Jesse James and his "Mission Impossible" ability to hide from the law, Jesse James remains mysterious after death. History says that James was buried in Kearney, Mo., after being gunned down on April 3, 1882, in his St. Joseph, Mo., home by fellow gang member Bob Ford. But not everyone buys that story. In Elvis style, there are those who believe Jesse James hid out and faked his death leaving some to question the DNA analysis of his remains. How can this be? According to an AP article, "Jesse James Museum in Wichita Claims Proof That Outlaw Faked Death," a Ron Pastore claimed he "… can prove through DNA tests that James died in Kansas in 1935."

"Pastore, who has opened the Jesse James Museum in Wichita, has had a Kansas man exhumed in his quest to prove that James faked his death and lived under the alias of Jeremiah M. James in Neodesha, where he was buried in 1935."

Bragg writes, "A Texas woman says James lived and died here, 30 miles south of Waco. He faked his death in Missouri, she says, and lived out his days here as a gentleman farmer. He died peacefully in 1943 and is buried under the name James L. Courtney in a family plot overlooking Deer Creek in Falls County."

The story most commonly heard is that it might have been one of his cousins that was shot in his place, and he was living underground with an assumed name. But the director of the Kearney museum, Elizabeth Gilliam Beckett, says, "Our (DNA) test was 99.7 percent accurate, but that 0.3 percent is what people are grasping to (sic).

"We have the real deal," says Beckett.

However, it makes it more problematic that the official story is that James did use an assumed name and was living under the name of Thomas Howard for a time in St. Joseph with his wife, Zerelda, and their two children. The body of "Thomas Howard" was buried as "Jesse James" after Howard's death in the front yard of the James family farm (Bragg).

And the remains were in poor condition. According to Anne C. Stone, Ph.D, James E. Strauss, L.L.M, and Mark Stone King, Ph.D., in the 2001 *Journal of Forensic Science* article, "Mitochondrial DNA Analysis of the Presumptive Remains of Jesse James," the remains of Jesse James from exhuming him in 1995 from Mount Olivet Cemetery in Kearney, Missouri, were "… poorly preserved, presumably due to wet and slightly acidic soil conditions, and insufficient DNA from analysis was obtained from two bone samples."

What did they rely on instead? Teeth and hair recovered in 1978 from the original burial site on the James farm. Using mitochondrial DNA analysis (mother's line) they did manage to match the teeth and the hair to Jesse James's sister. Do the mitochondrial results prove that the exhumed remains are those of Jesse James? Stone et al. say that there is a "remote" possibility that the remains are from "a different maternal relative," but "there is no scientific basis whatsoever for doubting that the exhumed remains are those of Jesse James . . . the burden of proof now shifts to those, who, for whatever reason, choose to still doubt the identification."

So we await the next chapter in the DNA mystery of Jesse James.

Jesse James remains part of our folklore of "romantic outlaws," and for some people, like Jon Watkins, he is distant kin.

September 24, 2013

It's a Wonderful Snowstorm

Genealogy is the grandfather of DNA ancestry testing, especially the latest state-of-the-art, autosomal tests. Some might wonder why not just focus on their family trees and what they know of their family history. While that is interesting, and definitely worth knowing as part of your family history, it is only hearing part of the story, like falling asleep in the middle of watching *It's a Wonderful Life*. It is time to wake up because there is a lot more to discover from your DNA beyond your genealogy. We have entered the Genomics Age.

However, there are still some true genealogy aficionados who are as yet unconvinced. It is an interesting hobby, but it not only is incomplete, it can also be rife with errors if one is not careful. How? It is only accurate if all the names have additional sourced information from archives.

Years ago, one actually went to archives and libraries and looked up the census or death records. Those records are now online, but all that takes work, and we live in a fast-paced world. As a result, in the world of genealogy, many share their family trees with others, but how many have made sure their information is correct? A lot of people have not and have incorrect information they pass on to others.

But how can one's genealogy be incomplete? No matter how careful you are, or how much time you take with a family tree, you cannot discover all that a DNA test can. Why not? Most people only concentrate on one or two ancestral lines at the most and often ignore the female side of much of their genealogy as it is difficult to discover much information if a maiden name is unknown. Also, there can be name changes, unknown adoptions or other non-paternity events. It seems that most genealogists believe all their ancestors were faithful, and while that is commendable, it might not be realistic. If you go back 50 generations, you probably have a non-paternity event somewhere. An adoption or some hanky-panky. One mistake like this can completely change the outcome of your known family history. Even the best genealogies like royalty have been proven to be fake and manipulated to be more prestigious than they are. None of the kings of England has a direct male descendant today. Many lines have died out. And many that are not royalty make assumptions because of what other people have "discovered" without doing real research because it takes time and effort. It is a fun game people have played for centuries, but how much of it is truth?

How can you find the truth? Haplotype tests are just one little step beyond genealogy as they still narrowly focus on the mother's or father's line. There is so much more to discover.

We are all more mixed than we realize. We are also unique in singular ways, like snowflakes.

October 25, 2013

All in the Family

Is it all in your genes? Sometimes it is, and sometimes it is not. Your ancestry is all in your genes. No matter what you do, you will always have the same family and set of ancestors, as Reese Witherspoon discovers in *Sweet Home Alabama.* There is nothing you can do to change that, and a DNA ancestry test will reflect your genetic connections to populations or people depending on the type of test you take.

However, genetic risk factors are not exactly the same. These are possibilities that depend on other epigenetic factors such as lifestyle and environment. Carl Zimmer, a famous geneticist, says as much.

We are just at the brink of the Genomics Age. DNA testing started with paternity testing and forensic testing back in the '90s and has now expanded to DNA testing for ancestry, advanced genetic screening for infants, and more. Perhaps, personal at-home genetics testing for medical risk factors needs more thought.

In concern for the consequences of public health, the FDA has pulled the plug on 23andMe. Is there a future for at-home medical DNA testing, or no?

November 22, 2013

9 DNA AND GENEALOGY

California's Mission Indians

Reclaiming a neglected part of California's past, historians Monday unveiled an immense data bank that for the first time chronicles the lives and deaths of more than 100,000 Indians in the Spanish missions of the 18th and 19th centuries.

In an eight-year effort, researchers at the Huntington Library in San Marino used handwritten records of baptisms, marriages and deaths at 21 Catholic missions and two other sites between 1769 and 1850 and created a cross-referenced computerized repository that is now open to public access.

The database should be a big boon to families researching their Southwest Indian ancestor who was adopted.

August 8, 2006

Basketmakers and Cannibals
Why You Don't Have (and Don't Want) Enemy Ancestors

When I first visited Chaco Canyon in northern New Mexico, I got a creepy feeling. Who were these people? The park staff, mostly Navajo and not descendants of the original inhabitants of the site, said they were the Anasazi, a Navajo term meaning "enemy ancestors." When I pressed for more answers, I learned only they belonged to the Chaco Culture.

They were Chacoans.

Anthropologists are used to calling people they don't understand a culture, but as I learned more of this one I don't think they had much of it. The culture of the Anasazi turns out to be one of head-hunting, terrorizing human slaves, drug use and cannibalism.

In a book subtitled *A History Forgotten*, George Franklin Feldman dispels the parlor concepts and sanitized history surrounding Native American practices and pieces together the frightful truth. It is the only book on its topic: *Cannibalism, Headhunting and Human Sacrifice in North America* (Hood: Chambersburg, 2008), and the author's work was an uphill battle against political correctness.

Forget about the Donner Party. When anthropologists and explorers first encountered the monuments and ruins of the people politely called the Ancestral Pueblo they found widespread evidence of a cannibalistic society that had gotten out of control and succumbed to its own inhumanity.

"The best documented indication that the Basketmakers were headhunters is ... Kinboko Canyon, evidence discovered by archeologist Samuel J. Guernsey of the Peabody Museum of Harvard in 1915, and reported in a 1919 publication of the Smithsonian Institution's Bureau of American Ethnology," writes Feldman in the chapter on the Anasazi. He goes on to describe this and other early excavations in the Four Corners area that were quickly hushed up and reburied in horror, including Battle Cave in Canyon del Muerto, now part of the Canyon de Chelly National Monument inside the Navajo Indian Reservation. We read now of flesh-stripping, bone crushing, roasting pits, and sliced off mastoids.

Around 950 A.D., 11 persons, including women and children, were killed and butchered, cooked, and eaten on Burnt Mesa in New Mexico north of the San Juan River. At a site near the Hopi villages in Arizona, a group of 30 individuals, 40 percent under the age of 18, were slaughtered and eaten. In a Colorado rock shelter, a large jar was found filled with splintered human bones. ...

The grisly record goes on and on. Feldman writes that by the year 2000 the number of such sites in the San Juan drainage where the Chaco Culture was centered had risen to 40 (p. 136).

To their credit, the Indians who were the Chacoans' food supply eventually overthrew their masters and left the area to settle far away, on the three Hopi mesas in Arizona and along the Rio Grande in new pueblos that survive today such as Taos and San Juan. We suggest that the aristocrats of Chaco may have been subject to degenerative neurological disorders like kuru or mad cow disease. But whether they self-destructed or were destroyed by their subject population, it doesn't seem to make a lot of sense to claim them as ancestors.

June 18, 2010

They Probably Always Talked Like That

One of the startling revelations by Stephen Oppenheimer is that a form of English was probably spoken from the beginning of the colonization of the British Isles. Just as genetic bedrock was laid down by the earliest inhabitants, to persist relatively unchanged through subsequent invasions by other peoples like the Romans, the English tongue has been dominant as the language of the land, admitting little admixture with Anglo-Saxon and Celtic. (See Stephen Oppenheimer, *The Origins of the British*, pp. 303ff.)

Pretty heady stuff, but Mick Harper, author of *The Secret History of the English Language* (Hoboken: Melville 2008), goes Oppenheimer one better by proposing that it was not proto-Anglo Saxon that the Ice Age inhabitants of Britain spoke but something very like Chaucer's pilgrims, only lacking, clearly, later invasive elements due to the Celts, Belgae, Romans and Normans.

Harper compares a sample of Old English (which we are taught is the same as "Anglo-Saxon") with Middle English and Modern English to show that Anglo-Saxon does not appear to be the same language as English—something all English graduate students suspect from the moment they are forced to read *Beowulf* for their comps. In the Anglo-Saxon epic (which survives in a single copy turning up in suspicious circumstances in Tudor England and is set in Sweden and never mentions England), "virtually every single word is *incomprehensible* except by translation," while in "the early English poetry of Chaucer and Piers Plowman ... virtually every single word is *comprehensible* except for spelling."

In case you do not believe it, here are the samples:

Nu scylun hergan hefaenricaes uard,
Metudaes maeti end his modgidanc,
Uerc uuldurfadur, sue he uundra gihuaes,
Eci dryctin, or astelidae.
(Caedmon, ca. 8th cent.)

A swerd and bokeler bar he by his side...
A whit cote and a blew hood wered he.
A bagpipe wel koude he blow and sowne,
And therwithal he brought us out of towne.
(Chaucer, The Prologue, 14th cent.)

You cannot say, or guess, for you know only
A heap of broken images, where the sun beats,
And the dead tree gives no shelter, the cricket no relief,
And the dry stone no sound of water.
(T. S. Eliot, 20th cent.)

Harper's comment is: "If Anglo-Saxon/English is one language, it's unique in the entire annals of languages on this our Earth, since it changes every goddamn word of itself" (p. 44). (Yes, he writes like that, too.)

The Anglo-Saxons were a small, obscure and illiterate tribe from, well, no one is quite sure, but perhaps northeast Germany, who arrived in waves after the Romans abandoned Britain in the fifth century, and who conquered most of the land and held it until the Danes and Norse (ca. 900) and Normans (1066) replaced them as rulers.

In Harper's view, they were just like the previous invaders, the Romans, Belgae and Celts, in having little effect on the language and customs of the populace. Just as there are only a handful of Celtic words in the English language, there was little impact on the linguistic bedrock of the kingdom the Anglo-Saxons carved out before they too had had their day. The fact that they left few monuments is unsurprising.

Which brings us to questions about the depth and breadth of Celtic heritage in Britain. If you are a Celtic fan (I'm not referring to the basketball team) you will not want to read *The Secret History of the English Language.* This book will disabuse you of many cherished notions. In Harper's view, the Celts were just one of the alternating foreign conquerors of the long-suffering English-speaking peoples. Their numbers were few, even on the Continent, and they left little genetic or cultural footprint except on the "Celtic fringe" where they were squeezed in their final days.

England has always been England. It's always spoken English. And France has always spoken French. "But that was in another country, and besides, the wench is dead. "We will have to save the French linguistic heresy for another post.

If you like the unusual and provocative ideas of M.J. Harper, who lives in London, check out the community of people who have bid farewell to the dunciad of academic research and unleashed their own personal pursuit of truth on a variety of intellectual topics at The Applied Epistemology Library. You can browse on the sly but must register (for free) to post your own comments and questions on threads.

June 30, 2010

Shot across the Bow of the Good Ship Mayflower
Jews and Muslims in Colonial America

More than eight years after starting the project, Elizabeth Caldwell Hirschman and Donald N. Yates won a book contract for their collaborative study of crypto-Jews and crypto-Muslims in the settlement of British North America. Titled *Jews and Muslims in British Colonial America, a Genealogical History*, the work will be published next year by McFarland, a leading U.S. publisher of scholarly, reference and academic books.

Among original investigations by the authors are genealogical studies of the West Country Gentlemen and others who proposed and promoted England's first colonies. From Sir John Hawkins (Sephardic Jewish surname Haquines, "physician" in Arabic) and Sir Francis Drake (whose family coat of arms bore a six-pointed star until it was air-brushed out by later historians) to Stephen Parmenius (a Jew from Turkish-held Hungary) and Captain John Smith, the principal players in England's colonization efforts are revealed to be far different from the white Anglo-Saxon Christian buccaneers American schoolchildren are taught about.

"England's reliance on Iberian Jews to promote its interests abroad goes back as far as the Tudors," according to the book.. "Henry VIII used Spanish Jewish lawyers to justify his divorce from Catherine of Aragon. One of them, an Italian banker tapped for his shrewdness and knowledge of international law, was the ancestor of Oliver Cromwell, Protector of the Commonwealth."

The book presents a series of colonial documents, contemporary firsthand accounts, records, portraits, family genealogies and ethnic DNA test results that fundamentally challenge the national storyline depicting America's first settlers as white, British and Christian. The authors postulate that many of the initial colonists were of Sephardic Jewish and Muslim Moorish ancestry, usually arriving as crypto-Jews or crypto-Muslims.

Names Ordinary and Illustrious

The footnotes in the study document origins and meanings of over 5,000 surnames previously assumed to be sturdy English ones of ancient bearing. The authors' research casts a sidelight on celebrated Jewish Americans who can trace back to colonial forebears. These range from the Massachusetts Kennedys to the Byrds of Virginia, from actors Johnny Depp and Adrien Brody to actresses Roseanne Barr and Gwyneth Paltrow, from writers Louise Glück and Neil Simon to politicians Barbara Boxer and Bernie Sanders and jurists Stephen Breyer and Elena Kagan.

"We hope that the remarkable stories of the men, women and families in *Jews and Muslims* will serve as a reminder of America's early diversity and stimulus for rewriting some of the inaccurate and injudicious portions in the country's history," said Yates.

Among the famous colonial figures discussed (and usually illustrated with a portrait) are William Byrd II, Patrick Henry, William Bradford, William Penn, George Mason, George Washington, Richard Lee II, Thomas Paine, Paul Revere, Peter Stuyvesant, Luis Gomez, Jacob Troxell, Anthony Ashley Cooper Lord Shaftesbury, Tench Tilghman, Christopher Gist, John Skeen, Sir Philip Sidney, Walter Raleigh, Humphrey Gilbert, Virginia Dare, Don Luis de Carvajal, Daniel Boone, William Cooper, the Salem Witches, Christopher Gist, Lord Saye and Sele, and various Lowells, Cabots, Lodges, Livingstones, Delanceys and

Roosevelts.

Chapter 2, "Sephardim in the New World" is a survey of Jews and crypto-Jews in North America, especially the Caribbean and Atlantic Islands. It includes autosomal DNA data proving the Melungeons are probably descended from Jews mixed with American Indians, Africans and Gypsies/Romani, as recently reported in this blog.

There are chapters and name-lists devoted to each of the original colonies. The book contains over 50 illustrations.

November 9, 2010

English navigators and explorers included many West Country gentlemen. Most were from intermarried crypto-Jewish families. New York Public Library.

DNA Studies Mature

Then: Genes of Old Testament Priests (Cohanim)
Now: Genetic Traces of Religions in Lebanese and Iranians

Then: Rare Genetic Disorders in Finnish Mitochondrial Haplotypes (U)
Now: Genome-Wide Association Studies in Saami

The whole business of direct-to-the-consumer DNA tests was founded upon the revelation in 1997 that Jewish men with the last name Cohen ("priest" in Hebrew) or something similar often preserved the genetic signature of Old Testament priests in the Y chromosome type handed down from father to son. Last year at long last, the so-called Cohen Modal Haplotype was completely pinned down and defined to everyone's satisfaction ("Does He or Doesn't He?"). Now similar genetic traces are being sought, and found, for other religions from the Middle East.

In response to customers asking whether being a Jew was a matter of ancestry or culture, genes or religion, I used to say, "Genes don't have religion, genes are older than religions, your DNA doesn't know what religion you are." But the increasingly adept methods of populations genetics are changing that pat response. The key tool is a program that uses advanced statistics to estimate population differentiations, BATWING. Standing for Bayesian Analysis of Trees With Internal Node Generation, this software can calculate the effective population sizes and growth rates from microsatellite data, assuming there was a split into several populations in the past. It is a little over 10 years old. The following article is likely to become a classic in this regard: "Influences of History, Geography, and Religion on Genetic Structure: the Maronites in Lebanon," by Marc Haber et al., *European Journal of Human Genetics* (2011) 19, 334–340; doi:10.1038/ejhg.2010.177; published online 1 December 2010.

Abstract. Cultural expansions, including of religions, frequently leave genetic traces of differentiation and in-migration. These expansions may be driven by complex doctrinal differentiation, together with major population migrations and gene flow. The aim of this study was to explore the genetic signature of the establishment of religious communities in a region where some of the most influential religions originated, using the Y chromosome as an informative male-lineage marker. A total of 3,139 samples were analyzed, including 647 Lebanese and Iranian samples newly genotyped for 28 binary markers and 19 short tandem repeats on the non-recombinant segment of the Y chromosome. Genetic organization was identified by geography and religion across Lebanon in the context of surrounding populations important in the expansions of the major sects of Lebanon, including Italy, Turkey, the Balkans, Syria, and Iran by employing principal component analysis, multidimensional scaling, and AMOVA. Timing of population differentiations was estimated using BATWING, in comparison with dates of historical religious events to determine if these differentiations could be caused by religious conversion, or rather, whether religious conversion was facilitated within already differentiated populations. Our analysis shows that the great religions in Lebanon were adopted within already distinguishable communities. Once religious affiliations were established, subsequent genetic signatures of the older differentiations were reinforced. Post-establishment differentiations are most plausibly explained by migrations of peoples seeking refuge to avoid the turmoil of major historical events.

Meanwhile in Autosomal DNA

A like expansion and intensification of research interests have also transformed the field of Finnish DNA. In the old days it was well appreciated, through the work of Finnila and others, that the people of Finland, Estonia, Sweden and neighboring regions in Russia had a peculiar genetic history. Strangely, at least on the basis of mitochondrial DNA, they were more closely related to the Berbers of North Africa than the neighboring Swedes, Poles, Lithuanians and Russians. Female haplogroups UK were associated with a risk of occipital stroke, migraine and other neuro-deficiencies. On another level, their unique genetic history was approached through the study of male haplogroup N, common among Laplanders and Saami.

The focus has now shifted from haplotyping and sex-linked genes to population genetics and autosomal DNA just as it has in consumer tests. After 10 years, an important autosomal study of the Saami has revolutionized the subject and shows promise of becoming the pilot to a new series of genome-wide disease association studies:

Jeroen R Huyghe et al. "A Genome-wide Analysis of Population Structure in the Finnish Saami with Implications for Genetic Association Studies." *European Journal of Human Genetics* (2011) 19, 347–352; doi:10.1038/ejhg.2010.179; published online 8 December 2010.

Abstract. The understanding of patterns of genetic variation within and among human populations is a prerequisite for successful genetic association mapping studies of complex diseases and traits. Some populations are more favorable for association mapping studies than others. The Saami from northern Scandinavia and the Kola Peninsula represent a population isolate that, among European populations, has been less extensively sampled, despite some early interest for association mapping studies. In this paper, we report the results of a first genome-wide SNP-based study of genetic population structure in the Finnish Saami. Using data from the HapMap and the human genome diversity project (HGDP-CEPH) and recently developed statistical methods, we studied individual genetic ancestry. We quantified genetic differentiation between the Saami population and the HGDP-CEPH populations by calculating pair-wise FST statistics and by characterizing identity-by-state sharing for pair-wise population comparisons. This study affirms an east Asian contribution to the predominantly European-derived Saami gene pool. Using model-based individual ancestry analysis, the median estimated percentage of the genome with east Asian ancestry was 6% (first and third quartiles: 5% and 8%, respectively). We found that genetic similarity between population pairs roughly correlated with geographic distance. Among the European HGDP-CEPH populations, FST was smallest for the comparison with the Russians (FST=0.0098), and estimates for the other population comparisons ranged from 0.0129 to 0.0263. Our analysis also revealed fine-scale substructure within the Finnish Saami and warns against the confounding effects of both hidden population structure and undocumented relatedness in genetic association studies of isolated populations.

The key to emerging triumphs of research here is the international HapMap project.

On two fronts—religious history and rare diseases—genetics has brought more advances in the past decade than in the previous century before that. That consumers can take part in these exciting developments by ordering an affordable autosomal analysis of their entire ancestry or confirming the paternity of their child with a simple test purchased at their local drugstore is a tribute to the present golden age of American science and industry.

February 17, 2011

Two New Population Studies

Two reports in the *European Journal of Human Genetics* underline how specific autosomal DNA can be in revealing the geographical structure among populations. One uses genome-wide data from the Illumina Human Hap300 project to predict the village of origin of a person's four grandparents given European origins. The other used genotyping from 3,367 individuals from seven different European, mostly British Isles populations to lay bare the detailed population structure and linkage disequilibrium patterns of Ireland, England, Scotland and Wales.

Both studies have Colm O'Dushlaine of Trinity College, Dublin as their lead author and highlight how genotyping with autosomal DNA dominates genetics today (and DNA testing), eclipsing in many respects the older style of tests known as sex-linked haplotyping or lineage analysis, which focused on the male Y chromosome and female mitochondrial lines only.

Colm O'Duschlaine et al., "Genes Predict Village of Origin in Rural Europe," *European Journal of Human Genetics* 18 (2010), 1269-1270.

Colm O'Duschlaine et al., "Population Structure and Genome-Wide Patterns of Variation in Ireland and Britain," *European Journal of Human Genetics* 18 (2010) 1248-1254.

April 14, 2011

A Country Built on People

Not everything you were told in school about the Pilgrims, George Washington and the other brave, white Christian Founding Fathers of America is true. In fact, according to Elizabeth Caldwell Hirschman and Donald N. Yates' new book, some of the familiar figures were not even Christian. Appearing last month, *Jews and Muslims in British Colonial America* offers a fresh perspective on the early American experience, with chapters and emigrant lists on all the original colonies, from Virginia to Georgia. Here are "tweetable" excerpts with some of the study's surprising Hebrew, Arabic, Turkish and Jewish names.

From the Introduction

"American genealogies rarely jump the Atlantic, and in Norman pedigrees there seems always to be a disconnect between the invasion of 1066 and sudden appearance of the family's supposed ancestor on, say, the Pipe Rolls from the reign of Henry II or Magna Carta of 1215. In America we sometimes have the ship's passenger list mentioning our ancestor but then a maddening interval of silence before a reliable nexus of birth, marriage and death records can be established. Oral traditions fill in the breach.

"The first president of Nigeria, Nnamdi Azikiwe, once remarked America 'is a country built more on people than on territory.' He saw Jews as the ingredient that gave Americans their characteristic distinctiveness and diversity, saying they 'will come from everywhere: from France, from Russia, from America, from Yemen. ... Their faith is their passport.'

"We have made an effort to bring to life some of the forgotten women who shaped the Colonial American experience. These include the ancient *Ur* mother of European royalty, the Occitan Jewish Queen Itta, Charlemagne's great-great-great-grandmother; Mary Lago, the Sephardic Jewish mother of the Quakers' founder George Fox; and Malea Cooper, the Jewish Mulatto wife of Daniel Boone's guide (whose Hebrew name is the same as one of President Barack Obama's daughters). Some of their stories will surprise readers, running counter to traditional historical accounts. Few people realize that one of the Salem witches mounted a legal defense using the family's professional connections and wealth. Sarah Town(e) Cloyes (or Clayes) fought her accusers and lived to become a founder of the town of Framingham, Massachusetts. The mother of Patrick Henry according to diarist William Byrd was 'a handsome woman of the Family of Esau' whose first husband in Aberdeen, Scotland, was 'of the Family of Saracens.' The Cherokee Beloved Woman Nancy Ward bore an Arabic name. One of the subthemes of the present study is the interaction between immigrants and American Indians. Readers will find Cherokee, Creek, Choctaw and other genealogies that reflect intermarriage between Jews or Muslims and America's indigenous inhabitants, believed in the thinking of the day to be Abraham's children.

"On a different level, an interlocking Presidential dynasty has been suggested with twelve patrician families at its core—Adams, Bayard, Breckbridge, Harrison, Kennedy, Lee, Livingston, Lodge, Randolph, Roosevelt, Taft and Tucker—most of which we will have recourse to mentioning in the chapters of this book.

"The footnotes in this study are intended not only to document origins and surname histories in unequivocal fashion but also to cast a sidelight on celebrated Jewish Americans who can trace back to colonial forebears and their relatives in European Jewry."

March 4, 2012

Melungeon Revelation
By Kari Carpenter

Thank you to Donald Panther-Yates and the DNA Consultants staff for the prompt return of my Melungeon DNA Fingerprint results. Several weeks ago, I had never even heard the word "Melungeon." In preparation for an upcoming genealogy research trip, I just happened to go on amazon.com and read the introduction to Dr. Yates' book: *Old World Roots of the Cherokee*. I was tremendously excited to see a description of the terms "Black Irish" and "Black Dutch." Up until this time, I had been unable to find a satisfactory definition of these terms. My maternal grandmother had always stated that she was "Black Irish." Another maternal great-grandmother reported her lineage as being "Black Dutch." None of my family members seemed to know what those terms meant.

Needless to say, I quickly ordered the above-mentioned book, as well as other books by N. Brent Kennedy, Elizabeth Caldwell Hirschman and Wayne Winkler. It took me no time at all to realize that much of the family tree that I have diligently put together (over the last six years) was/is profoundly Melungeon. If I have accurately understood the information I have rapaciously absorbed over the last few weeks, and responsibly documented my family lineage, the following facts and information support this conclusion.

1. After reading *Old World Roots of the Cherokee*, I realized that I am a distant Ramey/Reamy cousin of Teresa (Grimwood) Panther-Yates. My paternal grandmother was born a Reamy in Texas. Although I had known that the Reamys had been French Huguenots, I had no clue about their Sephardic ancestry until I read this book. Similar to the Grimwood/Rameys (in Chapter 9) my Reamy people also have a variety of unusual and distinctive first names: Othera, Vida, Vita, Orvia (male), Olive (male), Rozella, Oleta, etc.

2. The name of my fourth great-grandmother (paternal side) was Margaret Anna Goins. Her granddaughter, Margaret Ann England married my second great-grandfather, Olive Nathaniel Reamy. Margaret Anna Goins and her husband, William James Morris, resided in Tennessee and Alabama before moving to Texas after the Civil War (with their daughter's family).

3. My great-grandfather Orvia N. Reamy consistently reported that his grandmother was of Cherokee ancestry. (He did so delightedly and repeatedly because it upset many other family members who would have much rather maintained a complete silence regarding their "less-than-white" ancestry.) This woman would have been Elmina C. (Morris) England—daughter of Margaret Anna Goins. In addition to Goins being a major Melungeon surname, I believe that the inter-related Englands, Morrises, and Barnes of this branch of my family are also probably of Cherokee ancestry. Elmina Morris and husband Landy England gave their sons rather distinctive names of certain military generals prominent in the Creek/Redstick War.

4. My great-grandmother Jesse Alice Ketchand (my mother's father's mother) stated that she was Black Dutch. Family pictures show that Jesse Alice and her mother, Martha Ann Cammack, had Asian eye folds. This evidence and the family lore regarding Jesse Alice's grandfather William

Peterson Ketchand, strongly support Melungeon ancestry. It is said that sometime around 1834, "W.P." got into some kind of trouble in Virginia, whereupon he and his new bride rapidly left the area. W.P. subsequently made up a new surname: "Ketchand". (The only people you will find in the U.S. today with this particular surname are descendants of W.P.). I have a very strong hunch that my family is probably connected and/or related to "Sister Kitchen" and "them melungins" mentioned in the Stony Creek Primitive Baptist Church records of 1813. I have researched and found that after leaving Virginia, W.P. and his family lived for some time in Georgia before moving on to southeastern Arkansas, where he was a minister of a Primitive Baptist Church in Ashley Co., Arkansas. I have visited this church building—it looks exactly like those described by Elizabeth C. Hirschman in *Melungeons: The Last Lost Tribe in America*. W.P. and his son Jesse Enos Ketchand were listed as white on 1850 and 1860 U.S. Federal Census documents. However, in the 1870 Ashley County Arkansas U.S. Federal Census, Jesse Enos Ketchand and his family (including my great-grandmother Jesse Alice) were labeled "mulatto."

5. My mother's mother—Anna Laura Williams, stated that she was of "Black Irish" ancestry. I have traced her maternal line only as far back as Anna's great-grandmother: Cynthia/"Sinthy" Love in Lawrence County, Alabama, in the 1820s/1830s. Lawrence County seemed to be a gathering spot for a large mixed race community, and the surnames that "Sinthy" was linked/married to appear to have connections with the Chickasaw and Cherokee tribes (Love, Rodgers/Rogers, Carpenter). I also received a fairly rare mtDNA haplogroup (W6) from this woman.

6. My various immediate family members exhibit numerous examples of Anatolian bumps and ridges, Mongolian Blue Spots/Birthmarks, Asian Shovel Teeth, palatal torus and missing wisdom teeth/molars (one of my nephews has complete hypodontia; i.e. no adult molars).

7. Before my "Melungeon Revelation," I had researched that I was a descendant of Bacon's Rebellion co-leader James Crewes and his Pahmunkey/Powhatan "wife." Although colonial law did not allow his half-breed daughter Hannah any legal recognition or rights as his heir and daughter, James did his best to write a will bequeathing much of his estate to Hannah and her husband, Giles Carter, (shortly before he was hanged by Governor Berkeley for his part in the 1676 uprising). I have

two family images of their descendants (my second and third great-grandmothers) that show features of their Native American heritage.

Given all the above details, I expected that my DNA results would reveal a wonderfully diverse Melungeon-American background. Thanks again to DNA Consultants for both the expected and unexpected ethnic details. The Romani element was a bit of a surprise, and I'll admit that I was disappointed to see that I did not inherit a Native American marker from either parent. (I have no doubt that one of my siblings probably did.) These results will greatly aid me as I continue my genealogical research.

July 15, 2013

Shawn

My journey into DNA testing began with my desire to expand my known heritage, while clarifying debated Jewish ancestry. What I have found in return is that my ancestral paper trail only uncovers a small portion of the blood that runs through my veins.

My DNA Consultants results, for the most part, were quite surprising. My European matches were fairly consistent with my country origins on paper and surrounding areas. The major surprise, however, was that my No. 1 European match was Romani/Gypsy and my No. 10 match was Czech Republic. ...

Things became much more interesting with my World Population Matches. My scores (in order) were Romani/Gypsy, Middle Eastern, African, Iberian, Central European, African-American, Jewish, Mediterranean European, European American and Eastern European. I also came up with Native American admixture to top it off. These results are causing me to believe that there may be a line or more of family lineages that I have yet to tap.

I have received comments from others concerning my phenotype such as "I'm not sure what you are," "You don't look Irish" and "You must have some Black ancestry." Some have even just assumed I was Hispanic or Caucasian. Almost all acknowledge they see my Italian/Spanish phenotype, while a few also see slight Native American.

While my results provided insight into how diverse my blood really is, they also put an end to an age-old family debate as to our Jewish ethnicity. One of my relatives from a few generations past would passionately defend her position that our family line was indeed Jewish, while another family member would vigorously put forth his position that we were not Jewish. He would try to prove our non-Jewishness any time he could. I also had another family member in that family line say that he almost did not get hired for a job because the hirer thought he was Jewish. I always believed these accounts, especially since from as far back as I can remember I have found this side of my family (Italian and German) to phenotypically look Italian and/or Jewish.

Where does this leave me now? My results show my blood is much more than simply Italian, French, Irish and German. They confirm family testimony of Spanish/Portuguese/Iberian and Jewish ancestry. Perhaps more interestingly, my results leave me reassessing my ethnicity or multi-ethnic heritage, end years of family verbal passages or debates, and leave me with intriguing new ancestries that are waiting to be discovered.

July 22, 2013

Maria O'Connor commented on 23-Jul-2013

Shawn: Countries' frontiers are artificial. For example, there are people of Celtic heritage in northern Spain, northern Portugal, all over Ireland, all over England, all over Scotland, all over Wales, Southern Germany, northern France, Northern Italy, etc. All of them, even considering they come from different places, have the same Celtic DNA. So, if you have an ancestor from Spain or Portugal, that person could be of Celtic origin, or Mediterranean origin. If a person has Jewish Sefardi DNA, it could be from southern Spain, southern Portugal, North Africa, Middle East, etc. Also, in South America there are great numbers of people of European ancestry, including non-Hispanic, non-Portugue ancestry, like Irish, German, Italians, etc. It's quite complicated, due to ancient and new migrations.

Phyllis Starnes: Designer Genes

We interviewed Phyllis E. Starnes, assistant investigator, to find out what fascinates her about the field of DNA testing. Her story is the first in a series titled "Behind the Numbers" about the workers behind the scenes in our industry, from lab technicians to statisticians.

Interviewer: How did you first get interested in DNA?

PES: I went to the Melungeon Union in Kingsport (Tennessee, in 2002). Beth Hirschman had her "stalk," a diagram of her Melungeon family tree with all the names in her genealogy, many of which were also my surnames. I heard Dr. Yates speak at that meeting. They had their lines all pinpointed, thanks to DNA studies.

Interviewer: What was your next step after that?

PES: I came home and did a lot of genealogy research on the computer.

Interviewer: And then?

PES: The first year DNA Consultants opened for business, which was 10 years ago, I ordered a Y chromosome test for my husband, Billy. Other companies were offering the same product, but DNA Consultants was the only one to give you a full analysis and customized explanation of things. Then I ordered my own mitochondrial DNA test.

Interviewer: Any surprises?

PES: Billy's top matches for his male line, the Starnes surname line, were Macedonia and Albania. My mitochondrial mutations matched Native Americans. I became the first of the "Anomalous Cherokees" whose female lineages didn't fit in the traditional scheme of "Indians out of Asia." In fact, my Hypervariable Region 2 mutations matched only one other sample in the world, and that was Dr. Yates, who is Cherokee in his direct female line.

Interviewer: What did your husband and the rest of your family think?

PES: Some were excited, as I was, but most were just not interested. My kids thought the strong Native American matches were very interesting.

Interviewer: What other family members did you test?

PES: As soon as autosomal testing arrived, with the DNA Fingerprint Test, I did Billy and myself, of course, Julia, Kiely and Holli (our three daughters), our granddaughter Keely, my dad's sister and mother's sister, an uncle and his wife, a niece and a cousin.

Interviewer: What did you find out?

PES: Within the immediate family, it was obvious who got which ancestry and trait from whom, and how they all resonated. One of the big surprises was my father's side, which proved to have quite a bit of Native American and Iberian. The "First Peoples" gene came from his side and passed on down through our girls. On my mother's side, 11 out of 20 matches was India.

Interviewer: India!?

PES: Yes, it appears we were finally seeing the extensive Romani/Gypsy heritage in her family. People had always told me I was like a Gypsy, from my clothes and jewelry to my attitude and outlook. When Billy was in the Navy, I told him one day, "I'm tired of being a Gypsy." I said I wanted to settle down in one place.

Interviewer: Did you settle down?

PES: Yes, we've lived in a small town in East Tennessee for almost 40 years. We moved here in 1973.

Interviewer: Any other surprises in your DNA?

PES: If you were to chart our geographical matches, both in terms of autosomal DNA as well as the female and male lines, it would surround the Mediterranean. That's where familial Mediterranean fever comes in.

Interviewer: Who has FMF in your family?

PES: Billy, myself, Julia, Holli and a cousin. I'm sure others have it but it has not been diagnosed and they may call it instead fibromyalgia. Brent Kennedy (author of a book on Melungeons and their genetics) is a cousin many times over.

Interviewer: What do you enjoy about your job?

PES: It's like a holiday every day. With customers coming out of North Carolina or East Tennessee, I see a lot of the same matches and genealogy I have personally encountered in my own experience with DNA testing. I recognize a lot of genetic cousins.

Interviewer: When did you first hear the word "Melungeon"?

PES: I grew up in Southwest Virginia in the little town where the Stony Creek Church is located. The church minutes contain the first written instance of the word. The register is all of mine and Billy's ancestors, and part of Beth's (Elizabeth Hirschman, author of books on Melungeons).

Interviewer: What do you see in the future of DNA testing?

PES: I think we've only glimpsed the tip of the iceberg so far, even though it's been 10 years. We'll continue to have new knowledge, new products. I highly recommend our customized approach.

Interviewer: Any parting shots?

PES: I've worked in sales all my life—jewelry management and design, my own interior decorating shop, running my own hair salon—but I have found something to be truly excited about in DNA. Funny I couldn't get this excited about selling diamonds! If you think about it, your genes are the ultimate design for living.

November 20, 2012

DNA and the French Revolution

If you want a good mystery, look at DNA. DNA has all the best stories and tales of murder and intrigue. According to Sam Kean, author of *The Violinist's Thumb: And other Lost Tales of Love, War, and Genius as Told by Our Genetic Code,* " ... Somewhere in the tangle of strands are the answers to many historical mysteries about human beings that were once thought lost forever."

Kean says each one of us has " ... enough DNA to stretch from Pluto to the sun and back," and " ... every human activity leaves a forensic trace in our DNA" and the story that DNA tells, Kean says, " ... is a larger and more intricate tale of the rise of human beings on Earth: why we're one of nature's most absurd creatures, as well as its crowning glory." We just have to know how to read the story and solve the mystery. Like the one about the French Revolution and King Louis XVI.

Victor Hugo's classic novel about the French Revolution, *Les Miserables*, has been given a face-lift for a modern audience to ponder over popcorn in the theater. Or discuss at a local café or bookstore over cappuccino. But few of us would imagine that scientists have had any interest in the French Revolution. But we would be wrong. And it is a gruesome tale.

Phillipe Charlier, a forensic scientist, dubbed the "Indiana Jones of the Graveyards," according to the recent *Abroad in the Yard* article, "DNA Analysis Links Blood of Louis XVI, Beheaded in French Revolution, and Mummified Head of His Ancestor Henri IV," by Tom Martin Scroft, has linked blood stains in a decorated squash gourd to the mummified head of King Henri IV. There was once a handkerchief, according to the article, that had been "in the possession of an Italian family for over a century" in an "ornate calabash gourd" that had been "dipped in the (beheaded) blood of King Louis XVI" by a Maximilien Bourdalou.

According to an earlier *Discovery News* article by Jennifer Viegas, "Royal Blood May Be Hidden inside Decorated Gourd," the handkerchief "is now missing." Most certainly it has "decomposed" by now as David Blair suggests in his recent article, "Louis XVI Blood Mystery Solved." Viegas also says the ornately decorated gourd was "dated to 1793" and that the dried squash reads, "Maximilien Bourdaloue on January 21st, dipped his handkerchief in the blood of Louis XVI after his beheading." Why would he have done such a thing? For a bloody relic no doubt. What a coffee table conversation piece. Viegas quotes Carles Lalueza Fox, "lead author of the study and a researcher at Spain's Institute of Evolutionary Biology," as saying that the act was common: "In fact, many people went there to dip their handkerchiefs in the blood." How gruesome. But linking blood found in the Italian family's gourd to King Louis XVI? It took DNA analysis to validate that tale.

How was that done? In careful steps. First, according to Fox (quoted by Viegas) her team had to identify the "brownish substance" inside the squash as "dried blood." Later, Fox remembered that the King had "blue eyes" and he identified the genetic marker for the "blue eyes mutation." But that is a long way off from identifying it as the blood of King Louis XVI. The researchers also analyzed its mitochondrial profile and its Y-Chromosome profile and they found the "DNA profile (before they had a match) …was rare among Eurasians," which "suggest(ed) that it (might) derive from a royal bloodline." But Fox knew that they had to have "'someone'" for comparison. They first thought of the "(pickled) heart located in a royal French crypt thought to belong to the King's son, Louis XVII." It is beginning to sound like a tale from Edgar Allen Poe.

But they didn't use the heart after all. They discovered the mummified head of King Henri IV, who ruled France from 1589 until 1610, which had been "shuffled between private collections ever since it disappeared during the French Revolution," according to Marie Cheng's *AP* article, "Scientists ID Head of France's King Henry IV." (According to the article, "Henry IV was buried in the Basilica of Saint Denis near Paris, but during the frenzy of the French Revolution, the royal graves were dug up and revolutionaries chopped off Henry's head which was then snatched.") I don't suppose the head was in such great condition after all this shuffling about, but it still turned out to be useful.

With that mummified head, DNA analysis has "solved a mystery that has lasted for almost 220 years," according to Blair. He quotes a new study in the current issue of *Forensic Science International* as saying that the comparative analysis with the mummified head of King Henri IV confirms the connection by "... establish(ing) that Henri possessed a rare partial " 'Y' " chromosome" and Louis, a "direct male-line descendant, separated by seven generations," (had) this same Y chromosome. Along with "other (genetic) matches," the study concluded that "... historically speaking, this forensic DNA data would confirm the identity of the previous Louis XVI sample."

And you thought scientists were boring. Another DNA mystery solved.

December 31, 2012

Wanna Indian Card?

DNA testing for Native American ancestry is the best way to connect to one's past if the genealogy trail has come to an end. But enrolling in a federally recognized tribe is a lot more difficult than taking a cheek swab test. Perhaps there were family stories about having American Indian ancestry in your family history, but there was little way to prove this. Now, anyone can determine whether the family tales were tales or true with an easy cheek swab test. There are now known genetic markers for Native American ancestry that can be revealed in one's DNA testing results. Some autosomal DNA tests will even provide specific matches to Indian tribes like Apache or Cherokee, if the genetic match is strong enough, but that doesn't mean anyone is handing out an Indian card to you like candy. Why not?

For one thing, tribal guidelines aren't that simple. It is, generally, a more complicated process of finding someone you relate to that is in the tribe or on the federal census rolls, like the Dawes roll for the Cherokees. All tribes have different guidelines, but most of them want you to have also figured out your genealogy and how you connect to that specific tribe. However, some of them do accept DNA testing as part of the process though, generally again, DNA testing that shows a specific relationship to a tribal member (BIA, "A Guide to Tracing American Indian and Alaskan Native Ancestry'). There are many ways that can be proved, depending on the type of DNA testing that is done.

If you have family members in a tribe, or have formed some type of kinship to a tribe, like being adopted, pursuing tribal enrollment could be very important. However, some are looking to enroll because of the "myth of the monthly check," which they believe they will receive because of a racial identity alone. Nevertheless, these payments have been made according to specific judgments resulting from treaty obligations or settlements of claims, although there are few outstanding payments (BIA, "A Guide to Tracing American Indian and Alaskan Native Ancestry").

Unfortunately, DNA testing for Native American ancestry will most likely not put a check in the mail for you. However, what it can tell you is the truth. What could be more precious?

January 14, 2013

Skeletons in the Closet

I love genealogy, but it has its pitfalls. Those negatives eventually made me trust in the science of DNA testing more than genealogy. Why? People can lie. And why use old tools? We are now in the Genomics Age. It is time to go beyond interrogating your Hatfield and McCoy relatives. DNA doesn't lie.

I have had a love-hate relationship with genealogy for quite a while now. Genealogy too often has given me the silent treatment and not treated me as well as I had hoped. It was time for a new romance. I used to be a religious believer in the records of genealogy. I paid attention and played by the rules—interviewing family members, filling out family trees, going on genealogy forums, and researching for ancestors properly. My husband and I even trekked hundreds of miles to different library archives and courthouses in the days before public and vital records were assembled and published online.

None of it mattered. There were too many skeletons in my genealogy closet. There were ancestors who skipped across county lines and changed their names. One just disappeared. And no one was talking. One aunt hung up on me, and another told me to give up. It was frustrating trying to unlock my genealogy secrets. My father actually told me, "If you keep digging, you will find people you do not want to be related to."

But even if you do not have a murky ancestral past, genealogy is not a walk in the park.

Why not? First of all, there can be blind spots and unverified lines—like adoption—in our genealogy research and public records. If we just ignore this, or are unaware of it, that entire line of ancestry is incorrect. This can be multiplied if we then share our family tree. For instance, I have several great-grandmothers far back in some of my ancestral lines who are listed with only the first name. No further information is given or can be discovered. That is a blind spot in genealogy, but it does not mean that "Mary" did not have her own personal family history and story that has added to your story. And genealogy and sex do not mix. In fact, genealogy would have one think that not one woman in all of civilization ever cheated on her husband. However, the further back you follow any line in your family tree increases the odds of a "non-paternity" event of either adoption or an unknown dalliance. No one wants to believe their parents or grandparents had sex, but if they did not, you would not be here, and not all sex that has happened during history is just between the known marriage partners. I have a friend who was told quite late in life that her father was not her father. Why did she ask? Her DNA testing results did not make sense otherwise, and she confronted her mother with them. Soap operas aren't just on television. Her elderly mother confessed to an affair when her husband was at war.

Also, many people changed surnames when they immigrated to America because of Jewish or other exotic ancestry. That can mean that you do not really understand the ancestry of this line. I have two cases of this in my own family genealogy that I discovered. One was my grandmother's maiden name. It explains my DNA testing results with the Jewish markers as the original name is Jewish.

Again, genealogy can lie because people can lie. You can get incorrect information from your family about your genealogy as they may not wish to divulge any family secrets, and you can get other incorrect information from genealogy forums from researchers who have not done proper research and/or have unknown errors.

That is not to say genealogy cannot be fun. Just realize it is limited and can be flawed.

February 12, 2013

The Myth of Percentages

It is a delusion that one can take a Native American DNA test and get a card from the federal government. But it is ingrained in popular thinking, like the "fact" that Columbus discovered America. He didn't, of course. The original inhabitants beat him to the punch more than 10,000 years before. Ironically, Columbus believed he had landed in China.

How does one get federal Indian recognition and benefits? One might as well buy a lottery ticket. It is a complex and thorny issue. There are some 700 to 800 American Indian nations, and the requirements for qualifications from the BIA differ. But no federally recognized American Indian tribe will accept just a DNA test. For instance, the Cherokee Nation requires that one be accepted by one's community as American Indian, know of one's American Indian ancestry prior to being 18, and have found a direct ancestor on the federal government's final Indian census rolls, compiled and closed more than a hundred years ago. This not only means one needs to have done in-depth genealogy research but, most likely, have been born and raised on the reservation.

Then why take a Native American DNA test? It is important to know the story of one's ancestry. That story can be a meaningful chapter to pass down as part of one's family history. And DNA testing, especially new autosomal DNA testing based on forensic science, can tell you if you have Native American ancestry in your background. How? New tests provide a comprehensive analysis of all your ancestry in one simple, cheek-swab DNA test. It isn't just a test of one or two ancestral lines but about 1,000 or so.

There are some non-federally recognized state tribes that do accept DNA testing for qualification purposes. It is up to you to go to the tribe and find out their specific qualifications. No DNA testing company can do this for you.

However, an autosomal Native American DNA test will not tell you "how much Indian" you have. Why not? This is a very scientific approach to DNA testing for ancestry. And percentages are not scientific at all. If you are picking up your child's crayons, think of how each color is a blend of other colors. People are blends as well, so there is no "pure" population to use as a reference point. Portuguese or Hispanic means many different things to different people because they are so mixed, but all populations are actually mixed because of admixture and intermarriage between populations. Percentages are really just educated guesses—not real science.

Native American DNA testing is an important step, but you have to be in it for more than the money. The odds are against your winning that game.

March 28, 2013

10 DNA AND HEALTH

Tapestries of Illness
By Phyllis Starnes

Some trips we take are of the mind and not the body. Sometimes all you know about yourself, or think you know about yourself, changes or gets such additions as to make the picture different enough to feel like a change. It's amazing how the catalyst for such a moment can be something as simple as a clipping from a newspaper.

In 2002 my mother sent me a copy of a clipping from the Kingsport Times-News because it mentioned a friend from my school days. "I thought you and Billy might be interested in reading this …" she had written in blue ink across the article right under the title "Solving a puzzle to relieve the pain." She thought our only interest would be in the fact that the article featured someone that we'd known. I don't think that she realized that she had set my feet upon a path that would carry my family and me to the other side of the world and back again.

I found the article to be of interest for a very different reason than the one my mother had thought would interest me. It was my first introduction to Brent Kennedy and to the term familial Mediterranean fever (FMF).

For as long as I can remember I have never felt good. There are days that it's a challenge to put one foot in front of the other. My daughter says that some days you operate on the principal of "might as well …" as in, "I'm up so I might as well do what needs doing," she says. I read that article, and I heard every ache and every complaint I'd ever had in it.

I was driven to learn more, but the more I learned the more there was to know.

I bought Brent's book, *The Melungeons: The Resurrection of a Proud People*, because I wanted to know more about how he came to be diagnosed with FMF—a genetic disorder found most often in people of non-Ashkenazi Jewish, Armenian, Arab and Turkish backgrounds. As I was reading the book, I kept thinking that his family names sounded so familiar. A look at a family tree that had been kept up away from sight revealed that Brent's ancestors were the brothers, sisters and cousins to my own ancestors. If nothing else—I had found a new cousin.

My mystery illness might finally have a name! Armed with the newspaper clipping, my family tree, and Brent's book I returned to my primary physician. A few months earlier in a state of frustration at not being able to find a reason for all my aches and pains he had at last said, "Don't come back, Phyllis; there's nothing I can do for you." On the basis that I was related to Brent and had suffered the same symptoms, my doctor gave me a referral to Dr. Christopher Morris in Kingsport, Tennessee, who had been mentioned in the article. I think he figured that this might at last put a name to my condition or at least prove that it was "all in my head."

I found out in my appointment with Dr. Morris that diagnosing familial Mediterranean fever is often a process of eliminating anything else that might explain the symptoms. There has to be the genetic connection—at that point to Brent and some few other patients that had also been diagnosed—and then all the tests to eliminate other illnesses that have similar symptoms. Once the other possible culprits are eliminated then a trial of colchicine could be prescribed. A positive reaction to the medication is a positive indicator of FMF. I have been taking colchicine for six years now. There are still days that are challenging, but the episodes of joint pain, sore throat, low grade fevers, pleurisy, and unexplained abdominal pain are less frequent and shorter in duration than before my introduction to colchicine. My husband and two of three children are also taking colchicine. It has made life better for all of us. Only someone who has suffered a nameless condition can ever truly appreciate how it feels just to have a name for what ails you. It's not enough to know the what, though; I had to know the how. How could a family from southwest Virginia be diagnosed with a condition for which there was no connection to our known origins?

This was the start of my travels down paper trails, to county courthouses, family reunions with people I hardly knew or didn't know at all. The names keep multiplying as I go back through time tracking generations. Adams, Berry, Beverly, Bowling, Campbell, Carter, Caudill, Cole, Counts, Cox, Coxe, Fields, Freeman, Gibson, Hill, Jackson, Kiser, Lawson, Lucas, Mosely, Mullins, Osborn, Osborne, Powers, Pruitt, Rasnick, Robinson, Shephard, Short, Stallard, Stanley, Stewart, Taylor, Tipton, Turner, Williams, Williamson and others.

My family has lived by and played in and on the banks of the Clinch River and Stony Creek for generations—the same Stony Creek where the old Primitive Baptist church of the same name once stood—the same Stony Creek church that is often put forth as the place where the earliest known written reference to the word "melungins" occurred in 1813. The listing of membership from that time is a "Who's Who" of mine and my husband's family trees. The original church building was washed away by floods a long time ago. My husband's family and my mother were members at the Pine Grove church that replaced it. My mother sometimes took my daughters to services there when they stayed with her during summer breaks. It was still a one-room church house with a curtain to pull for Sunday school, and an outhouse. Baptisms were still performed in the deep spot right behind it in Stony Creek—the same deep spot that served as a swimming hole on Saturdays.

Our family trees have long and deep roots; I am still following threads here and there trying to sort them out. The roots and branches are intertwined so that in some places it's hard to tell where one tree ends and the other begins. I have reached some ends that I can no longer follow. I am sometimes like a squirrel whose weight bears down on the branch too much so that I must leap to another one for a while. My search has led me to counties outside of Scott County, Virginia such as Lee, Wise, Russell, Dickenson, Patrick, Louisa, and Cumberland, Virginia. I have crossed the border to Tennessee to Hawkins, Hancock, Grainger, Greene, Claiborne, Sullivan and Washington Counties. I've made trips to Letcher, Perry, and Clay counties Kentucky chasing down a name or a possible ancestor. I have found that at some point the paper trail runs thin—fires, floods, and politics can be blamed for records being lost or destroyed. In all of the trails that I can follow this way, I still haven't found out how a Southeastern Appalachian family could possibly have a Mediterranean medical disorder.

My husband and I attended the Fourth Union hosted by the Melungeon Heritage Association (MHA) in Kingsport, Tennessee, in 2002 right after I had first visited with Dr. Morris. I had only been taking colchicine for a few weeks at that point. This was the event where the preliminary results of the first Melungeon DNA project were announced by Dr. Kevin Jones. It is unfortunate that project wasn't able to be followed to its end due to Dr. Jones' illness, but this event did have some lasting effects for my family in a personal way. I met Cousin Brent and found cousins, Beth (Caldwell Hirschman), Donald (Panther-Yates), and Nancy (Sparks Morrison) among others; although, at the time I didn't realize that they were my cousins. Their research has added to that of my own, and we have all been able to fill in pieces of the puzzles that we have been working to put together individually. Subsequent MHA events for many years have been like extended family reunions.

The first time I ever saw Donald Panther-Yates, before he'd even spoken, I told my husband, "I don't know who he is, but I want to pick his brain." Once I heard him speak at Fourth Union—his topic, "Jewish Indian: Who'd've Thunk It!"—I knew he was right. If you've ever had an epiphany—a moment when you *know* the truth when you've heard it, this was one of those moments. I had never thought about it. It can only happen if you are willing to accept the possibilities—I was willing to accept this new aspect as to who my ancestors might have been: Jewish, Muslim, Christian, Native American, Mediterranean, Middle Eastern, Romani, Northern African, as well as northern European. I went from being a white, middle class Appalachian American to multiracial. It was a huge leap, and yet I am so much richer for having made it. My family doesn't look any different from everyone I grew up with in Fort Blackmore, Virginia.

At one point in history, though, a Virginia state official apparently thought that families with my ancestors' surnames were different enough to warrant being treated differently. Did Walter Plecker cause some members of my family to "whitewash" the family lore? At what point in time did my Native American ancestors decide not to be Native American? Why did my Romani ancestors decide that they weren't Roma? When did my Mediterranean forefathers and mothers decide that they had to be the same as their neighbors? The truth is lost irretrievably from the prospective of paper records and my family lore has long since forgotten the answers to any of these questions. I now had more clues about our origins, but how to prove it? Science provided me with another avenue to follow. Where a door closes, often a window opens.

The new and upcoming thing in genealogy for a few years now has been the use of genetic and DNA technologies. Once I accepted that the paper trail was never going to go back far enough to find my family's origins or to explain our medical condition, I decided that this new path was the only one that might possibly provide the links. Paper trails can definitely be lies; blood trails don't lie but neither can they always be taken at face value and some results only create more questions.

We chose to support Donald's new venture, DNA Consultants, and ordered the mtDNA test for myself and both the Y-DNA and mtDNA test for my husband. My husband's Y-DNA results showed an Albanian or Macedonian modal (R1b). His mtDNA results (T*) had an exact match in Iceland. My mtDNA came back U2e, which is normally considered a European haplogroup, but I have no HVS1 matches and my HVS2 region matches that of a documented Cherokee woman. Answers? Yes. Questions? Yes.

I have continued to chase down paper trails, but my thirst for knowledge has not been quenched. A new type of DNA testing has become feasible for public consumption—autosomal testing has become affordable. DNA Consultants offers a DNA ethnotyping test that matches CoDIS markers to contemporary populations. It is the same test used to match evidence to victims or suspects in crimes. It is the same test that is used to prove maternity and paternity. The numbers alone won't give you much information; the profile and map provided with the test answer many questions. They also gave me new topics to research –I've found it interesting that many of our individual matches can be connected through historical events and shared borders. I have, at this point, had a few of my maternal and paternal family members tested as well as my husband, my three daughters and one grandchild. My family's common matches show that we are, indeed, of northern European descent—with many members having matches in the United Kingdom, France, several Scandinavian countries and the Iberian Peninsula. However, this wasn't the whole story as the history books would have us believe.

We also had common matches in Morocco, Tunisia, Syria, Saudi Arabia, Turkey, Greece, and Afghanistan to name a few Middle Eastern matches. Every family member had at least two or more of these countries on their matches, and the matches occurred for at least three or more members. There are also matches in several scattered countries including northern India that would indicate a Romani or Gypsy ancestry. There were even some Australian Aboriginal matches and some Sub-Saharan African matches.

What may be the most recent additions to the family history are our numerous Native American matches—the top or second match of many family members being a Lumbee match. There are other Native American matches all over North, Central, and South America. None of these can be interpreted as a specific tribal affiliation.

I have at least proven that it's possible for me to suffer from familial Mediterranean fever. I have the ethnic backgrounds most often affected in my genetic makeup. If for no other reason, the decision to have my DNA tested was worth knowing/proving this to be the case. There is one thing that anyone thinking about taking this path should consider—it may disprove as easily as it may prove any stories of origin that your family may have. I do not think that I would have gotten as much out of it, if I hadn't been willing to accept all the possibilities.

Will I ever track down which ancestors came from where exactly? Most likely not, because I think many of the ancestors that you could say came from here or there are probably hundreds of years back. I believe that most of the mixing of various ethnicities happened before my ancestors came to the Americas. I now believe that I have ancestors that were here prior to my ancestors that I know came before or during the Revolutionary War. I am more firmly than ever an Appalachian American. And I am also Native American, northern European, Mediterranean, Middle Eastern, and Romani (Gypsy). And yet since my ancestors chose (perhaps freely, perhaps out of necessity) to become Anglo-American—I am none of these things. I sometimes feel like I am an "orphan of the world." I know that many people from the cultures and heritages that are a part of me wouldn't accept me as a part of them.

It seems to be human nature to want to label and categorize things—including people. It is most definitely human nature to want to belong and to be accepted. I think the greatest challenge in being a multiracial person in a time when society is supposedly more enlightened is resisting pressure to choose to acknowledge one facet of your heritage over the others in order to wear a certain label. That pressure comes from those that are uncomfortable with you defying their attempt to put you in one category or keep you out of another. It seems to me that if I reject one part of myself to satisfy anyone's desire to label me or my own longing to belong then I repeat history. I start down the same slippery slope that some ancestor slid down long ago. I hope not to repeat the decision that I consider in the present to have been a mistake in the past.

It is my hope that one day the world will be able to accept the diversity that makes up the foundations of America. It is my hope that one day the average American will recognize that our history is much deeper than what we thought it was. I think that groups like the Melungeon Heritage Association are the best way to start. If we can hold to the belief that the purpose of the MHA is to celebrate the richness of culture and the diversity of heritage that is Appalachia—if we can keep to the belief that we are "one people, all colors," then we stand the best chance of finding belonging and acceptance for how we are different. We will finally be able to be true to ourselves.

The beauty of Appalachia's people is that we are much like a tapestry, a kilim or a quilt—it takes many threads of many colors to make a whole.

September 15, 2003

Children of Parents with Alzheimer's Want to Know If They Have the Gene

Genetic screening for the apolipoprotein E gene associated with Alzheimer's disease appears to be beneficial and welcome to nonsymptomatic adult children of people with the disease, according to an article by Kerri Wachter in this month's *Adult Psychiatry*. Dr. Robert Green made the announcement at an international conference by the Alzheimer's Association.

Dr. Green said he was surprised by the results of his study, which determined that knowledge of having the gene, not having the gene and not wanting to be screened for it made no difference in levels of depression or anxiety.

Revealing such genetic risk information to asymptomatic patients is not usually done, because the gene is only a marker for susceptibility, not for the disease itself.

Another reason for not sharing knowledge of the gene's presence is that there is no treatment at present, and it was feared that screening might increase people's anxiety.

Happily, Dr. Green's study showed that people want to know and can seemingly deal with the results of that screening.

March 31, 2006

Linkage Disequilibrium Lower in African-Americans?
Referencing *Genes and Immunity* 6 (2005) 723-727

Linkage disequilibrium across the human genome is generally lower in West Africans and African-Americans than Europeans. However, in one region of the DNA rich in immune genes, scientitists have found significantly more examples of apparent positive selection in West African and African-American populations.

July 28, 2005

My Genome. So What?

According to an editorial in *Nature* with this title, recent additions to the growing database of personal genomes dictate that research be undertaken on how people are using the new genetic clues about their health.

November 6, 2008

Putting the Test to a Test

In the last blog post, we responded to the call of *Nature* in "Genetics without Borders." In this, we examine the second of three editorials in this week's issue concerning regulation of DNA testing companies: "Putting DNA to the Test."

It should be pointed out at the beginning that the wrath of the editors descends in unequal fury on commercial enterprises. They are not as irate at ancestry companies as "personal genomics providers." They seem to have in mind mostly concerns like 23andMe, which promises for $399.00 to sequence your personal genome and give you "access to all health, disease, and trait reports" maintained by its staff, together with "all ancestry features and raw data download." To the editors of *Nature*, this is akin to practicing medicine and genetic counseling on the Internet. Spit into a cup, discover your personal DNA sequences and get an email when a medical article mentions your nucleotide position.

The presumptuous and condescending attitude of the editors is evident in their first paragraph (italics added): "The availability of affordable, direct-to-consumer genetic tests has *mushroomed*, leaving regulation lagging behind. Dozens of companies now offer *inexpensive* (elsewhere: *cheap*) home kits that allow people to *spit* into tubes, send the samples for DNA analysis and receive a report that *allegedly* details their ancestry or their *possible* susceptibility to *a long list* of disorders that have been linked—*often tenuously*—to particular genes. But the value of these tests remains *debatable*, which is why (*bad predication in our grammar book*) the *industry* needs a strong set of quality standards and codes of conduct to protect both *its consumers* and its own *credibility*."

Aside from poor writing (which seems to be a requirement for an advanced degree in the natural sciences today), there are numerous examples of logical fallacies in this and the rest of the article. Perhaps Ludwig Wittgenstein was right. What cannot be put into words cannot be thought. What can only be poorly expressed is poorly thought.

It is unclear whether the regulators would extend the same benevolent protection to the academic researchers who also consume genomics laboratory services. A case can be made that even *their* understanding is not always perfect and up-to-date. Elsewhere in the same issue of *Nature,* are warnings to fellow scientists who make exaggerated claims about their research. The editors also reprove wayward brethren who seek to dip more than once in the immortalizing waters of the Pierian springs, by submitting their work to multiple journals, often under different guises or false pretenses.

The world of science has so much dirty underwear of its own, it is surprising it wishes to examine that of others. Credibility seems to be in short supply everywhere.

Without dissecting what is a mess of snips to start with, let us draw attention to one scenario the would-be regulators raise. "Customers," they predict, "will frequently receive results telling them only that they face the ambiguous possibility of a somewhat elevated risk of a little-understood disorder. ... If the ambiguous, slightly elevated risk relates to a frightening condition such as breast cancer, some individuals might feel compelled to undertake drastic and perhaps needless measures, such as prophylactic mastectomy (surgical removal of a breast to avoid cancer)."

I read this horrific statement to a friend of mine over the phone, who said she had been in that exact situation. Doctors found a lump in her breast. Knowing that ancestry testing had placed her in a category of predisposition to developing breast cancer, she underwent, after due deliberation, "prophylactic mastectomy." "I was thankful I took the DNA test," she said, "because it gave me information that helped me evaluate my risks." She says she is sure that if she had not taken the step she did, she would have breast cancer today.

It is arrogant of scientists to think they must protect people from information. This is the stance of a totalitarian state that controls and censures the information consumed by the populace, or of a state religion such as that which ruled supreme during the Middle Ages. It was attempted with disastrous results in so-called "activist era" of the Federal Trade Commission during the 1960s and 1970s under Commissioner Mary Gardiner Jones. An institutional ideology of this sort assumes that consumers require protection from scientific information that they may misinterpret and that may lead to personal or social distress. For example, misplaced information like this might lead historically disadvantaged communities to increase their distrust of the scientific establishment ... as though the scientific establishment didn't do enough in that direction!

Another of the editors' arguments against releasing genetic information to the populace is that genetic information is always evolving and may not be complete. Quoadusque, we may well ask with Cicero. When will it be complete? Or complete enough? And who is to make that judgement?

Instead of mad, speculative and needless worry about consumers who are supposedly ignorant and defenseless, why don't we let reason and the unimpeded flow of information take their course? Those two forces educated, up to a point, the scientists who are now trying to second-guess the rest of us. While it may have taught them a lot of facts, it did little apparently to sharpen their powers of philosophical reflection.

October 8, 2009

To Do or Not to Do DNA?

Mad Hatter's Tea Party at California Colleges

American education is in such a state of decline and confusion that the following new program, with all its pros and cons, seems tantamount to a mad hatter's tea party. We reproduce a description of it from *Nature* in all its carefully nuanced and agonizing detail. We suggest that rather than fretting over whether DNA testing companies might use predatory marketing on teenagers or students be pressured into making purchases and be psychologically damaged by DNA results, the school authorities worry about real threats like the fast-food poisons served up in the cafeteria franchises on their campuses. Or overpriced and watered down textbooks. Or alcohol in dorms. Or date rape. Or just about anything else.

A DNA Education (*Nature* 465: 845-46 16 June 2010)

"Taking personal genetic testing into the classroom brings ethical and legal sensitivities to the fore. Although personalized genetic testing is still very new and controversial, its increasing use in health care seems inevitable—a trend that makes it essential to give consumers and physicians a better education in the technology's strengths and weaknesses.

"That was the rationale behind an announcement made last month by the College of Letters and Science at the University of California, Berkeley. This year, instead of sending its incoming students a book for later discussion in class, the college will send them a kit to swab their DNA. If they so choose, students can send in their sample to be analysed for three common gene variants that indicate how an individual metabolizes alcohol, lactose (found in dairy products) and folic acid, a vitamin common in leafy green vegetables....

"A telling contrast in approach has been provided by Stanford

University in Palo Alto, California, which announced a similar course designed for medical students shortly after Berkeley announced its program. Recognizing the potential for controversy from the outset, Stanford officials first appointed a task force of basic scientists, clinicians, legal professionals, genetic counselors, ethicists and students who spent a year designing precautions against coercion and conflicts of interest by the institution, and working out access to counseling.

"That said, the Berkeley and Stanford programs are both still experimental ... these two programmes are only the beginning of a long conversation that needs to happen on campuses worldwide."

June 16, 2009

Officious or Official?

Council of Europe Adopts Protocol on Genetic Testing for Health Purposes

In this report by Laurence Lwoff in the *European Journal of Human Genetics* (2009) 17, 1374–1377, first published online July 1, it is noted that the Council of Europe has weighed in on one of the most controversial areas of DNA testing, whole-genome sequencing and SNP testing to find genetic predisposition to disease for individual customers. Recent editorials in *Nature* have called for similar measures in the United States, which is home to 23andMe and other companies offering such services.

So far, no regulatory proposals have been aimed at genetic ancestry testing, only medical and health-related screening. One of the warnings often raised in the public discussion on genetic testing for health purposes, however, is that results may confuse and unnecessarily alarm consumers—a criticism that could apply equally to ancestry services. Another is that commercial research scientists and business operators may jump the gun with findings and peddle bad science, although critics admit that the state of knowledge on nearly every topic of interest to geneticists and medical researchers is in a constant state of flux. A finding about a gene for Alzheimer's will be trumpeted in the pages of a major journal one week only to be updated or withdrawn in the next.

This being the case, one wonders when discoveries will ever be fit to be commercialized or made available to the public. Should science only serve scientists?

We have always maintained that the would-be regulators underestimate moderately educated people's ability to understand emerging science. They overestimate commercial companies' disregard for professional practices and responsible communications. Most of the measures under discussion will have the effect of denying people access to valuable information. Regulation will also hamper growth in a direct-to-the-consumer business with unimaginable promise for society at large. A home paternity test purchased at the corner drugstore may make all the difference in the life of a family. Discovery of varied ancestry through a DNA test can be an important factor in furthering a consumer's interest in other peoples and countries, in history, and ultimately in tolerance of others. DNA testing can help bring peace of mind, but it can also help bring peace in the world.

Many, if not most, of the innovative contributions to society have come from non-specialists. The scientific establishment is not oriented toward practical applications of knowledge. The Croatian engineer Nikola Tesla dropped out of college and never received any formal training. Driven entirely by his natural aptitude, he patented some of the most important inventions in the birth of commercial electricity, including alternating current (AC) electric power systems and the AC motor. He died penniless at the age of 86 in 1943. No government program or university gave him any support or assistance.

Whatever else the Council of Europe deliberated about, we hope they were not cynical or self-important enough to discount the possibility there may be many more popular scientists like Tesla in Europe's future. Science and technology are increasingly becoming a way of life for millions of people around the world who do not happen to have an advanced degree. It is a positive sign that consumers are so eager to take responsibility for their own health they will use the latest innovations from genomics to gain knowledge and control. Scientists should be glad they have such an impact. They should not squander the respect they enjoy in our eyes with pedantic discussions.

October 22, 2009

Future Shock or Future Letdown?
Does deCODE's Bankruptcy Signal False Promise of Genetic Medicine?

New York Times reporter and DNA book author Nicholas Wade raised an interesting question in his report on the bankruptcy last month of Iceland's deCODE Genetics, which attempted to make it possible for an ordinary consumer to buy the latest applicable information on the connection between their personal genes and their personal disease risk. The article was titled "A Genetics Company Fails, Its Research Too Complex."

In the November 17 edition in the Science and Technology section, Wade wrote, "The company's demise suggests that the medical promise of the human genome may take much longer to be fulfilled than its sponsors had hoped." But there may be more to the story." The discovery that major diseases do not have any simple genetic pattern of causation has dealt a serious setback to the gene-hunting field as a whole," he added.

Signs of the deflation in the field of "gene hunting" over the past 10 years since the Human Genome Project was completed and the second phase of the HGP was announced as focusing on the "conquest" of disease are:

- Discovery that genes are not found in continuous sequences or segments or even on the same chromosome.
- Realization no DNA can be considered "junk DNA" and even "non-coding" loci have at least place-holder functions and hence their values are not neutral.
- Greater respect for the role of environment in inheritance, including the nano-environments within the cell where DNA is stored and replicates.
- Jumping the gun on numerous claims concerning genome-wide association studies in scientific journals like *Nature* and *Science*, and subsequent retractions by editors and authors.
- Ever-increasing sample sizes with ever-increasing lack of robustness for the data and clarity for conclusions.
- A push for extending genetic surveys to rare and under-represented populations, with few surprises in the analysis of the

implications for medical research or consequent benefit for public health.

- Diminishing returns on research investment (ROI) on nearly every front.
- Not a single viable gene-therapy product ever introduced.
- Realization that only very rare genes are discoverable and selection usually takes care of them and extinguishes them over time; hence the bulk of medical research funds goes toward the rarest of cases and not widespread disease such as cancer or diabetes.

Harvard biology professor Richard Lewontin maintained as long ago as the 1960s, and continued to warn even on the eve of the completion of HGP I in 2000, that gene hunting was essentially a scientific fetish with little true power or efficacy. In 1992, he wrote "The Dream of the Human Genome" as a review article in response to *The Code of Codes: Scientific and Social Issues in the Human Genome Project*, edited by Daniel J. Kevles and Leroy Hood, and seven other recently published books on the subject of genetics and medicine. The essay was reprinted in Lewontin's own book *It Ain't Necessarily So* (second edition, New York Review, 2001).

I think it is time to elevate gene hunting to the danger of something beyond a harmless fetish for the members of a narrow profession or scientific sect. Its waste and failures have taken on the proportions of a national form of folly and collective denial. While huge expense and sensational efforts continue to be thrown away on the molecular biology revolution, the need to renovate our neglected infrastructure and reform political mechanisms goes unanswered. Resources that might be better allocated keep dwindling. The supposedly most-advanced society in history turns a blind eye on such relatively easy measures of public health as universal health care and uncontaminated chemical-free food and water supplies. While geneticists continue to cackle about inch-sized strides in their progress toward scaling the distant peaks of genetic medicine we are slipping into the abyss of logical disconnects.

December 6, 2009

The New Science of Epigenetics

The sequencing of the human genome capped off the 20th century's tireless search for genetic causes for all diseases. But epigenetics is the hot new science now. Dr. Anne Marie Fine, a Scottsdale physician, certainly thinks so. Dr. Fine spoke in Paris recently on Epigenetics and Beauty and next month will present a paper called "Dining at the Epigenetic Cafe" in Monte Carlo, Monaco, at the largest European physicians' anti-aging conference. In June she will present a paper titled "Epigenetics and the Autosomal DNA of Human Populations: Clinical Perspectives and Personal Genome Tests" at the University of British Colombia, Canada, with Donald Yates, principal investigator at DNA Consultants, along with participating in a 90 minute colloquium on epigenetics, autosomal DNA and ethnic identity. Clearly, epigenetics is stealing the show!

From the Fine Center for Natural Medicine News, here is how Dr. Fine describes epigenetics and its promise:

"Epi" literally means "above" so epigenetics are the influences from above that affect the DNA. Epigenetics refers to modifications to DNA and chromatin, the protein scaffolding that surrounds the DNA, that persist from one cell division to the next, despite a lack of change in the underlying DNA sequence. So, the "epigenome" refers to the interface between the environment and the genome. This is the basis behind the new science of epigenetics— how the environment affects the cellular DNA. Cells are bathed continuously in a sea of changing environmental conditions. This means the epigenome is dynamic and responsive to environmental signals especially during development, but also throughout life. It is becoming increasingly apparent that stress, environmental chemicals, and nutrient deficiencies are some of the biggest factors that promote epigenetic changes to the DNA. In addition, some of these changes in gene expression persist long after the exposure has stopped. What this means is that these changes can transcend generations.

Researchers at the University of Pittsburgh stated in the journal *Medical Hypotheses* in 2009:

It is becoming clear that a wide variety of common illnesses, behaviors, and other health conditions may have at least a partial epigenetic etiology, including cancer, respiratory, cardiovascular, reproductive, and autoimmune diseases, neurological disorders such as Parkinson's, Alzheimer's, and other cognitive dysfunctions, psychiatric illnesses, obesity and diabetes, infertility and sexual dysfunction. Effectors of epigenetic changes include many agents, such as heavy metals, pesticides, tobacco smoke, polycyclic aromatic hydrocarbons, hormones, radioactivity, viruses, bacteria, basic nutrients, and the social environment, including maternal care. It has even been suggested that our thoughts and emotions can induce epigenetic changes.

"Incredibly, only about 2 percent of diseases can be attributed to locked-in single gene mutations," says Dr. Fine. Most disease occurs as a complex interaction between genetic susceptibility and the environment. This means, while there are genetic predispositions, there are environmental triggers that actually start the disease, but also environmental factors that protect against developing the disease. The key is to understand which factors promote disease, and avoid them, and which protect, and seek them out. Our genetic makeup doesn't necessarily determine our biological fate. "Genes may load the gun, but environment pulls the trigger," says Dr. Fine.

James Watson once said that the double helix contains a library of detailed information about all generations of our ancestry "if only we could read it." Combining epigenetics and the advances in autosomal DNA tests, we are *beginning* to read the whole of human medical, evolutionary and ethnic history, at least in outline form.

February 14, 2012

Tommy Dionisio commented on 16-May-2012
Very well said. High time we began looking closer at the environmental factors associated with disease. The more we understand, the greater our knowledge, the more empowered we become to exercise prophylactic exclusion of many of the harmful chemicals we expose our genome to in the products we eat, apply to our skin and inhale on a daily basis.

How Good Is Direct-to-Consumer Genetic Screening?
Not Very, According to Study

Direct-to-consumer Genetic Testing Services: What are the Medical Benefits?
Thierry Frebourg
European Journal of Human Genetics 20 (2012) 483; advance online publication, January 4, 2012; 10.1038/ejhg.2011.229.

It is clear that consumers, and many geneticists and medical professionals, underestimate the complexity of genetically determined diseases and their risk levels as measured by genomic testing. The question is whether it is ethically sound to sell consumers packaged DNA tests that could exaggerate their risk for say, heart disease, or render a false negative result.

In one study, DTC testing on average handed out very slight risk factors across the board, lower than those known to be in the general population. Another study, the first of its kind, performs an experiment comparing traditional genetic screening by counselors to "insta-testing" by Navigenics. It looks like the technology of DTC has a long way to go, while our understanding itself of genetically inherited disorders is still in a rudimentary stage of development. Rather than exploring new sites and new tests the emphasis needs to be on interpreting the studies and data we already have.

April 19, 2012

The Sins of Science

Science, it seems, has been "the new religion" for a long time. It has always had its apostates and heretics, even its unremarkable and quotidian sinners.

In an article titled "Disgrace," Charles Gross, a professor of psychology and neuroscience at Princeton University, reviews the whole matter of contemporary and historical scientific misconduct (*The Nation*, Jan. 9/16, 2012, pp. 25-32). He finds nothing new in the shocking case of Harvard's Marc Hauser, who was exposed two years ago for scientific misconduct, in of all fields, the biological basis of morality and genetic inheritance of doing evil. Hauser apparently was guilty of the very venial sin of fudging facts. The three ways to do that, all frowned upon, are by fabrication (making data up), falsification (altering or selecting data, cherry picking) and sheer plagiarism (which all but entering freshmen understand).

In 1830, computer science pioneer Charles Babbage published a book in which he distinguished "several species of impositions that have been practised in science ... hoaxing, forging, trimming, and cooking."

Gross classifies the Piltdown man as an example of hoaxing. This fossil combining parts of an ape and human skull was discovered in 1911 and not discredited until the 1950s. Most hoaxes are intended to poke fun at the public's credulousness, but the Piltdown hoax was undertaken by well-meaning British imperialists who hoped their construction would fill an awkward gap in the record. Like God, if the missing link did not exist, we should have to invent one. Pip pip for the Royal Society!

Babbage believed that forging was uncommon. Rarely are results completely counterfeited and pulled out of thin air.

"Trimming" is probably a form of scientific misconduct that few scientists confess to their most exacting monitors such as the National Science Foundation but rather quietly cover up in bland hypocrisy. It consists of "eliminating outliers to make results look more accurate, while keeping the average the same." Who has not committed that little white sin? Let him who is without self-assurance cast the first chad.

"Cooking," on the other hand, the purposive selection and distortion of data, might be a real concern for all of us.

Gross goes on to inspect the career of Harvard's "war crimes professor" Richard Herrnstein, who became a co-author after his death of the book *The Bell Curve* about racial differences in intelligence. It is not a very pretty kettle of fish.

Charles Darwin essentially stole the idea of natural selection from Alfred Russel Wallace, the father of biogeography, did he not, and if he didn't, certainly failed to credit some of his predecessors in his rush to fame and self-glorification.

In genetics, we are reminded that the saintly Gregor Mendel probably falsified the suspiciously exact 1:3 ratio he "observed" in comparing pure dominant with hybrid peas (p. 26).

Alarmingly, we learn that "the modal scientific miscreant is a bright and ambitious young man at an elite institution," just the sort of role model worshiped by the popular press.

Maybe our society should be examining a few of science's feet of clay rather than pompously setting more laurels on the heads of its exalted heroes.

April 21, 2012

Undying Memories

The next time you experience déjà vu think about this. It might be more than a trick of the brain. Scientists have recently confirmed that genes can pass down the memories of our ancestors to us.

It sounds like something out of "Lord of the Rings," but medical researchers at the University in Cambridge have determined with experiments that DNA might pass down some genetic markers from generations ago (David Cornish, Wired Science, "Study: Genes Could Retain 'Memory' When Passed to Offspring").

This means that if you have a phobia about boarding an airplane it might be because your great-grandfather had a fear of flying. It also might mean that if you have a certain condition or disease, it could be because someone in your family history had that disease. We have had the latter idea of genetic inheritance being linked to disease for some time. That is why, with women, doctors might ask who in their family had breast cancer. Doctors even know that it is a greater risk if it is on the mother's side of the family.

Until now, scientists did not understand this process of passing epigenetic markers on, piggy-back style, through generations. Epigenetics was the mystery person in the room no one wanted to question or talk about, whether out of politeness or ignorance.

First, what do epigenetic markers do? They are like the managers in a factory telling the cells to either "use or ignore a particular gene." For this reason, although it should not be ignored, it is not the entire story if you discover with genetic testing that you, or someone you know, have a genetically high risk for a certain disease like breast cancer. It is possible to get lucky and have an epigenetic marker that is telling a gene to just sleep a few decades and not turn into cancer (Carl Zimmer, "Hope,

Hype, and Genetic Breakthroughs").

But how are epigenetic markers passed down through generations? Most of them are erased from one generation to the next (Cornish). However, James Gallagher, in his article, "'Memories Pass Between Generations,'" says that traumatic events can alter DNA and affect both the "brains and behavior of subsequent generations." Reputable publications now suggest that genetic memory is not science fiction. Genetic memory is stored in your DNA. Both genetic memory and epigenetics are now considered mainstream science.

December 6, 2013

Carried Away on the Red Balloon

We are now in the DNA age. What can DNA testing reveal today, and what will it reveal tomorrow? Genetics testing is an exciting field as it has just now become accessible and affordable, and the science is advancing at an unprecedented level. How can anyone not be in awe of the cutting-edge advancements in this field? After all, DNA testing is fast becoming a more important tool for understanding one's health conditions.

However, it is as important to be recognizant of possible limitations in this field as it is to understand what it can tell us. Genetic testing results can tell us a great deal. And we should be aware of this important resource to acquire what knowledge we can about ourselves, but we need some perspective. Otherwise, our outlook is one more akin to pure blind and deterministic faith than a scientific one. We should not be like the boy in the movie, *Red Balloon,* where we get carried away into the clouds because we think DNA testing can do anything we desire. We need to come down to Earth once more and analyze what it can and cannot do. To go too far down the path of thinking DNA testing for medical purposes provides all answers can be as much of a problem as shunning genetic testing science altogether out of unfounded fears. Genetic testing is an innovative tool. But it is still a scientific tool—not a religion.

And if genetic testing is a tool, aren't there other tools in the toolbox? Many have pondered about the role that epigenetics and the environment play on the genetic level. Surely it is not a life sentence to get a certain gene for cancer, and there are other factors like this to consider, at least in some cases. The formula then may not always be so simplistic: If one has X gene, one contracts Y disease in a pure deterministic fashion. Instead, perhaps we should view genetic markers as a pattern of guideposts warning us of greater health risks.

This is important and is not to make light of what genetic testing can do, because we could all use such warnings. In other words, even if we are unable to establish individualized medicine based on some futuristic, genetic testing system on the level that many hope for in the foreseeable future, that is no reason not to embrace what genetic testing can do for us today. And it has already saved lives and solved mystery diseases according to Marilyn Marchione's *Time* article, "Gene Scans Solve Mystery Diseases in Kids, Adults." That is plenty enough for me for now.

October 2, 2013

DNA Vaccines

What do you first think of when you think of DNA? DNA tests for ancestry, paternity, and forensics come to my mind first. Drugs and vaccines were not on my list, but they should have been. In the future, DNA will help advance drug delivery. We are living just at the dawn of the Genomics Age.

It sounds like a scene from the science-fiction show, *Continuum,* but it isn't. Nanoscale-sized DNA "cages" have been shown in recent clinical trials to be effective for a more optimal delivery of drugs. These miniscule DNA cubes are made from actual strands of DNA. The DNA cube carries cargo like a UPS box carries the phone you just bought, and it is carried to your door. Just like any package, it is shipped to its source. But like a Trojan horse, opening it means it will attack diseased cells by dispensing medicine. How does it open? The cancer cells, or other diseased source, trigger the release of the DNA cargo full of medicine. ("DNA 'Cages' May Aid Drug Delivery" *Science Daily).* Like the soldiers surrounding the Trojan horse, it seems, they have to see what is inside, and it is their undoing.

Why use DNA to deliver drugs? According to Thomas Edwardson, a McGill doctoral student, "DNA structures can be built with great precision, they are biodegradable, and their size, shape, and properties can be easily tuned" ("DNA Cages"). They are, after all, building blocks, and it seems that we can use them to build in other ways that we never thought possible.

In fact, the future is now. Though they are not yet on the market, vaccines that are currently being developed use DNA ("Biologicals" *World Health Organization*). What could the future look like? It could be that in the near future you will get a genetic test that shows you have cancer, and then you get a DNA-delivered drug or vaccine to help cure it.

September 3, 2013

Born to be Vulnerable

Little doubt exists that genetic testing and sequencing of infants is an extremely innovative and valuable tool. The amount of knowledge that we can garner from this cutting-edge technology is far beyond what we have been able to do up to this point with a medical diagnosis. It reminds sci-fi fans of the scans on *Star Trek, The Next Generation.* Truly we are living in the future. But are we ready for it?

The NIH, National Institutes of Health, quotes Alan E. Guttmacher, M.D., director of NICHD, as saying, "Genomic sequencing has [the] potential to diagnose a vast array of disorders and conditions at the very start of life, but the ability to decipher an individual's genetic code rapidly also brings with it a host of clinical and ethical issues, which is why it is important that ... the trio of technical, clinical, and ethical aspects of genomics research ... (is part of the research for treating newborns)." Now, we just have to decide how to use this tool.

If DNA testing is used to diagnose a baby with a disease or disorder at the start of life, and the infant can be better treated and evaluated from the start of life that would be a great advantage for all concerned. We could save lives. We could catch rare genetic conditions earlier. That is the desire of those in the genetic testing field, and the hope many have discussed is that this will become common for these reasons.

However, what if like *Gattaca*, a child is found with genetic testing to have a high probability of having a heart problem or other disorder or disease later in life? Should the parents be told? Should they later tell the child? Would they treat a child differently like the parents in *Gattaca*? It seems that many parents do tend to overreact if their child's DNA testing results show high probabilities of genetic disorders or diseases. Rob Stein in his *NPR* article, "Genome Sequencing For Babies Brings Knowledge and Conflicts," quotes bioethicist Mark Rothstein of the University of Louisville as saying these test results can sometimes lead to the so-called "vulnerable child syndrome" where a child is not allowed to have a "normal childhood" including "… riding bikes and climbing trees." Perhaps, some children would not even be wanted.

However, is this a problem with the science of genetic testing? There have always been overprotective parents. Sadly, there have always been unwanted children. It seems that what is needed is more education about genomic sequencing and possibly guidelines for all involved in the process, including the parents.

October 18, 2013

Lazy Genes

Lifestyle choices play a part, but DNA tests now show it isn't just how much fast food you eat that causes obesity. Some claim they cannot lose weight because of their genes. And many stigmatize people who are larger than average, like the Republican hopeful for the U.S. 2016 election, Chris Christie, because they assume it is easy to control weight. However, we are all unique. Now science and genetic testing is backing them up and coming to the rescue. Genetic testing shows that we have unique genetic variations, and genetics plays a larger role than many formerly thought when it comes to how much we weigh (Jacque Wilson, *CNN*, "5 Cool Things DNA Testing Can Do").

How do our genes differ? They differ in many ways, so this is just one more reason that it might be good to get to know your genes and your genetic profile with genetic testing. Scientists have determined with genetic testing that some people's genetic profile may predispose them and their children more to obesity (Wilson).

Also, genes regulate about 80 percent of our body fat (Alexander Sifferlin, *Time*, "What Mice Can Tell Us about Obesity and Genetics"). Why should that concern us? Genetic testing on mice shows that some of our genes are better at the job than others. In other words, some genes are better at turning calories into energy than others. Some sit down on the job.

How can we use this information to lose weight? It might be all the rage in the future, but there is no diet plan according to your genetic profile. "While there is a genetic component to obesity, our understanding of it is limited," says CNN diet and fitness expert Dr. Melina Jampolis (Wilson).

What can I take from these findings then? Certainly, one should just not use it is one more excuse to eat whatever one wants and sit on the couch instead of exercising. Exercise, like it or not, seems to be the more important factor. "In the case of the mice, it looks like the metabolism and activity is way more important than the amount of food consumed," says Jake Luis, a part of the team doing genetic testing on the mice, and a professor of medicine and human genetics and of microbiology (Sifferlin).

It looks as if we will have to wait for a specially selected diet based on our DNA testing results, but we might not like it anyway. For now, it looks as if we will have to keep putting on our running shoes to make up for our lazy genes.

April 25, 2013

Reversing Aging

What if you could reset your DNA biological age and become ever younger like Benjamin Button in the recent film with Brad Pitt? Sound too much like science fiction? Actually, scientists have just discovered the DNA biological body clock. Now, they are set to work on how to reset it to reverse aging. Of course, we are not talking about actually reversing one's actual age but using genetic testing and genomic science to reverse the age of one's cells and organs.

Is this possible? Not now, but scientists think this discovery brings them a step closer to finding this DNA Fountain of Youth. Why? First of all, they have revealed through genetic testing that not all of our organs age at the same rate. Tissue within and surrounding diseased organs may be decades older than healthy tissue found in the same person. (Ian Sample, *The Guardian,* "Scientists Discover DNA Body Clock)

Now, they just have to figure out how to slow it down. Actually, they are on the way to doing that using genetic testing and stem-cell research.

Steve Horvath, professor of genetics and biostatistics at the University of California in Los Angeles, in the journal, *Genome Biology,* said they managed to reset cells back to zero by "converting adult cells into stem cells, which can grow into any tissue in the body" (Sample). Ever said you wanted a new body? We might be just around the corner of doing just that and living longer lives.

Perhaps, not far in the future, we will go to the doctor to have genetic testing done to see which old or diseased parts of our body we need to reset. Puts a whole new meaning on the phrase "recreating your life."

October 21, 2013

Genes Are Not Contagious, Though Genetic Testing May Be

With any new science, there must be guidelines. With the advent of the first automobiles, there were no speeding signs until we discovered that driving fast is equally fun and dangerous. There are now emerging new and more sensible guidelines for genetic testing for children and newborns. The panel composed of the American Academy of Pediatrics and the American College of Medical Genetics and Genomics recommends genetic testing for childhood diseases for children and newborns. Diseases that wait to strike in adulthood—like Alzheimer's—are only to be tested depending on the parents' wishes and what is in the best interest of the child, according to Denise Mann in her *US News & World Report* article, "New Guidelines Issued for Genetic Screening."

There are exceptions. For instance, it might be reasonable to do genetic testing if there is a family history of a rare, genetic disease, even an "adult-onset" one, Mann quotes Dr. Joyce Fox, a medical genetics doctor at North Shore Hospital in Manhasset, N.Y. Mann also quotes a Dr. Helaine Ross, pediatrician at the University of Chicago as saying, "The focus should be on the education of families, counseling them and helping them make decisions that focus on the child's best interest."

What is not in the child's best interest when it comes to genetic testing? Genetic testing done out of placating the adult's morbid curiosity is not in the interest of the child. However, the panel did leave it up to the parents to do further genetic testing, and "what is in the best interest of the child" could be interpreted differently. It is conceivable that some parents could, as a consequence of genetic screening, treat a child like

Vincent was treated in *Gattaca* by being overly protective. However, there have been plenty of overprotective parents without genetic testing. Still, this could be problematic for some children and families.

My best friend died of leukemia when I was a child and though my mother was brilliant, she somehow became convinced I could get it as well and demanded my pediatrician do genetic testing to look at my white blood-cell count. It was normal. I am not sure what she would have done with this tool. Nevertheless, it is an incalculable scientific breakthrough to have such an advanced level of genetic testing. We now have genetic tests for more diseases and can provide a better treatment plan and diagnosis for many children and even save lives. We didn't stop making cars because some people speed.

October 28, 2013

11 DNA AND MOM

Haplogroups J and K Have Low Susceptibility to Parkinson's Disease

Confirming previous studies, a new article finds that K and its related haplogroup J—both common Jewish lineages—have a lower risk than other haplogroups for Parkinson's disease. D. Ghezzi et al., "Mitochondrial DNA haplogroup K Is Associated with a Lower Risk of Parkinson's Disease in Italians," *European Journal of Human Genetics* 13 (2005) 748-752.

April 14, 2005

Lineages of Indigenous Mexican Populations

Characterization of mtDNA Haplogroups in 14 Mexican Indigenous Populations
R. I. Peñaloza-Espinosa et al.
Human Biology 79/3 (Jun 2007) 313-20

Abstract. In this descriptive study we investigated the genetic structure of 513 Mexican indigenous subjects grouped in 14 populations (Mixteca-Alta, Mixteca-Baja, Otomi, Purpecha, Tzeltal, Tarahumara, Huichol, Nahua-Atocpan, Nahua-Xochimilco, Nahua-Zitlala, Nahua-Chilacachapa, Nahua-Ixhuatlancillo, Nahua-Necoxtla, and Nahua-Coyolillo) based on mtDNA haplogroups. These communities are geographically and culturally isolated; parents and grandparents were born in the community. Our data show that 98.6% of the mtDNA was distributed in haplogroups A1, A2, B1, B2, C1, C2, D1, and D2. Haplotype X6 was present in the Tarahumara (1/53) and Huichol (3/15), and haplotype L was present in the Nahua-Coyolillo (3/38). The first two principal components accounted for 95.9% of the total variation in the sample. The mtDNA haplogroup frequencies in the Purpecha and Zitlala were intermediate to cluster 1 (Otomi, Nahua-Ixhuatlancillo, Nahua-Xochimilco, Mixteca-Baja, and Tzeltal) and cluster 2 (Nahua-Necoxtla, Nahua-Atocpan, and Nahua-Chilacachapa). The Huichol, Tarahumara, Mixteca-Alta, and Nahua-Coyolillo were separated from the rest of the populations. According to these findings, the distribution of mtDNA haplogroups found in Mexican indigenous groups is similar to other Amerindian haplogroups, except for the African haplogroup found in one population.

July 7, 2007

Another "Anomalous" Cherokee Descendant

When I first got my mitochondrial results about four years ago, I learned that I matched no one in the world! Indeed, I wondered if I might be from Mars. Why didn't I match someone?

Later, I discovered that my type U5b fell into the Cherokee "anomalous" category, which meant that my family stories of Cherokee descent were, as I knew, not just stories. And this was in a strict matrilineal fashion though these were not the stereotypical Native Americans envisaged by standard genetic testing and hidebound anthropologists.

My matrilineal line goes back to Elizabeth (Eliza) Ann Culver, who died Dec. 8, 1847, in Hancock County, Georgia. By a sleight of hand very common at the time, she was whitewashed and made into a Marylander like her presumed father Levin Ellis, but she was, as her mitochondrial DNA suggests, in reality probably a Lower Cherokee woman. The Elati language she spoke is now extinct, as are apparently most of the Elati clans. Witness my lack of matches anywhere.

Hoping to find a match as this study proceeds into Phase II!

My Direct Female Line

> Elizabeth Ann Ellis Culver born 1775 prob. in Lower Cherokee
>   country in Georgia
> Martha Matilda Culver Grace born 1807 in Georgia
> Martha Roberta Grace Meares Brackin born Feb. 19, 1847, in
>   Georgia
> Mollie Brackin Box born 1867 in Georgia
> Luta Mae Box Newberry born about 1890 in Newville, Henry
>   County, Alabama
> Nina Jo Newberry—my mother

My mother had Cherokee ancestry from both parents. She was French/Tuscarora from my grandfather, Anderson Colbert Newberry, who descends from French Prevatts who settled in Robeson County, N.C. and then intermarried with Tuscaroras and Cherokees.

February 12, 2010

On the Trail of Spider Woman

Thoughts about the origin of mitochondrial haplogroup B and Mother Earth symbolism among the Hopi, Zuni, Hohokam, Fremont Indians and others.

I got a holiday present from my wife of an unusual little book titled *On the Trail of Spider Woman: Petroglyphs, Pictographs, and Myths of the Southwest,* by Carol Patterson-Rudolph (Santa Fe: Ancient City Press, 1997). Putting this intriguing study together with a travel book by David Hatcher Childress, my son and I took a four-day road trip into the homeland of the Indians credited with having the first civilization in the Southwest, a settled town life marked by desert agriculture, canals, pottery, baskets, ballcourts, plazas and adobe pueblos, pithouses and kivas. Previous occupants of the area were non-sedentary hunter-gatherers considered to be "paleo-Indians."

We visited Painted Rock Petroglyph Site outside Gila Bend, in the middle of nowhere, and ended the trip in the barren sands outside Phoenix, where we began it, visiting the Hohokam Pima National

Archeological Monument, also known as Snaketown. The former is little seen, and the latter cannot be seen, because the Pima (now Gila River Indians) had the ballcourt and other ruins reburied by backhoes in the 1970s. The caretakers of this declared national treasure decided not to open it to anyone to view or visit because of its "sensitive" nature. There are no signs, no roads, nothing left above ground.

At Painted Rock, the first mystery we pondered was why it was called "painted" rock when there is no paint. Petroglyphs are produced by pecking away the dark desert varnish to make a negative image on the underlying lighter rock. We wondered if it had anything to do with the Paint People, or Phoenicians, Kanawah Indians of the East Coast or Cherokee and Saponi Paint Clans.

The second mystery was the abundance of snake imagery. Famously, snakes in Indian tradition stand for boats and water. We noticed a Corn Cross, the symbol of the Feathered Serpent or Quetzlcoatl religion, supposedly introduced into Mexico from both the East and the West by white, bearded strangers in ships, who brought rule by laws and numerous arts of civilization and banned human sacrifice.

The third thing we remarked upon were the many Great Goddess or Earth Mother or birthing/fertility symbols. Such places were probably shrines where women came to be blessed and get married and give birth. Sun Park in Hopiland has numerous hemispherical carvings about 2 inches wide where people ground out minerals to eat. These cupmarks or cupoles at petroglyph sites puzzled archeologists until an important article in a scholarly journal clarified their meaning as part of the worldwide phenomenon of pica (pronounced "pie-ka"), "the desire to ingest nonfood substances such as rock powder, clay, chalk, dirt, and other material by some humans, especially pregnant women" (Kevin L. Callahan, "Pica, Geophagy, and Rock-Art in the Eastern United States," in *The Rock-Art of Eastern North America,* ed. Carol Diaz-Granados and James R. Duncan, Tuscaloosa: U of Alabama P, 2004, p. 65).

In general, petroglyphs are neglected both by archeology and anthropology. Their study is a no man's land. Sun Park has a birthing cave and birthing stone. Canyon de Chelly has the most photographed Mother Earth rock formation in the world, Spider Rock, a chthonic monument discussed on page 83 of the Spider Woman book by Patterson-Rudolph.

Who Were the Hohokam?

At Sun Park, there were clear images of horses, riders, people praying, spirals, axis mundi (center of the earth) symbols like the iron butterfly and cross, labyrinths, bilobed axes, irrigation plans, horned beasts, felines, palaces or villages and warriors with spears and shields. We searched in vain for anomalous depictions of whales, elephants and deep water fish, found at other similar sites, but the sun was sinking and we did not have time to make a thorough inspection of the motifs. As we returned to Third Mesa we did find a famous petroglyph of a whale at Old Oraibi.

The name Ho-ho-kam is usually explained as meaning "Those Who Are Departed," but such an etymology is more gloss than literal translation of its meaning and origin. Like many words in the Hopi, Zuni, Pima and Uto-Aztecan languages, it actually appears to be composed of South Semitic elements. In Ancient Egyptian, it means "Sea Peoples" or "Foreigners." The historic Sea Peoples who came from Asia Minor and points northerly once threatened to conquer the Egyptian empire. The Philistines and Phoenicians are related to them. They were remarkable for their feather bonnets and, like their relatives the Cretans (whose language also came from Asia Minor), for a long-protracted continuance of Mother Goddess worship down into the Bronze Age.

Haplogroup B is the signature lineage of certain Indians in North America. Its ultimate source is Southeast Asia (not Mongolia, as has been suggested for the other three classic Native American haplogroups A, C and D), whence it took multiple circum- or trans-Pacific migratory routes to the Americas (Eschleman et al. 2004). It has high frequencies in Polynesia, which was settled from Southeast Asia, and among the Western Indians of the U.S. such as the Hopi, Zuni (77%), Anasazi (78%), Yuman, and Jemez Pueblo (89%). It is also found in frequencies approaching 70% in the Cherokee and Chickasaw.

We believe Spider Woman is simply an aspect of the Stone Age Great Goddess worshiped by those who came from Southeast Asia through Polynesia and helped colonize the American Southwest. She is the same as the Earth Mother. As in other cultures, she was replaced by sky and sun deities and male hierarchies. But her religion seems to have persisted in the Hohokam, Cherokee and Hopi tribes in a similar fashion to the survival of her cult in the Cretans, Phoenicians and Sea Peoples.

According to Hopi and other traditions, at the end of the last age, the Mother Goddess ceased to be the leader of the people in their wanderings and went back "under the sea" *to the east and west* whence she and they had emerged. We can only infer from this that Spider Woman, as she was called in Asia and the Pacific, and the Great Goddess, as she was known in the Old World of the Middle East, relinquished her role as supreme deity to the new male pantheons and withdrew across the Pacific and Atlantic Oceans to the distant origins of civilization outside the Americas. Ironically, her memory survived better among the Indian nations than in the war-torn empires and materialistic cultures that dominate world history elsewhere. Indian societies today exhibit rare examples of matriarchy as opposed to patriarchy.

December 31, 2010

Haplogroup N

In Europe, Asia Minor and American Southwest and Now in the Cherokee

Haplogroup N1a came to the fore in genetics literature when Wolfgang Haak et al.'s studies on 7,500-year-old skeletons in Central Europe revealed that 25 percent of the Neolithic European population might have belonged to this lineage. The skeletons were found to be members of the Linear Pottery Culture (LBK ware), which is credited with being the first farming culture in Central Europe.

The study was a major development in the debate on the origin of European populations, since Haak et al. argued that:

> The discovery of mitochondrial type N1a in Central European Neolithic skeletons at a high frequency enabled us to answer the question of whether the modern population is maternally descended from the early farmers instead of addressing the traditional question of the origin of early European farmers.

Two competing scenarios exist for the spread of the Neolithic from the Near East to Europe:

- Demic diffusion (in which farming is brought by farmers), for example Renfrew's Neolithic Discontinuity Anatolian hypothesis
- Cultural diffusion (in which farming is spread by the passage of ideas), which is the assumption in Alinei's Paleolithic Continuity Theory.

The study's authors concluded: "Our finding lends weight to a proposed Paleolithic ancestry for modern Europeans."

N currently appears in only 0.18 to 0.2 percent of regional populations. It is widely distributed throughout Eurasia and Northern Africa and is divided into the European, Central Asian, and African/South Asian branches based on specific genetic markers. Exact origins and migration patterns of this haplogroup are still unknown and a subject of some debate.

Although not one of the classic Native American lineages (A, B, C, D, and X—Schurr), N has been identified in the ancient Southwest in the Fremont Culture centered in Utah. It is one of the Middle Eastern lineages that appear in the Cherokee and other Indians; see DNA Consultants Blog, "Anomalous Mitochondrial DNA Lineages in the Cherokee". Most investigators attribute this phenomenon to recent European admixture. But such haplotypes if only instanced in North America without exact Old World matches could just as well be considered Native American.

It has been suggested that N is also characteristic of the Sea Peoples, who may have traveled to the American Southwest in antiquity.

January 1, 2011

Haplogroup B and Water Clan Symbols
Native Hawaiians and Native Americans

In a previous post, "On the Trail of Spider Woman," we suggested that petroglyphs in Arizona and Utah with female goddess symbolism and birthing ceremonies were connected with the Hohokam ("Sea Peoples") and other Indians who followed in their wake, corresponding to archeology and anthropology's Basketmaker Culture. In this and a series of posts over the next few months, we will show pictures of "emergence" petroglyphs from Hawaii, New Guinea, California, Hopi, Zuni, Pima,

Papago, Fremont, Zuni, Mimbres, Palavayu and Eastern Woodlands cultural sites that support our thesis. We believe them to be the footsteps and stepping stones of female haplogroup B and its associated lineages.

Mitochondrial haplogroup B does not have as its dissemination center Mongolia or Siberia or Central Asia but Southeast Asia, specifically Taiwan and Indonesia, and is characteristic, in contrast with Indian groups emphasizing A, C and D, of the Pueblo Indians and some Southeastern Indians such as the Cherokee, Chickasaw and Choctaw. It entered the Americas in successive waves, some of them seaborne, over many millennia.

The first picture comes from the western coast of the island of Hawaii. It is considered one of the oldest religious shrines in the Hawaiian Islands. It shows a stick figure carved into a rock set in the ground. As we will see, this is a typical "emergence" figure marking the arrival of a people in a new phase of existence.

The symbolism is of a female mother figure giving birth, her progeny here depicted by the tail-like extension coming from between her legs. There are thousands of variations of this tribal or clan mother iconography scattered over Asia and the Americas (but not apparently found in Europe or Africa).

The Hawaiians considered the western coast of the Big Island their place of emergence. According to their legends, their people came from the sea from the southwest and were noted for their ability to twist plants and fibers into ropes. Their capital was hence called Hilo (twisted, plaited).

On account of their subtlety in these arts they adopted the hula (twist) dance as their national dance. Its original purpose was as a fertility ritual to increase population. (Johannes C. Andersen, *Myths and Legends of the Polynesians*, Tokyo: Tuttle, 1969.) The main song sung during the enactment of the hula was called "The Water of Kane, or Waters of Life."

The Hawaiian Mother symbol seems to be connected with a certain clan. As is often the case, the head of the female figure is differentiated to show which clan. This one has horns and could represent a dragonfly. This insect recurs in American Indian petroglyphs where it is associated with the Water Clan and fertility rites.

To read the Hawaiian petroglyph properly we might say, "Here is the spot where the Head Mother of the Water Clan emerged and gave birth to her people." It is likely (although no legends are preserved regarding its use) that women made offerings here to become fertile, attract husbands and be delivered of healthy children. In similar ceremonial sites, such figures mark an actual birthing stone where women squatted to give birth, attended by midwives and clan mothers.

To show the physical resemblance of the Hawaiian design to American Indian symbols we will reproduce thumbprints below from different traditions. They will be linked together and explicated in subsequent posts in this series.

January 14, 2011

Hidebound History

Her profession is her religion
Her sin her lifelessness
          —Bob Dylan

Will the archeological establishment's obtuseness about the religion of the Great Goddess ever falter? In an article titled "Pieces of a Bronze Age Puzzle" in the current issue of *Archaeology Magazine* (Sept/Oct 2011, p. 15), Jessica Woodard discusses the "enigma" of thousands of broken Cycladic figurines from the tiny, uninhabited island of Keros near Naxos. Summarizing the decades long work of Cambridge archeologist Colin Renfrew, she dates the site to 2800 to 2300 BCE and (are you ready for this) speculates there was a lot of "social activity as well as ritual activity ... relating to beliefs about life, death, and perhaps the hereafter."

This is tantamount to saying that the deliberately broken figurines were broken by people, human beings who lived a long time ago, on purpose. But what kind of rituals and "beliefs"? The word "religion" is mentioned nowhere. Evidently, since archeologists profess no religion themselves they cannot detect it in any of the people whose graves and relics they dig up.

Greek mythology tells how Venus, the eldest of the Fates, was born at sea and stepped ashore on several islands, where her cult continued, notably at Cythera, Crete, Naxos and Cyprus. All the "enigmatic" broken figures clearly relate to the worship of the Mother Goddess.

Marija Gimbutas covers the featureless face, arms crossed over breasts and other unmistakable signs of the Goddess or Magna Mater in her voluminous writings, including *The Language of the Goddess*. We suggest if Colin Renfrew cannot bring himself to read Gimbutas he at least dip into Pausanias, the second century C.E. author of a guidebook to Greece in 10 volumes. There he will find many descriptions of these votive offerings to the Goddess.

Archeologists may also want to acquire at least a bowing acquaintance with Riane Eisler's *The Chalice and the Blade.* Both Gimbutas and Eisler describe three invasions of the warriors of the steppes with their male gods following the year 3000 B.C.E. that spelled an end to the long period of female-based life-celebrating religion in the Middle East and Old Europe. Only the Minoans, Etruscans and certain other peoples from Asia Minor and the Greek Islands were able to retain the Mother Goddess in the new mostly male pantheon, which was focused more on death than rebirth.

The only puzzling part of the Keros Hoard is how archeologists could overlook its abundant testimony to the Mother Goddess religion.

August 15, 2011

Emergence Petroglyphs

Emergence petroglyphs as featured on previous blog posts about the Hopi, Sea Peoples, Hohokam, Fremont Indians and Cherokee ("Haplogroup B and Water Clan Symbols") have also turned up now in Patagonia in southern Chile, on the tip of the South American continent's Pacific Coast. They were identified in Hawaii already. These findings suggest the stick figure of a woman giving birth, or emergence petroglyph, is Pacific-wide and confined to that hemisphere, not instanced in Europe, the Middle East or Africa.

February 5, 2011

The Eternal Female

Among the explosive stories in Sam Kean's excellent "biography" of DNA, *The Violinist's Thumb* (Little, Brown, 2012), is a chapter on what makes mammals mammals. The short answer is a placenta, and Kean begins by discussing a rare case of mother and daughter coming down with a hereditary form of leukemia, a sort of "simultaneous cancer."

The chapter is titled "Love and Atavisms," and it zeroes in on the role of the major histocompatibility complex in the origin of mammals and selection of sex partners. MHC is the factor that makes some potential mates smell better than others to females (through pheromones). In one experiment, women were given T-shirts males slept in to smell, and they identified the "wildest" smelling as the most attractive. Same smelling T-shirts were spurned, perhaps for the same reason members of our own nuclear family are not sexy-seeming to us. (In places with little genetic variety, like Utah, it was shown that pheromone sensors were inoperative.)

"In humans, MHC is located on the shorter arm of chromosome six (p. 169)." In the case of the infant born with her mother's congenital proneness to leukemia, the cancer cells had deleted her MHC. She developed acute lymphoblastic leukemia, eventually being successfully treated for it, although her mother died of it in the same time frame.

"We don't think of cancer as a transmissible disease," Kean writes. "Twins can nevertheless pass cancer to each other in the womb; transplanted organs can pass cancer to the organ recipient; and mothers can indeed pass cancer to their unborn children, despite the defenses of the placenta (p. 176)."

But now comes the truly amazing revelation.

"Other scientists have painstakingly determined that most, if not all, of us harbor thousands of clandestine cells from our mothers, stowaways from our fetal days that burrowed into our vital organs. Every mother has almost certainly secreted away a few memento cells from each of her children inside her, too.

Such discoveries are opening up fascinating new facets of our biology; as one scientist wondered, "What constitutes our psychological self if our brains are not entirely our own? More personally, these findings show that even after the death of a mother or child, cells from one can live on in the other. It's another facet of the mother-child connection that makes mammals special." (pp. 176-77) Thus, female DNA lives on down the ages, while male DNA shows a fragile character and is doomed to experience dead ends. Siblings may get different "clandestine" cells from the same mother and develop different heteroplasmies, accounting, in part, for their different personalities, abilities and family traits.

August 10, 2012

Were Solutreans in Ancient America Egyptians?

Yes, according to Bill Tiffee, whose article on Solutreans in America will appear in volume 29 of the series Epigraphic Society Occasional Papers. Titled "Were Ancient Egyptians the Solutreans Who First Settled America?" the new study, he says, "looks at the possibility that the Solutreans who first settled America were from Egypt, and that the genetic marker X is found in the highest concentrations among the Druze (who migrated from Egypt 1,000 years ago) and the descendants of the Moundbuilder Native groups including the Sioux and Algonquin and possibly the Cherokee."

We have previously suggested that the Cherokee incorporate both Greek and Egyptian DNA. Chapter 3 of Donald Yates' new book *Old World Roots of the Cherokee* is devoted to the DNA story of the so-called "anomalous" Cherokee lines, including haplogroups T and X.

Several prominent scholars have argued that Europeans known to archeologists as the Solutreans of France and Spain around 18,000 years ago were the first to settle the Americas. Tiffee examines the similarities between Solutrean and Clovis or Paleo-Indian stone technology and reconstructs the Solutrean culture in Egypt beginning 24,000 years ago (p. 119). He links ancient Egyptians with genetic marker E-M78, mitochondrial haplogroup X, Tula and the Spiro Complex mounds in Oklahoma, among other North American sites. He also discusses the Great Flood of about 10,000 years ago, the legends surrounding Osiris and the rise of agriculture in southern Turkey (Gobekli Tepe).

"Perhaps," he concludes, "Egyptologists need to rethink their paradigms of ancient Egypt. And perhaps modern Native American descendants of the Moundbuilders, the Algonquin groups, Sioux,

Cherokee, Chickasaw (and other Native cultures closely related to mound-building) need to reconsider where their most ancient ancestors came from (129)."

In DNA Consultants' Cherokee DNA study, "Anomalous Mitochondrial DNA Lineages in the Cherokee," as well as numerous blog posts since 2009, it was reported that haplogroups U, T, K, J, N, X and L are found in Cherokee descendants in frequencies mirroring those of Egypt.

June 19, 2012

Real People's DNA Stories: James Shoemaker

Bible Studies, DNA Tests, Mother's Nursing-Home Confessions Lead to New Life

Until he took an autosomal ancestry test, James T. Shoemaker had little concept of his heritage. He assumed he was just an average white European American like his Appalachian neighbors.

Although raised in a Pentecostal Church, Shoemaker always felt a strong pull toward Jewish culture. So, last year, he went to his doctor and asked for a DNA test. "I wanted to see if there were any Jewish lines in my ancestry," he said.

He ended up taking a DNA Fingerprint Plus, a complete analysis of one's genetic ancestry that includes ethnic markers and megapopulation admixture matches.

Fast forward from that first eye-opener, and, today, the 53-year-old Waynesboro, Pa., resident is halfway through a conversion process to Judaism at B'nai Abraham, a Reform congregation in Hagerstown, Md., where he is being mentored by youngish Rabbi Ari Plost.

"I got all three ancestral markers for Jewish I, II and III," Shoemaker recalls, "so I went to see my mother, Jacqueline Rose, at the nursing home in Hagerstown, and she admitted, 'Well, yeah, my parents, uh, they were both Jewish."

It was the first he had heard of it. "Mom never said a word about having Jewish ancestors. It turned my life around."

The fact that he got a "double dose" of Jewish alleles in his marker results confirmed the truth of his mother's admission that both she and his father came from Jewish families.

Shoemaker next took a Premium Male DNA Ancestry Test to determine whether his father's Y chromosome line was perhaps Jewish. The results were delivered to him in mid-November.

His particular haplotype did indeed match several other Jewish men, including those with the surnames Brown, Hendrix, Shepard, Getz, Phillips, Lewetag and Sequeira. "The subject's specific male haplotype (surname line) probably came from Southwest Germany or the Low Lands, to judge from the modal matches and patterns of distribution," according to the report.

As for the surname itself, the Surname History section (included in every Premium Male report, cost $325), had some valuable clues for Shoemaker's genealogy.

"Shoemaker is probably a translation of the Dutch or German equivalent Schuhmacher or Shumacher meaning "shoemaker." It is noted as a Jewish family name in Southwest Germany and the Saarland in France, including Lörrach in Baden (Lars Menk, *A Dictionary of German Jewish Surnames*, Bergenfield: Avotaynu, 2005, pp. 673-74). It could also come from Schuster, a more common Jewish German surname (p. 675)."

A Mason since 1990, and flirting at one time with Messianic Judaism, Shoemaker feels as though his earlier attempts to connect with his Jewish heritage were blind and unguided without the hard testimony of DNA.

"All these things I've been interested in with my studies and religious life now fall into place," he said. "I'm finding out why."

What lies in the future? This Pesach, Shoemaker will have an official bar mitzvah, complete with ritual bath and reading from the Torah. He then plans to attend Hebrew Union College in Cincinnati. "What I am really looking forward to," he says, "is making *aliyah* to the Land of Israel."

November 16, 2013

The Junk Bonds of Genetics

The tripartite Asian Model of the peopling of the Americas through "Beringia" was re-asserted with "the most comprehensive survey of genetic diversity in Native Americans so far" in a study published in *Nature* this week, "Reconstructing Native American Population History," by Harvard's David Reich et al. If ever there was a blue chip study, this is it. Only it is more like junk bonds in which no one should put stock.

If you read the fine print of this new issue from the Ivy League anthropological establishment, you may discover:

* Although the authors claim to go beyond examining single loci on the mitochondrial genome or Y chromosome and to analyze instead 364,470 SNPs, they are still stuck on the same biased samples. In one of their feats of prestidigitation, they statistically filter out "West-Eurasian-related and sub-Saharan African related ancestry in many Native Americans" (p. 371). They ignore anything that does not support their preconceived conclusions.
* Anthropologists have always insisted on the Bering Land bridge. Geneticists start with anthropologists' assumptions and test their model. Guess what? After enough manipulations you can make it work!
* Whole genome sequencing was adopted because it has become most economical, but half the samples were just adopted into the new study after doctoring the pre-existing data. These biased data (Pima, Inuit etc.) were not reliable when they were collected (as far back as the 1990s), and have only been improved through statistical voodoo. The new Indians' samples (heavily geared to Mexico, Central America and northern South America) were probably subjected to SNP investigation out of interest in biodiversity and possible medical applications anyway. The motives of investigators who mostly belong to medical faculties are tainted.

Here's the conclusion:

> Our analyses show that the great majority of Native American populations—from Canada to the southern tip of Chile—derive their ancestry from a homogeneous "First American" ancestral population, presumably the one that crossed the Bering Strait more than 15,000 years ago. We also document at least two additional streams of Asian gene flow into America, allowing us to reject the view that all present-day Native Americans stem from a single

migration wave, and supporting the more-complex scenarios proposed by some other studies. In particular, the three distinct Asian lineages we detect—"First American", "Eskimo–Aleut" and a separate one in the Na-Dene-speaking Chipewyan—are consistent with a three-wave model proposed mostly on the basis of dental morphology and a controversial interpretation of the linguistic data.

So, we're back to Greenberg and other discredited believers in the linguistic explanation of human diversity, something they used to call racism. Maybe that's because culturally inferior American Indians make such great subjects for grant getting in the first place. Especially if they are safely dead, on a reservation, or far away and helpless and completely extraneous to our society.

August 19, 2012

Haplogroup T among the Cherokee
A Surprising Middle Eastern Component

Haplogroup T (named Tara by Bryan Sykes in *The Seven Daughters of Eve*) is usually not seen as a Native American lineage. But it is discussed as such in Donald Yates' *Old World Roots of the Cherokee*, where it takes its rightful place among other Middle Eastern haplogroups like U, J and X. Moreover, several geneticists have drawn attention to its prevalence in New World Jewish and crypto-Jewish populations.

The following comes from Chapter 3, "DNA," pp. 55-57, and discusses some living examples of "Taras" who verified their Native American genealogies with a DNA test from DNA Consultants in 2007-2009, as reported in "Anomalous Mitochondrial DNA Lineages in the Cherokee":

Maternal lineage T arose in Mesopotamia approximately 10,000 to 12,000 years ago. It spread northward through the Caucasus and west from Anatolia into Europe. It shares a common source with haplogroup J in the parent haplogroup JT. Ancient people bearing haplogroup T and J are viewed by geneticists as some of the first farmers, introducing agriculture to Europe with the Neolithic Revolution. Europe's previous substrate emphasized older haplogroups U and N. The T lineage includes about 10% of modern Europeans. The closer one goes to its origin in the Fertile Crescent the more prevalent it is.

All T's in the Cherokee project are unmatched in Old World populations. They do, however, in some cases, match each other. Such kinship indicates we are looking at members of the same definite group, with the same set of clan mothers as their ancestors. So let us briefly introduce some of these descendants of Middle Eastern-originating Cherokee lines.

Jonlyn L. Roberts, had a puzzling, but typical genealogy that led her to embark on a lifelong quest for answers. Her mother, Zella, was adopted by the George and Mary Hand family of Hand County, South Dakota, in 1901. Little information was passed down, but piecing together clues from her childhood, Roberts believes that her mother's original family might have come from the Red Lake Ojibwe Indian Reservation or one of the North or South Dakota reservations. At any rate, her mtDNA haplotype is a unique form of T, one with certain distinctive variations in common with others in the study.

Another T in the study fully matched four other people, all born in the United States. One of these noted their ancestor as being Birdie Burns, born 1889 in Arkansas, the daughter of Alice Cook, a Cherokee.

Gail Lynn Dean (T) is the wife of another participant, whose type belongs to anomalous U. Both she and her husband claim Cherokee ancestries.

T participant Linda Burckhalter is the great-great-granddaughter of Sully Firebush, the daughter of a Cherokee chief. Sully married Solomon Sutton, stowaway son of a London merchant, in what would seem to be another variation of "Jewish trader marries chief's daughter."

Two cases of T represent descent in separate lines from the historically documented Gentry sisters, Elizabeth and Nancy, daughters of Tyree Gentry, who moved to Arkansas in 1817. The tested descendants are aunts or cousins of Patrick Pynes, a non-registered Cherokee and professor of American Indian studies. Learning of the results of the study, Pynes commented, "The possible connections to Egyptian heritage among these Cherokee descendants are especially interesting. We have a photograph of one of the women in this T* line (a granddaughter of Nancy Gentry, I think), and she is wearing an Ankh necklace. We all thought that was kind of strange. As far as I know, the Gentrys were

Methodist Episcopalians."

Three participants with T previously unknown to each other, and living in different parts of the country, turn out to be very close cousins descended from the same Cherokee ancestress. Their mitochondrial mutations exactly and fully match. Two claim Melungeon ancestry—a Yates male-linked cousin of the author and a relative of Phyllis Starnes (U, matching the author's). The third has adoption in the family, so the female ancestry is unknown.

A case of rare T5, Cheryl, took not only the mitochondrial test but also our CoDIS-marker-based ethnic population test, DNA Fingerprint, to validate "Cherokee or Jewish ancestry" from her mother. The results of the DNA Fingerprint Test show Ashkenazi Jewish in the No. 1 position, followed by assorted American Indian matches. Cheryl says that she is exploring returning to Judaism, but that in the remote Texas town where her family lives there are few avenues or resources to pursue.

As tabulated in Appendix A, our small survey shows a great deal of diversity and relatedness. It includes more than a few participants who discovered they share the same Cherokee ancestry, maybe even the same clan. Unlike a random sample of the U.S. population, they exhibit a mix that turns the conventional numbers on their head. Haplogroup H, instead of an expected 50% dominant position, is one of the smallest, with only 7.7%. Haplogroup U, an older lineage representing the Stone Age colonization of Europe before the ascendency of H, contributes 25% of the total number. Haplogroup X, marked by an exiguous presence elsewhere, attains a frequency in the Cherokee more than tenfold that of Eurasia or rest of Native America.

Yet the most startling statistic concerns T haplotypes now verified in the Cherokee. At 27%, they constitute the leading anomalous haplogroup not corresponding to the types A, B, C, or D. Several of them evidently stem from the same Cherokee family or clan, although they have been scattered from their original home by historical circumstances. So much consistency in the findings reinforces the conclusion that this is an accurate cross-section of a population, not a random collection of DNA test subjects. No such mix could result from post-1492 European gene flow into the Cherokee Nation. To dismiss the evidence as admixture would mean that there was a large influx of Middle Eastern-born women selectively marrying Cherokee men in historical times, something not even suggested by historical records. Mitochondrial DNA can only come from mothers; it cannot be imported into a country by men.

If not from Siberia, Mongolia or Asia, where do our anomalous, non-Amerindian-appearing lineages come from? The level of haplogroup T in the Cherokee mirrors the percentage for Egypt, one of the only countries where T attains a major showing among the other types. In Egypt, T is three times what it is in Europe. Haplogroup U in our sample is about the same as the Middle East in general. Its frequency is similar to that of Turkey and Greece.

September 14, 2012

Christie Hill-Troglin commented on 15-Nov-2013
My third great-grandmother (via my paternal lineage) is Sully Firebrush. I would love to have any additional information that you would be willing to share.

Barry Vann commented on 24-Jan-2014
Hi, I am one of the folks whose family on both my mom and dad's sides insist that there was a Cherokee ancestor. On my maternal side, my grandmother claimed her maternal grandmother drew some sort of an Indian claim (presumably a cash payment) in the 1920s. Her married name was Mary Crisp, and she lived in Graham County, North Carolina. Our mtDNA is T2e, and my autosomal results (FTDNA and Geno 2) tell me that I am 17% Middle Eastern. I have no known connection to the Middle East. Is it possible that Middle East can be confused for Native American? Thanks, Barry

My Grandfather Was a Cherokee

My Cherokee grandfather told me we had Cherokee Indian ancestry when I was just a child. We were sitting on the front porch in Alabama eating corn pone and watching fireflies.

Many have a family history of American Indian ancestry but have been unable to trace this in their genealogy. Can Native American DNA testing confirm Native American heritage? It can though there are several scenarios and DNA testing options. DNA testing for Native American ancestry is an innovative science but not magic, and there are different advantages and limitations with various methods.

For instance, if one is fairly certain that one's maternal line is American Indian, then a mitochondrial DNA test can be taken to test this line. The maternal line of ancestry follows the mother's line of ancestry (one's mother, and her mother, and her mother, and hers, and so on). What could one discover with those DNA testing results? One could discover if the Eve, the origin of one's maternal line, is American Indian. This is an excellent method and one of the most common types of DNA testing for American Indian ancestry because it can delve into deep ancestry. If one has a family history of the maternal line being American Indian, this is the test to find out if that is the case. However, it will not work if one's American Indian ancestry skips from the maternal to the paternal line. In other words, if one's mother had American Indian ancestry but received this ancestry from her father.

If that is the situation, or if one is not sure, there are other DNA testing options as well. Autosomal DNA testing based on forensic science can test genetic markers on both the mother's and the father's side. However, the test is not able to tell whether American Indian ancestry is from the mother's or father's side.

May 8, 2013

12 DNA AND NATIONALITY

Vikings Left Little Genetic Trace in Ireland

The Scale and Nature of Viking Settlement in Ireland from Y-chromosome Admixture Analysis
Brian McEvoy, Claire Brady, Laoise T. Moore and Daniel G. Bradley
Smurfit Institute of Genetics, Trinity College, Dublin, Ireland
European Journal of Human Genetics 14 (2006) 1288-1294; doi:10.1038/sj.ejhg.5201709; published online 6 September 2006

You may want to consider returning your Viking Certificate. According to a new study of Y chromosome lineages among Irish males with surnames seeming to indicate Norse ancestry, the Vikings did not leave as much of an imprint on the Irish population as many people would like to think. After all, they only stayed for a couple of centuries in most cases before going back to Norway. That's not much time to leave a sizable genetic legacy, even if you are a Viking.

Abstract. The Vikings (or Norse) played a prominent role in Irish history, but despite this, their genetic legacy in Ireland, which may provide insights into the nature and scale of their immigration, is largely unexplored. Irish surnames, some of which are thought to have Norse roots, are paternally inherited in a similar manner to Y-chromosomes. The correspondence of Scandinavian patrilineal ancestry in a cohort of Irish men bearing surnames of putative Norse origin was examined using both slow mutating unique event polymorphisms and relatively rapidly changing short tandem repeat Y-chromosome markers. Irish and Scandinavian admixture proportions were explored for both systems using six different admixture estimators, allowing a parallel investigation of the impact of method and marker type in Y-chromosome admixture analysis. Admixture proportion estimates in the putative Norse surname group were highly consistent and detected little trace of Scandinavian ancestry. In addition, there is scant evidence of Scandinavian Y-chromosome introgression in a general Irish population sample. Although conclusions are largely dependent on the accurate identification of Norse surnames, the findings are consistent with a relatively small number of Norse settlers (and descendants) migrating to Ireland during the Viking period (ca. A.D. 800-1200) suggesting that Norse colonial settlements might have been largely composed of indigenous Irish. This observation adds to previous genetic studies that point to a flexible Viking settlement approach across North Atlantic Europe.

September 6, 2006

Myths of British History
Prospect, October 2006

Everything you know about British and Irish ancestry is wrong, according to geneticist Stephen Oppenheimer. Our English ancestors were Basques, not Celts. The Celts were not wiped out by the Anglo-Saxons. In fact, neither had much impact on the genetic stock of these islands. And a form of English, not Celtic, was spoken in most places before the Roman invasion. Phylogenetic studies have overturned many of the commonly held beliefs about British history.

October 31, 2006

China Revamps Surname System

China has decided to open the door to a million new surname combinations to better distinguish individuals in its society. Previously, it has always been said there were only a hundred surnames to choose from, with Li heading the list. The news was mentioned in the current issue of *Nature* (vol. 448, no. 7153) with an article titled "The 'Hundred Surnames' of China Run into Thousands," by Dafeng Hui (p. 553). Below is a previous news announcement on the changes:

New Naming Rules in Mainland China
Wen Hua Du
The Epoch Times
Jun 29, 2007

China is proposing a new naming regulation that would require newborn children to adopt either or both parents' surnames as their surname. The plans are currently being discussed at a grass-roots level. To date, there has been no official regulation on naming.

According to the *Yangzi Evening News*, there are 1,601 surnames in China. If the policy goes into effect, approximately 1,280,000 new surnames will be created. This will reduce the amount of duplicate names and help to more clearly identify children's blood relationships.

The regulation states that new names must contain between two to six characters and must not damage the country's dignity, infringe on local folk customs, or have a negative impact on society. The names must use the simplified version of Chinese characters if they exist, and must not use eliminated variant characters, self-created characters, foreign names, or pinyin or Arabic numerals.

To prevent frequent name changes, the regulation would permit citizens 18 years and over to change their name (both surname and given name) only once. Those that change their names with fake certifications would be fined 800 yuan (approximately $105).

Chinese Naming Customs

Research shows that the Chinese naming convention stems from a matriarchal society where people formed clans around mothers. To differentiate clans, they used the mother's surname as the clan title. These surnames commonly contained the character component (radical) that stood for "woman."

In addition to this, the following factors are believed to have influenced the creation and use of Chinese surnames:

1. A forefather's home state or kingdom, such as the states of Zhao, Song, Qin, and Wu.
2. A forefather's title, such as the ancient official titles 'Sima' and 'Situ'.
3. A forefather's rank of nobility, such as 'Wang' and 'Hou'.
4. A forefather's given name.
5. An occupation, such as a pottery maker using the word for 'pottery' as a surname.
6. A place name or physical description of the location's scenery such as Dongguo, Ximen, Chi, and Liu.
7. Worship of animals like horses, cows, sheep, and dragons. Elements from these characters were mixed into surnames.

The book *One Hundred Surnames*, written during the Song Dynasty, includes about 500 surnames, of which 60 are compound surnames. According to statistics, there were about 5,000 surnames in ancient documents. Now, there are about 200 common surnames.

Ancient people's names were more complex than modern people's. Besides surnames and given names, additional words were used to identify one's position in the family hierarchy. People sharing the same hierarchy were required to have one identical word in their names.

People who were cultured or had social status had additional names such as 'Zi' (a courtesy name traditionally given to Chinese males at the age of 20, marking their coming of age), and 'Hao' (a self-selected courtesy name, usually referred to as the pseudonym).

For example, the Song Dynasty author Su Shi's surname was 'Su' and his given name was 'Shi'. In addition to this, he had 'Zizhan' as his 'Zi' name, and 'Dongpo Jushi' as his 'Hao' name. The Tang Dynasty poet Li Bai lived in Shichuan's Qinglian County. He thus often used an additional name 'Qinglian Jushi' (as his Hao name) to signify: "Person who lives in Qinglian."

Traditionally, many people also believed that names foretold one's future. Extensive research went into choosing a name based on date of birth and the corresponding eight characters of the Chinese horoscope.

August 1, 2007

Population Geneticists, Forensic Scientists Construct Genetic Map of Europe

A review by Nicholas Wade in the Science section of *The New York Times*, August 13, 2008, summarizes the recent work by Dr. Manfred Kayser and others published in *Current Biology* revealing the genetic relationship between 23 populations in Europe.

With few exceptions, the DNA landscape with its clusters and clines and genetic distances follows the political and linguistic boundaries of European countries. Finns were found to be very different from everybody else, and the Italian population was shown to be extremely stable over the ages. Who would ever want to leave Italy anyway once they arrived?

August 13, 2008

Elizabeth Rex

Like Egyptian Queen Hatshepsut, who wore the royal beard in portraits, Queen Elizabeth I was careful to refer to herself as a "prince," not "princess," and sign her name *Elizabeth R.* (the *R* standing for Rex, "king"). We recently watched, courtesy of Roku, all six episodes of the 1970-71 BBC mini-series "Elizabeth R.," starring the definitive Glenda Jackson. Elizabeth I, as one professor argues, posthumously founded the movie industry, beginning with a performance by Sarah Bernhardt.

We suggest she is also a key figure in understanding the crypto-Jewish strain in English royalty and aristocratic genealogies.

Consider, for instance:

• Elizabeth is not—or was not—a common Anglo-Saxon name. It is the Greek translation of the Hebrew name Elisheva, meaning "God's promise," "oath of God," or "I am God's daughter." Many Elizabeths in England and Scotland owe their naming directly or indirectly to St. Elizabeth of Hungary. Queen Margaret, the daughter of Edmund the Exile and granddaughter of Edmund Ironsides, initiated a flood of return Scotsmen to the British Isles in the 11th century. Cognates are Eliza, Elzina (Arabic) and many others.

- Elizabeth called Sir Francis Walsingham, her chief foreign policy adviser, "little Moor." He probably had Sephardic Jewish antecedents, accounting for his knowledge of Spanish and secret spy network.
- Elizabeth was a Boleyn (also spelling Bullen, Bullent, Bulleyn, Bullin, Bullon, Bollen, Boulin, Boullen, Bullan, Bullant, Bulein, Bolen, Bollyng, Bolling, etc.). Authoritative surname sources derive this name from Boulogne, a Norman town, or a Middle English word meaning "excessive drinking." It may also have come from Bolognia in Italy.
- The Tudors, who were Welsh, joined with the Stewarts, who were Scottish. Both families have many Jewish connections and attributes.

Elizabeth's counselors and favorites included: Francis Drake (whose Y chromosome line matches Ashkenazi Jews), John Hawkins (Sephardic surname), John Dee (professor of Hebrew and Cabala), John Frobisher (Forbes, Scottish Jewish), Admiral Sir Charles Howard, Walter Raleigh, Francis Bacon and others of their ilk. These mapmakers, privateers and promoters are discussed in Chapter 1 of the book *Jews and Muslims in British Colonial America.*

November 28, 2010

When Ireland Was Jewish

The royal mound cemetery at Taillten, modern Telltown, in County Meath, houses the burials of numerous kings and nobles from early Ireland. These begin with Ollamh Fodhla, whose death occurred in 1277 B.C.E., and run to just before Conchobar Mac Ness, who died in A.D. 33 according to the *Annals of Tighernach,* written in Old Irish and Latin in the early Middle Ages.

Pronounced "CON ah war," Conchobar is the first of the name Connor or O'Connor in Irish annals. His mother was Queen Ness, and his nephew Cuchulain, the famous hero of the Ulster cycle of stories. "Our oldest and most trustworthy authorities state that Taillten ceased to be used as a cemetery on the death of Conchobhor," wrote Irish antiquarian (the term used before "archeologist") William F. Wakeman in *The Handbook of*

Irish Antiquities in 1891, drawing on field reports dating back to 1848 (London: Studio, 1995).

What made Conchobar's burial unusual was that, unlike the previous kings of Ulster entombed at Taillten, he "wished that he should be carried to a place between Slea and the sea, with his face to the east, on account of the Faith which he had embraced" (p. 94).

It seems obvious that Conchobar converted from pagan druidism to a new religion, one that emphasized burial facing east. The new religion could not have been Christianity, although Irish myths claim, anachronistically, that Conchobar died upon being told by druids that Jesus had just been killed "by the Jews." Christianity was not widespread until the fourth century of the Common Era. Jews, like Christians, are buried facing east.   According to Celtic tradition, Conchobar was one of the two men who believed in God in Ireland before the coming of the Faith, Morann being the other man. Such a statement can only mean that Conchobar and his adviser Morann were monotheists or Jews.

In Conchobar's day, Judaism attracted millions of converts. Eventually, between one-tenth and one-quarter of the inhabitants within the limit of the Roman Empire professed Judaism. Often it was a syncretistic form combined with other rites and beliefs.

The story of a king converting to Judaism with many of his subjects and descendants following him is a phenomenon documented from more than one historical time point associated with mass conversions. The Hellenized Hasmonean dynasty established under Judas Maccabeus ruled the Kingdom of Israel for over a hundred years, spreading Jewish religion to many others in the Middle East. Bulan was a Khazar king who led the conversion of the Khazars to Judaism. The Babylonian prince Machir, also known as Todros, Theodore, Theodorich, Dietrich, William and by other names, converted most of the inhabitants of the kingdom of Narbonne/Septimania in southern France.

It is likely the Irish high king Conchobar inspired many of his people to accept Judaism. The introduction and early spread of Christianity in Ireland could have been facilitated by the pre-existence of Jewish institutions in the country.

King Ollamh Fodhla's Throne at Taillten, covered with tifinagh (North African) inscriptions. Conwell, On the Cemetery of Taillten (1879).

November 21, 2010

Should the DNA Marketspace Be Regulated?

In a paper to be delivered at the American Marketing Association's meeting in Washington in June, Elizabeth C. Hirschman estimates that the number of people who have purchased a DNA test now exceeds 1.5 million. Her work suggests that the value of the market (excluding paternity testing) in 2011 will reach nearly $150 million in sales.

That seems like too big an industry to escape government oversight, and it's true that several scientists have targeted the direct-to-the-consumer DNA testing business for criticism, particularly personal genomics companies like 23andMe.

Before another academic grant gets written to send out another industry questionnaire, however, marketing professionals and public-policy analysts ought to have a look at Hirschman's new case study, destined, we think, to become a classic.

"Altruistic Economics and Consumer Cooperatives in the DNA Marketspace" sketches a vibrant picture of DNA test takers busy following up on their results in social-networking sites like DNA Communities and even joining in the design process for product improvements by the leaders in the industry. No unhappy campers there!

The proof of the pudding is in the eating. Rather than mount yet another policy-making roundtable, would-be regulators should just order some of the DNA tests available from today's leading companies and judge for themselves how accurate or valuable or harmful they are. That makes a lot more sense than writing another food review for a restaurant they do not intend to patronize, or for a cuisine that is not to their taste.

The AMA's Marketing and Public Policy Conference is the premier national and international event for marketing academics, public-policy makers, and marketing practitioners interested in social and public policy.

Another point made by the paper is that "The industry has completed the introduction, early growth stages and consolidation phase of its life cycle. ... It is a mature field facing few new technology thresholds, and it is still very much confined to the United States, Canada and England."

All that having been said, it may be too late to regulate the industry. It seems to be doing fine all by itself. Like the pharmaceutical and computer industries, the DNA marketspace is an American phenomenon we should all just basically let thrive and be proud of.

March 8, 2011

Sailing to Byzantium

Claudio Ottoni et al., "Mitochondrial Analysis of a Byzantine Population Reveals the Differential Impact of Multiple Historical Events in South Anatolia," *European Journal of Human Genetics* 19 (2011) 571-76.

Abstract. The archaeological site of Sagalassos is in Southwest Turkey, in the western part of the Taurus mountain range. Human occupation of its territory is attested from the late 12th millennium B.P. up to the 13th century A.D. By analysing the mtDNA variation in 85 skeletons from Sagalassos dated to the 11th–13th century A.D., this study attempts to reconstruct the genetic signature potentially left in this region of Anatolia by the many civilizations, which succeeded one another over the centuries until the mid-Byzantine period (13th century). Authentic ancient DNA data were determined from the control region and some SNPs in the coding region of the mtDNA in 53 individuals. Comparative analyses with up to 157 modern populations allowed us to reconstruct the origin of the mid-Byzantine people still dwelling in dispersed hamlets in Sagalassos, and to detect the maternal contribution of their potential ancestors. By integrating the genetic data with historical and archaeological information, we were able to attest in Sagalassos a significant maternal genetic signature of Balkan/Greek populations, as well as ancient Persians and populations from the Italian peninsula. Some contribution from the Levant has been also detected, whereas no contribution from Central Asian population could be ascertained.

A recent comparison of medieval mitochondrial DNA from a Byzantine cemetery with modern populations in Southwest Turkey shows what we have assumed in our population analyses house in our database atDNA. The integration of historical with archaeological information proves that the little South Anatolian town of Sagalassos has a clearly structured Balkan/Greek maternal population with some ancient Persians and Italians in the mix but no Central Asian (Turkic) contributions discernible.

The inference is that when the Turks conquered Anatolia and eventually took control of the Byzantine capital (modern-day Constantinople) they remained largely a ruling class with little penetration into the ancient settlements scattered through Turkey. Even though the general populace accepted their conquerors' religion, Islam, their bedrock DNA did not significantly alter, at least not in the female lines. The male lines may tell a different story.

April 21, 2011

Two Days Too Late

Nancy J. Cooper et al. v. The Choctaw Nation is one of the classic botched cases in the annals of the Dawes Commission, the Federal government's attempt to deal a death blow to tribal sovereignty at the close of the 19th century. I had heard rumors about my Cooper relatives and how they were kicked out of the Choctaw Nation. But I never knew the whole story until recently.

J.W. Howell mentions the case in a textbook studied today in law schools. John Cooper, our ancestor, was a Choctaw chief who owned a plantation near Linden, Tennessee. The family was seated at the dinner table one evening when a vigilante mob broke in. They were told at gunpoint to leave, their possessions forfeit.

The men swam their horses across the Mississippi River at Memphis and left the women encamped under the willows on the other side while they went back to try to recover some of their cattle. John and Nancy's old mother, who was in her 80s, died before they returned, empty-handed. The party proceeded to Indian Territory.

In 1896, the family encouraged Nancy, blind, unmarried and no longer able to care for herself, to enroll with the Choctaw Nation. They and a large group of kinsmen won roll numbers. But they were all stricken from the rolls by an adverse decision of the Choctaw-Chickasaw citizenship court a couple of years later. More than a hundred of them joined in a class-action suit.

"We're still fighting it," says Pam Kahler of Vian, Oklahoma. "My husband and I talked to the BIA in Muskogee and found out about the old ruling. They told us the reason it was overturned was because the people named in the court ruling were not living in the Choctaw area when they were added to the Dawes rolls." They, in fact, were living in the Chickasaw area of Duncan, Comanche area, Stephens County.

Aunt Artie Meecie was told that the family was "too poor to be on the rolls."

In February and March 1907 matters came to a head. The attorney general of the United States declared the lower courts out of line and ordered that hundreds of Choctaw Coopers, Browns and others were, after all, entitled to enrollment.

The only trouble was that the attorney general's decision of March 4, 1907, did not reach the department until March 6, 1907, two days after the rolls were closed by operation of law. There was then no authority in the Office of the Secretary of the Interior, under the law, to enroll them.

Nancy Cooper was laid in a pauper's grave. Not only was the family too poor to be Indian, it was two days too late.

June 22, 2011

Sorbs Probably Not "the" Core Ashkenazi Jewish Population

It used to be thought that the Sorbs, a medieval East German Slavic population, formed the core of a group of Jews who ended up as the dominant element of Ashkenaz, Eastern Jews in Poland, Lithuania, Ukraine, Silesia and Russia. No longer.

According to an article published in *European Journal of Human Genetics* (2011) 19, 995-1001, "Genetic Variation in the Sorbs of Eastern Germany in the Context of Broader European Genetic Diversity," by Veeramah et al., the Sorb "population isolate" is not that isolated but was "less than that observed for the Sardinians and French Basque," true isolates. Sorbs are, in reality, part of the larger population of West Slavs, including Poles and Czechs, and scarcely distinguishable from them.

Histories of Judaism will have to revise their accounts of the genesis of Ashkenazim accordingly. The story of European Jews is more diversified and dynamic than most history or coffee-table type books allow.

For an example of a research article on Ashkenazi roots overemphasizing Sorbs, see Behar et al. "Multiple Origins of Ashkenazi Levites: Y Chromosome Evidence for Both Near Eastern and European Ancestries," *American Journal of Human Genetics* 73:768–779, 2003, who theorizes they were a leading constituent of Levites.

August 18, 2011

Why Italians Live So Long

The Genetic Component of Human Longevity: Analysis of the Survival Advantage of Parents and Siblings of Italian Nonagenarians
Alberto Montesanto, Valeria Latorre, Marco Giordano, Cinzia Martino, Filippo Domma and Giuseppe Passarino
European Journal of Human Genetics 19 (2011) 882–886;
doi:10.1038/ejhg.2011.40; published online 16 March 2011.

Abstract. Many epidemiological studies have shown that parents, siblings and offspring of long-lived subjects have a significant survival advantage when compared with the general population. However, how much of this reported advantage is due to common genetic factors or to a shared environment remains to be resolved. We reconstructed 202 families of nonagenarians from a population of southern Italy. To estimate the familiarity of human longevity, we compared survival data of parents and siblings of long-lived subjects to that of appropriate Italian birth cohorts. Then, to estimate the genetic component of longevity while minimizing the variability due to environment factors, we compared the survival functions of nonagenarians' siblings with those of their spouses (intrafamily control group). We found that both parents and siblings of the probands had a significant survival advantage over their Italian birth cohort counterparts. On the other hand, although a substantial survival advantage was observed in male siblings of probands with respect to the male intrafamily control group, female siblings did not show a similar advantage. In addition, we observed that the presence of a male nonagenarian in a family significantly decreased the instant mortality rate throughout lifetime for all the siblings; in the case of a female nonagenarian such an advantage persisted only for her male siblings. The methodological approach used here allowed us to distinguish the effects of environmental and genetic factors on human longevity. Our results suggest that genetic factors in males have a higher impact than in females on attaining longevity.

We just returned from a long trip through Italy and were struck by Italians' apparent immunity to all the forces of aging that plague Americans and other members of the First World. "Italian men," said Paolo, our driver, "smoke, drink, womanize and curse all day and live to a hundred." Maybe the answers why are in this new report on Italian longevity.

August 5, 2011

Skeletons of the English Peoples
British Bones Push Back Date for "First Anatomically Modern Human" in Northwestern Europe: A Missing Link from Kent's Cavern in Devonshire

A prehistoric maxilla (upper jawbone) fragment was discovered in the cavern during a 1927 excavation by the Torquay Natural History Society, and named *Kents Cavern 4*. The specimen is on display at the Torquay Museum.

Although previous radiocarbon dating suggested the bone was about 35,000 years old, a new study in *Nature* redates it securely to 44.2-41.5

kyr. The article by Tom Higham et al., "The Earliest Evidence for Anatomically Modern Humans in Northwestern Europe," also claims that on the basis of dental comparisons it is "human" rather than "Neanderthal."

The Kent's Cavern fragment "therefore represents the oldest known anatomically modern human fossil in northwestern Europe, fills a key gap between the earliest dated Aurignacian remains and the earliest human skeletal remains, and demonstrates the wide and rapid dispersal of early modern humans across Europe more than 40 kyr ago."

A related article in the same issue of *Nature* is "Early Dispersal of Modern Humans in Europe and Implications for Neanderthal Behavior," by Stefano Benazzi et al. It attempts to place the so-called Cavallo fossil from southern Italy in a time frame of about 44,000 years ago, thus suggesting a "rapid dispersal of modern humans across the continent before the Aurignacian and the disappearance of Neanderthals."

Neither study considers that the evidence they are examining may be the result of hybridization between "humans" and "Neanderthals." As most geneticists, the authors have rigid categories and do not consider that our definitions of species and sub-species and transitions in technocomplexes and traits are in flux as new discoveries are made.

One man's Mede may be another man's Persian, and we note that the "fossil race" is not devoid of scientific jingoism pitting one country's news-making finds against another's. So far, England seems to be winning.

However, the British still have to live down Piltdown Man, a fraud of biblical proportions that fooled the world for almost half a century until the 1950s. The Piltdown hoax is, perhaps, the most-famous paleontological hoax ever. It has been prominent for two reasons: the attention paid to the issue of human evolution, and the length of time that elapsed from its discovery to its full exposure as a forgery combining the lower jawbone of an orangutan with the skull of a fully developed modern human.

The editors sum up the two new studies by writing, "The reanalysis of findings from two archaeological sites calls for a reassessment of when modern humans settled in Europe, and of Neanderthal cultural achievements." We wish that the paleontological community would think more out of the box and reassess how, when and where "humans" and "Neanderthals" interbred.

November 23, 2011

When Wales Was Jewish

Short Answer: Pre-Roman Times

As is well-known, haplogroup E1b1b1 accounts for approximately 18 percent to 20 percent of Ashkenazi and 8.6 percent to 30 percent of Sephardic Y-chromosomes. This North African type appears to be one of the major founding lineages of the Jewish population.[1]

In Britain, this quintessential Jewish type (together with J, another telltale sign of Middle Eastern roots) is absent or negligible in many towns and regions but reported in elevated frequencies in Wales (Llanidloes 7 percent, Llangefni 5 percent), the Midlands (Southwell, Nottinghamshire 12 prcent, Uttoxeter 8 percent), Faversham in Kent (9 percent), Dorchester in the West Country with historic harbors (7 percent), Midhurst in West Sussex commanding ancient sea-ports (5 percent) and the Channel Islands, always an important crossroads of influences (5 percent).[2] Bryan Sykes' survey of paternal clans in England and Wales confirms significant traces of the E haplogroup that he dubs Eshu in southern England (4.9 percent) and Wales (3.1 percent).[3] It reaches its highest point in Britain in Abergele, Wales (nearly 40 percent), an anomaly that has been attributed to Roman soldiers of Balkan origin but may have alternative and more complex explanations. See our blog post "Right Pew, Wrong Church," arguing that the footprint of E in Britain is attributable to North African influence, not the descendants of Roman legionnaires from the Balkans.

In 2011, Llangefni and Wrexham in North Wales became the focus of a call for local men to provide samples of their unusual DNA. A team of scientists led by Andy Grierson and Robert Johnston from the University

---

[1] A. Nebel et al, "The Y Chromosome Pool of Jews as Part of the Genetic Landscape of the Middle East", *American Journal of Human Genetics* 69.5(2001) 1095–1112.

[2] C. Capelli et al, "A Y Chromosome Census of the British Isles," *Current Biology* 13 (2003) 979–984.

[3] Bryan Sykes, *Saxons, Vikings and Celts* (Norton: 2007) 206, 290.

of Sheffield hoped to link the migration of men from the Mediterranean to the copper mined at Parys Mountain on Anglesey and on the Great Orme promontory nearby. A preliminary analysis of 500 participants showed 30 percent of the men carried E1b1b, compared to 1 percent of men elsewhere in the United Kingdom.[4]

Significantly, Welsh tradition associates the Iron Age hilltop town on Conwy Mountain known as Castell Caer Seion with a settlement of ancient Jews. This site overlooks Conwy Bay on the north coast of Wales and lies on the ancient road between Prestatyn in Denbighshire and Bangor in Gwynedd opposite Anglesey. In the Black Book of Caermarthen, the Welsh national bard Taliesin casually remarks in the persona of the battling hero,

> When I return from Caer Seon,
> From contending with Jews,
> I will come to the city of Lleu and Gwidion.[5][v]

Lleu and Gwidion are the names of two other legendary figures; they are believed to be historical and to have lived in the early centuries of the Common Era or anterior to it.

It is hard to avoid the thought that the hilly area to the west of the town of Conwy, in North Wales was once inhabited by Jews.

April 2, 2012

Stephen Blevins commented on 03-Apr-2012
My DNA is E1b1b1, my most distant ancestor is William Blevins (Longhunter) from the area you mentioned. My autosomal DNA places my ancestors in the Orkney islands of Scotland. I'm convinced that a tribe of Jews migrated from Israel to north to Scandinavia or Denmark

---

[4] "'Extraordinary' Genetic Make-up of North-east Wales Men," BBC News North East Wales, article retrieved Jan. 2012 at http://www.bbc.co.uk/news/uk-wales-north-east-wales-14173910. On Dienekes' Anthropology Blog there is speculation about whether the main sub-clade involved is Balkan or North African E; posts and comments retrieved Jan. 2012 at http://dienekes.blogspot.com/2011/07/eastern-mediterranean-marker-in.html.
[5] William F. Skene, *The Four Ancient Books of Wales* (Edinburgh, 1868, republished 2007 by Forgotten Books) 206.

and may have been a part of the invasion by Vikings to Scotland before they were found in Wales as Poweys in the Northern Mountains. Blevins comes from Blethyn meaning little wolf or (Hero) look up Ap Blethyn of Gwynedd.

Blevins Descendant commented on 12-Apr-2012
I was always told the Blevins came from Wales, but in checking this story out I was unable to verify it, nor could I find any substantiation of the etymology from Bleddyn ("son of wolf"). There is not a single Blevins in the Welsh census records, although the name is found sparsely in Cheshire, Lancashire and other northern English counties. "Formby, Wales" is actually Formby in Merseyside in Lancashaire. The -dd-element in the Welsh name Bleddyn cannot be twisted into a -v-. So, go figure.

Paul commented on 28-Apr-2012
My mother is a descendant of Henry Cook I of Devon. His ascendants were among the first settlers of Massachusetts and Connecticut. A great-uncle, Lemuel Cook, was the oldest surviving Revolutionary War veteran when he died at 106 years of age. We recently had my mother's autosomal DNA analyzed and found strong population matches from the Balkans (Croatia, Bosnia, Macedonia, Serbia, etc.)—which was very unexpected. There was also prominent representation from Spain and Portugal—not so unexpected. In my own 18 marker test, I had one Jewish III marker, though I can't say from whom. There is no known Judaism on either side. Sounds like your article might be describing the early Cooks. Interesting ...

Katarina Cadieux commented on 13-Dec-2013
Well, the language of the Welsh (Cymri) alone is very Hebraic. Here are some examples: Anudon (Welsh)/Aen Adon (Hebrew) (without God); Yni all sy-dda(Welsh)/Ani El Saddai (Hebrew) (I am almighty God); Llai iachu yngwyddd achau ni (Welsh)/Loa yichei neged acheinu (Hebrew) ("Let him not live before our brethren"); An annos (welsh)/ain ones (Hebrew) (None did compel). The amazing thing to me is how similar the words look and sound, the English is the meaning for both

Welsh and Hebrew, their meaning are the same. The Welsh are a very ancient people even their name for themselves in their language has Crimea roots that many Hebrew tribes migrated to.

Dafydd Gwilym W. Gates commented on 17-Feb-2014
Katarina Cadieux (13 Dec 2013) wrote some examples to show how Welsh had parallels in Hebrew. I'm a first language Welsh speaker and couldn't make sense of the Welsh examples, I'm afraid, it wasn't Welsh. So sorry, Dafydd

Where Do I Come From?

As soon as Euro DNA was released last month, I quickly studied my new list of European nationalities where I have significant ancestral lines according to DNA Consultants' new autosomal population analysis. I had come to know and accept, of course, the usual suspects, compiled from the 24 populations available from ENFSI (European Network of Forensic Science Institutes). But the new list represented 71 populations and far surpassed ENFSI or any other database in commercial use. It had, for instance, the first European comparisons for countries like Hungary, Lithuania, Malta and Iceland. So, how would my familiar matches—Scotland, Ireland, England, Belgium and the rest—shake out in the new oracle?

Some of the top matches—above British Isles or Northern European ancestry—were Central European. Here were the top 20:

Rank	European Population Matches
1	Slovakia—Saris (n=848)
2	Finland (n = 469)
3	Slovakia—Zemplin (n=558)
4	Netherlands (n = 231)
5	Slovakia—Spis (n=296)
6	Romanian - Transylvanian - Szekler (n = 257)
7	Romanian - Transylvanian - Csango (n = 220)
8	Scotland/Dundee (n = 228)
9	Switzerland (n = 200)

10   England/Wales (n = 437)
11   Ireland (n = 300)
12   Italy (n =103)
13   Denmark (n = 156)
14   Romanian (n=243)
15   Swedish (n = 311)
16   Serbian - Serbia / Vojvodina / Montenegro (n = 100)
17   Icelandic (n=151)
18   Estonia (n = 150)
19   Romanian - Transylvania/Banat (n = 219)
20   Norwegian (n=1000)

Slovakia? Romania? To be sure, I had always had a fascination with both countries. But why were these matches so prominent? Admittedly, the results of an autosomal ancestry test are cumulative and combinatory. While they *do* reflect all your ancestry, as no other test can, you are cautioned not to use the matches to try to pinpoint lines in your family genealogy. There is always a temptation to over-interpret.

My European admixture results from AncestryByDNA had yielded a confirmatory result: 20 percent Southeast Europe. That struck me at the time as odd. Yet, Hungarian was now one of my top metapopulation results, too. (Remember, Hungarian data did not figure into ENFSI because Hungary is not in the European Union.)

The Scottish (my grandmother was a McDonald) made sense, as did all the other matches from what I knew through years of paper genealogy research. But I was unaware of any strong Central European lines.

Sizemore Research: Pitfalls of Genetic Genealogy

Then I recalled the Sizemores. My great-great-grandmother was a Sizemore, and they were multiply connected with my Coopers, my mother's maiden name. Could the Central European effect in my Euro result be from the Sizemores?

Much ink—or at least many keystrokes—has been expended on the Sizemore controversy. There are pitched battles on genealogy forums and edit wars in cyberspace. One armed camp has them down as Melungeons and admixed Cherokees with crypto-Jewish strains. Another holds it as an article of faith that the Sizemores were a lily-white old Virginia British family and the surname comes from something like Sigismund (think Goetterdaemmerung). Y chromosome DNA shows ambiguous conclusions: You can visit the advertisement page sponsored by Family Tree DNA.

Alan Lerwick, a Salt Lake City genealogist, upset the apple cart some years ago by linking America's Sizemores to Michael Sismore, buried in the Flemish cemetery of the Collegiate Church of St. Katherine by the Tower in London in 1684. That was the same parish as my Coopers lived in. Then and now, it is the most Jewish section of London. Sizemore is not a British surname before the 16th century. It was clear to me long ago that neither my Sizemores nor my Coopers were Mama Bear, Papa Bear families. Spurred by my Euro DNA test results, I dug into my subscription at Ancestry.com and learned that Michael Sismore was recorded as being born as Michael Seasmer in Ashwell, an important village in north Hertfordshire, November 1, 1620. His parents were Edward Seasmer and Betterissa (a form of Beatrice). New information! Alert the list moderators and surname project guardians!

Seasmer is undoubtedly the same as Zizmer, an old Central European Jewish surname adduced in multiple families in Israel, Romania, Czechoslovakia, Germany, Austria, Russia, Moldavia and the United States. Edward and Michael are favored first-names in the U.S. branches. The Hebrew letters, which can be viewed on numerous burials in Israel, are (in reverse order, right to left) RMZZ. Cooper is a similar Jewish surname, common in Russia and Lithuania and Israel as well as the British Isles and the U.S. In fact, my father's surname, Yates, is a Hebrew anagram common in the same countries, meaning "Righteous Convert."

Hertfordshire was an important center for British Jewry, mentioned in the works of Hyamson, Jacobs and others. A good hypothesis to explain the transformation of Michael's name from Seasmer to Sismore and thence to Sizemore is this. His grandfather, a Zizmer, came to England in the time of Elizabeth, perhaps via the Low Lands, possibly as a soldier or cloth merchant. This could account for Michael Sizemore's burial in the Flemish cemetery of St Katharine's by the Tower, usually reserved for foreigners. It also explains the predilection in descendants for such names as Ephraim, Michael, Edward, William, John, Richard, James, George, Hiram, Isaac, Samuel, Solomon and Henry. And why girls were named Lillie, Lydia, Louisa, Naomi, Pharaba, Rebecca, Sarah and Vitula. The last name (also found in my wife's grandmother's name) was a Jewish amulet name. It meant "old woman" in Latin and was given to a child to augur a long life.

Zismer took the form of Cismar, Cismarik, Zhesmer, Zizmor, Ziesmer, Zausmer, Cismaru and Tzismaro—all amply attested in the records of European Jewry, including Jewish Gen's Holocaust Database, with the records of over 2 million victims and survivors of the Nazi genocide of World War II. I am proud of my Jewish heritage through my great-grandmother and through my halfblood Cherokee Indian mother Bessie Cooper Yates.

Thank you for indulging me in this genealogical excursus on a family mystery. Like a restaurant owner, I would be to blame if I didn't eat in my own establishment.

July 28, 2013

13 DNA AND MONEY

Breakthrough of the Year: Human Genetic Variation

Elizabeth Pennisi
Science 318/5858 (21 December 2007) 1842-43

"Equipped with faster, cheaper technologies for sequencing DNA and assessing variation in genomes on scales ranging from one to millions of bases, researchers are finding out how truly different we are from one another."

Genetics News Highlight of 2007
According to the Editors of *Nature*
First whole human genome decoded

James Watson, co-discoverer of the structure of DNA, and genomics pioneer Craig Venter announced that their full genomes had been sequenced. The achievements were the first in an anticipated wave of personal-genome sequencing and crucial steps toward personalized medicines tailored to an individual's genetic makeup. Watson's genome, announced in June, was analysed by Connecticut company 454 Life Sciences' new rapid-sequencing technique. Venter's, published in September, was the first fully sequenced diploid genome—detailing DNA inherited from both parents—and revealed that human genetic variation is greater than previously thought.

In November, Google-backed Californian biotech firm 23andMe

launched a $1,000 personal-genome service; the same month that Icelandic deCODE genetics offered DNA testing for disease-linked genes for the same price. And 2007 saw a splurge of research papers from genome-wide disease-association studies, including diabetes and cancer.

December 24, 2007

From the Human Genome to the Personal Genome

Innovations in DNA sequencing and genotyping are opening doors for personal genomics. Science writer Nathan Blow in *Nature* 449/7162:627 explores these technological advances and their implications.

Related articles review the state of the science of personal genomics, including economic feasibility and emerging technology that suggest it may soon be possible to sequence your own entire genome. Such an event could be the precursor to hundreds of gene-screening tests within the reach of consumers, inaugurating a whole new era in DNA testing for ordinary people.

According to Harvard biology professor Richard Lewontin, because the Human Genome Project was completed by 20 or more research institutes and much of it was "fill-in-the-blanks" computer guesswork, the Human Genome we now have is an artificial one corresponding to no living individual. Only by sequencing real human subjects and describing the vast variety of human genetics will medical science be able to achieve its aim of discovering the genetic basis of disease. This has been heralded as Phase II of the Human Genome Project, although it has attracted little funding (unlike Phase I).

October 15, 2007

Economies of Sale
High Density SNP Testing Is Transferred from Medicine to Forensics

An article titled "DNA Has Nowhere to Hide" in October's *Genetics* (published by *Nature* magazine) describes how the technology of high-density SNP genotyping that has opened up vast horizons in genome-wide association studies is now transforming forensic science.

The new techniques can detect trace quantities of a suspect's DNA in a complex mixture containing hundreds of other individuals, substituting a rifle for what was previously a shotgun approach to crime-scene evidence. High-density SNP testing uses allele profile frequencies—somewhat as a DNA fingerprint and population statistics are used to infer ethnicity in our DNA Fingerprint Test.

October 6, 2008

Google Wants to Do What?

Google Wants to Index Your DNA, Too
Marjorie Backman
Business Week Online, 4/21/2008, p.7.

Google is investing in genetic-screening companies, notably Navigenics, backed by start-up venture capital firm Kleiner, Perkins, Caufield & Byers. Navigenics conducts genetic tests to determine a person's vulnerability to 18 diseases. Google has pumped $4.4 million in 23andMe.

April 21, 2008

Iceland's deCODE Defunct

Nature 462/401
Published online 23 November 2009

In a report by Erika Check Hayden, the journal *Nature* gloats that the innovative personal-genomics company deCODE Genetics has gone out of business, leaving the disposition of valuable genetic data unclear.

"After struggling financially for years, the genomics company deCODE, based in Reykjavik, Iceland, filed for bankruptcy on 16 November," wrote Hayden, who follows the genealogy-and-genetics business beat for *Nature*. "The question now is whether other companies looking to commercialize genomics will follow the same path."

DNAPrint of Sarasota, Florida, went down that path last February without even an obit in scientific journals.

But according to Kari Stefansson, deCODE's CEO, the fate of the data never was in play since it belonged to individuals who had their DNA tested at their own expense with the service lab of deCODE. The lab, Islensk Erfdagreining, continues to operate today "under the same data and privacy protections as ever, rooted in the Icelandic community and within a tried and tested regulatory environment," wrote Stefansson in a comment on the online report by *Nature*.

Such an accidentally-on-purpose misunderstanding is more than sloppy science journalism or bad science. It reveals the fundamental hostility of academic geneticists and related disciplines to commercializing or even popularizing DNA. Geneticists should stop thinking they are doing God's work. They should give up the illusion that the great generality of humankind can only understand, profit from and benefit from their work if they, the scientific intelligentsia, condescend to allow it, controlling the conditions and goals of its use.

November 28, 2009

The $1,000 DNA Test

"Complete Genomics is about to release fast, cheap sequencing into a competitive market."
Erika Check Hayden
Published online 6 October 2008 in *Nature*

The era of the $1,000 genome has arrived. That's the claim from Complete Genomics, a young company based in Mountain View, California, that revealed today that it plans to sequence 1,000 human genomes next year—and 20,000 in 2010.

October 6, 2008

15th Anniversary of New Genome Sequencing

At a time when it seemed that American science had bitten off more than it could chew with the Human Genome Project, Craig Venter and his innovative company published "A New Strategy for Genome

Sequencing." Appearing in the journal *Nature* in 1996, the Venter multicenter approach bypassed laborious gene mapping and allowed the HGP to meet its goal of full sequence information on the human genome in 2000.

"In the race to sequence the human genome," write the editors of *Nature*'s DNA Technologies Milestones, "research groups had to choose between the random whole genome shotgun sequencing approach or the more ordered map-based sequencing approach."

The choice of randomness versus order was present from 1982, but the Venter strategy was resisted for many years. Finally, in 1996 it was accepted and given an equal emphasis with the more-orthodox approach.

After a standoff between the two groups of scientists, "a showdown ensued, with the biotechnology firm Celera Genomics wielding whole-genome shotgun sequencing and the International Human Genome Sequencing Consortium wielding map-based sequencing. Yet when the dust settled, it was a draw—both groups published their initial drafts of the human genome concurrently in 2001."

The maverick technology helped make high throughput genomic sequencing at commercial labs an economy reality and gave birth to a range of new DNA tests within the reach of ordinary consumers like you and me. Today, 15 years later, those interested in autosomal ancestry testing and personal genomics have biologist and entrepreneur Craig Venter and his irascible persistence as a scientific pioneer to thank.

November 21, 2011

Gazing into the Crystal Ball
Genome-Sequencing Anniversary

The Golden Age of Human Population Genetics
Molly Przeworski
Science 331/6017 (4 February 2011) 547
doi: 10.1126/science.1202571

The first draft of the genome provided the road map for the past decade of research in human genetics, allowing for the design of platforms that have been used to query variation in populations worldwide and helping to drive down the cost of sequencing by several orders of magnitude.

Within years, tens of thousands of complete genome sequences will be available from humans and from extinct hominids, as well as from thousands of other species. Given the human mutation rate, we will soon know of variation among individuals at almost all sites in the genome.

For population genetics, this ushers in a previously unimaginable opportunity to reconstruct the entire genealogical and mutational history of humans and pushes us against the limits of what we will be able to infer about the evolutionary and genetic forces that affected every region of the genome. Why are disease mutations present in human populations? What is the genetic basis of our cognitive and physiological adaptations? What was the sequence of demographic events that led to the colonization of the globe by modern humans? Stay tuned, and before long, we should know as much as genetic data alone can tell us.

Yes, we've heard exalted claims before, like 10 years ago, when the next phase of the Human Genome Project was to be devoted to the "conquest" of disease. How many diseases have been conquered in 10 years, after billions of research dollars? Guess. None.

And as far as population genetics goes, the whole story of "classic" Darwinian evolution seems to be unraveling before our eyes with every passing month (except of course in textbooks and the creationist opposition, where it never changes). If we can't be sure about evolution, how can we decide what is true about early human migrations?

February 5, 2011

Ancestry Painting and Global Similarity

We were surprised to see what DNA testing companies are calling their autosomal products these days. Ours is the DNA Fingerprint family of products, but 23andMe calls their entry "Ancestry Painting and Global Similarity" and "Personal Genome Service." Others offer "Genetic Ancestry Analysis," "Family Finders," and "Ancestral Origins."

Before the introduction of DNA fingerprinting for ancestry purposes, DNA testing was limited to the father's male line or mother's mitochondrial lineage. Newer autosomal tests can be taken by a male or female. They analyze all your lines at once, not just the two traditional ones of genetic genealogy. Autosomal DNA is the great equalizer, but it's not being marketed very adroitly.

A beginner's class titled "Ancestry Tracing with an Autosomal DNA Test: Conceptions and Misconceptions" will be presented by DNA Consultants principal investigator Donald N. Yates, Ph.D., at the upcoming Arizona Family History Expos on January 22, 2011. It will provide an overview of autosomal DNA tests that are capable of yielding a more-complete picture of your family tree and its roots. Covered are the science and history of DNA fingerprinting, what markers are delineated, the databases used for finding matches, and methods and strategies for interpreting your results, including follow-up websites, social networking and readings.

December 4, 2010

A Biography of DNA

Book Review: *The Language of Life by Francis S. Collins*
Nature 463 (21 January 2010) 298-299

Abdallah S. Daar in reviewing this new book by National Institutes of Health director Francis Collins maintains "we have entered the era of rapid, inexpensive genetic testing and genome sequencing" and must simply come to terms with the phenomenon of personal genomics and consumer genetics. In the next decade, he predicts, the price-tag for sequencing a human genome will drop to a few hundred dollars. The cost for the Human Genome Project was about $3 billion over 13 years.

Collins' book is *The Language of Life: DNA and the Revolution in Personalized Medicine* (Harper/Profile: 2010).

February 3, 2010

Yes, We Have No Patents

Can someone patent your DNA? Not anymore. It sounds Kafkaesque, but this practice has been business as usual in certain companies for three decades according to Jesse Holland in his AP article, "Court says Human Genes Cannot Be Patented."

This month the Supreme Court unanimously voted to deny patents on unaltered human DNA. The action invalidates more than 5,000 patents.

Yet, the court opted to approve patents on genetically altered DNA. Just what is the difference, and what does it mean for consumers?

Justice Clarence Thomas wrote that "a naturally occurring DNA segment ... is not patent eligible merely because it has been isolated" ("Supreme Court Bars Patents on Unaltered DNA" by Dianna Stafford). In other words, finding something in nature is not the same as inventing something.

According to Holland the court has said, "Laws of nature, natural phenomena and abstract ideas are not patentable."

Imagine that the first person that found a banana got a patent on it. Forget banana cream pie, banana pudding, or banana and peanut-butter sandwiches. Bananas would be exorbitantly expensive. Why? No one else would be allowed to grow bananas because that person would have a monopoly. It would be, "Yes, We Have No Bananas," for real.

Remember Angelina Jolie? We weren't just shocked that she had a preventative, double mastectomy by choice. We were shocked at the sticker price for the test that determined that she had a faulty gene and had inherited a higher genetic predisposition for breast cancer. Some $3,000. The test was too expensive for most women to even consider. And why? According to Holland, the company sells the only BRCA (diagnostic) gene test because it holds the patent and can keep the cost prohibitive." This means "their patents allowed them to have a monopoly on genetic testing for hereditary breast and ovarian cancer."

In other words, no one else could sell you this test. This company is not the only one that has such a patent, but it has been the deciding one for this case. Jason Koebler quotes Larry Brody of the National Institutes of Health's genome-technology branch as saying "the ruling will make the field of diagnostics, at least for breast cancer, much more open."

As the case was brought on by the American Civil Liberties Union, he also quotes Sandra Park, of the ACLU's Women's Rights Project as saying, " Today, the court struck down a major barrier to patient care and medical innovation ... because of this ruling, patients will have greater access to genetic testing and scientists can engage in research on these genes without fear of being sued."

And according to Stafford, biotech-research companies will continue to thrive as the court has ruled that so-called complementary DNA, or "cDNA," which is manipulated in laboratories, "is patent eligible because it is not naturally occurring." She quotes Meeta Patnaik, a life-science diagnostics consultant working with Kansas City-based MRIGlobal and other clients on the greater importance of synthetic DNA for biomedical research:

In the long run, patentability of synthetic or cDNA may be more important for the continuation for biomedical research ... Myriad actually was one of the few with patents on actual genes with no manipulation ... Most other patents are method or process patents that include manipulation of the genes. Advances in technology already were enabling other manufacturers and developers to overcome these patent issues, so the impact is not as significant as it could have been.

In other words, the future has arrived. And, yes, it has more genetic-testing choices for consumers and patients.

June 21, 2013

The Remaining 10 Percent
Human Genome Was Sequenced, Right?

Well, not completely. According to Larry Moran, a professor in the department of Biochemistry at the University of Toronto, "We can say that only 90 percent of the human genome has been sequenced and the remaining 10 percent falls into 357 gaps scattered throughout the genome."

Read all the numbers at Moran's blog "Sandwalk—Strolling with a Biochemist."

February 6, 2012

DNA: Storage Medium of the Future
One strand would hold libraries of digital information

The next decade's version of Facebook, Twitter, or Pandora could be digitally encoded on DNA. How? The next app? A card in one's wallet? Who knows?

This is not a new idea. It has been suggested before that entire libraries of information could be encoded on DNA and others have tried it without success. But no one thought it was possible until, according to an article in *Nature* by Monya Baker, "DNA Data Storage Breaks Records," reported that a team of Harvard researchers encoded a book on DNA and said there was plenty of room for more. A lot more. The possibilities could be limitless.

Cells die and replicate, so using cells is not a very secure system. The team, led by George Church, a synthetic biologist from the Medical School in Boston, devised a new system using ink to convert the data onto glass chips instead of cells. It worked. Of course, we're a long way from understanding this completely. And there are problems. It isn't feasible at the moment to encode an entire library onto DNA. It is far too costly. You can't use DNA to drive your computer instead of your hard drive. But it could happen. At first, many thought no one would want to drive a car, fly in an airplane, or even use a computer. In a similar manner, this could be the first glimpse of ushering in a new era just as all those were. How exciting!

But what would be the advantages? Durability and security of limitless information. Imagine what we would know if we could still go to the Library of Alexandria. Even today, beyond the everyday virus, there are talks of cyberwars and cyberthreats that could potentially shut down vast data and threaten the security not just of your personal computer but of nations. What if we no longer had to worry about those things? How would that change our world?

Also, if DNA can store this much information, what is being stored on it now that we don't even know about? What libraries of information are on our own DNA that we have yet to read? What will it tell us about ourselves in the future that we have yet to discover?

December 27, 2013

An Autosomal DNA Test Reflects How Unique You Are

It is great to be a part of any team, whether it is playing football for the Dallas Cowboys, or just working with your co-workers at work, but you are also an individual. In the same way, you may have patriotic pride in knowing you are an American, or lay claim to roots from any other country, but that isn't the whole story.

For instance, you are more than just an American if you come from

America. America has long been known as the melting pot of nations. In fact, 99.9 percent of us with the exception of some members of isolated tribes, like the Hopi and the Seminole, have ancestors from all around the globe. An autosomal DNA test will reflect how unique an ancestry you have and show you your top matches to specific regions. What objections could there be to that? Here are some answers to the most common ones:

"We are all related anyway if you go back to Adam." This argument is only half true and is too simplistic an analysis. It reflects the idea that we are all mixed, and while that is mostly true, the implication that we all have the same ancestry is not true. Some people may have no ancestry at all from a certain region. Also, even if two people both have Spanish-Portuguese connections, one may have a stronger genetic connection than someone else.

"I already know who my ancestors are." That is wonderful that you already have some knowledge of your ancestry. It is a great place to begin, but it isn't all you can discover. Why not? There are always gaps in our knowledge when it comes to genealogy whether from skeletons in the closet or typos on a census or unknown adoptions. Most people only know about one or two lines very well. Even if your data is perfect, an autosomal DNA test can test for some 100 lines of ancestry in a comprehensive manner on both sides of your family and can uncover hidden lines and deep ancestral markers. There are always surprises.

"I already had a DNA test." A lot of people do not understand that there are different types of DNA tests. They tell you different things, are based on different science, and cannot be compared. If you took either a Y-DNA (Male) test or a Mitochondrial (Female) test, those tests only look at those lines. Also, the information you get depends on the size of a company's academic database/s. If you go to a large library in a big city, you can check out more books than if you go to a rural library in a small town.

"I don't care about my ancestry." I find this the oddest objection to getting a DNA test. How could you not care who your ancestors were? Their story is part of your story. It might be why your family had certain traditions, for instance. And that is a story you can hand down and share with others.

March 7, 2012

Replace Shoe Leather with DNA

What do you know about your family history? Most people do not know much beyond their grandparents. If you want to get started discovering your personal genealogy, where do you begin? You begin with careful research.

My husband and I got started researching our family genealogy the old-fashioned way. We drove to archives and libraries. Sometimes, we drove for hundreds of miles to only arrive an hour or so before the library closed. We nearly went blind looking at microfiche to run down some of our Native American and other ancestors hopping from one county line to the next and giving the census taker a slightly different name each year. Most people look for royalty. I was beginning to think some of our ancestors were outlaws.

This was long before Native American DNA testing or any direct-to-consumer DNA testing for ancestry became possible.

However, genealogy is a good place to begin if you want to become familiar with your ancestors. First, find an online genealogy program. Depending on what you want, some are free and some are not. Fill in your parents' names and what you know in a family tree and print it if you wish. Next, call up or go interview as many relatives, especially elderly relatives, as you can. It might be good to record them. Make sure you ask about specific dates and places and think of significant life events when it comes to genealogy: weddings, funerals, military records, adoptions, etc. Don't forget to ask them stories about your family history. Ask them if there are any secrets hidden in your genealogy. Of course, it doesn't mean they will tell you, which is why you may want to finally do some DNA testing. Why? DNA, unlike people, doesn't lie.

After you have all this information, you can start to work on an online family tree. Fill in what you know and slowly begin work on the rest. Just do so carefully. Do not copy unsourced genealogy information into your family tree. This is a common error. Just because someone published their family tree, does not mean all their genealogy information is correct. Every ancestor in your family tree should have a record, like a census record or military record, attached.

No matter how hard you work on your genealogy though, you cannot find everything because of adoptions, hidden lines, missing records or misinformation like name changes. For these reasons, some of your information might be wrong, or you may have overlooked something. Consequently, you might want to look at what is called Genetic Genealogy or DNA testing for ancestry to provide more clues to your ancestry like deep ancestral markers or hidden genetic connections. The male and female DNA tests (Y-chromosome and mitochondrial tests) will give you information only about those particular lines and link you with cousins. An autosomal DNA test based on forensic science will provide a comprehensive analysis of your ancestry matching you not to people but populations.

What better hobby than to discover your family's story?

May 8, 2012

Cooking the Data

Don't get me wrong. Fudge? I love fudge. And I can make some dream fudge because I come from a line of Southern belles who made fudge so divine that when these ladies served it and a few other Southern delicacies to General William Tecumseh Sherman in their antebellum home on their rose-dappled china, he decided not to burn down Savannah, Georgia, after all. But fudging? I hate fudging. And professional genealogists and scientists are not immune. Fudging facts has become a national pastime.

And not everyone who claims to be a professional is one. According to Kimberly Powell in her *About* article, "5 Family History Scams to Avoid," some so-called "professional" family historians are making a career out of recreational genealogy when they aren't qualified:

> It is relatively easy for an amateur family historian to set up shop and charge money for tracing family trees. This is absolutely acceptable as long as the genealogist in question does not misrepresent their abilities or training ... Just because a genealogist doesn't have professional certification doesn't mean they don't know what they are doing. Professional genealogists are not usually licensed ... However, there have unfortunately been cases where people have been easily misled ... There have even been cases when so-called genealogists have "faked"

genealogical data to produce family histories for their clients.

Just that happened to me. A so-called "professional genealogist" called me up wanting to put my family tree in his genealogy book as he was writing about a line of mine. My people in a book! Then he said, "I really couldn't figure out how you were related to William, so I just made the assumption you were." Trouble is I really couldn't figure out how I was related to Irish William either, and I don't think I am.

As it turns out, this line is mixed with Native American and Melungeon ancestry. And almost all the males were named John for some reason. As soon as I had located one John, he had packed up his gear and mules and moved across the county line with all his family.

I learned much later that in the early days of this country, you had to play a game of "chicken" with the law. Why? The law might be more on your side in one county than another. And you had to keep up with it. He would go back and forth across state and county lines with slight misspellings of his name. Finally, he seems to disappear as far as the census records are concerned. I have had several different endings for the last John in this line play in my head—left with the Natives, took a ship to South America, got killed by a mountain lion, and so on. To this day, people will say to me, "Aren't you a cousin of so and so?" because they assume I am related to William. I just say, "If you just want to fudge the facts."

Initially, I was shocked about the assumptions the genealogist made but found out later this is common. Many people just connect dots. Don't facts matter anymore? They make the assumption that because someone else said so and so is a relative that it is true without checking the data themselves. Many download family trees instantly thinking they don't need to do any real work. Fudging will do. What a world of Wonderland we live in.

And the problem with surnames is largely ignored in articles like "What's in a Name? Y Chromosomes, and the Genetic Genealogy Revolution," by Turi E. King & Mark A. Jobling (*Trends in Genetics* V25 I8 August 2009 351-360). King and Jobling even state that if you share a paternal surname with someone, then you have a far greater likelihood of being related to that person ignoring much data. This ignores the well-known fact that many immigrants changed their names once they arrived to cover up Jewish or exotic ancestry and "Americanized" their names. It ignores the also well-known fact that some slaves took names of their masters. The craze for surname searches also ignores a rather glaring fact: Not all these lines are the same with the same origin. In fact, I only have one line that has only one immigrant. This means anyone in the United States that has that surname is related to me. But that is rare.

Unfortunately, faking the data is not just for genealogy. Science is not without their fudgers either. In an article titled "Disgrace," Charles Gross, a professor of psychology and neuroscience at Princeton University, reviews the whole subject of contemporary and historical scientific misconduct (*The Nation*, Jan. 9/16, 2012, pp. 25-32). He finds nothing new in the shocking case of Harvard's Marc Hauser, who was exposed two years ago for scientific misconduct, in of all fields, the biological basis of morality and genetic inheritance of doing evil.

Hauser apparently was guilty of the sin of fudging facts. The three ways to do that, all frowned upon, are by fabrication (making data up), falsification (altering or selecting data, cherry picking) and sheer plagiarism (which all but entering freshmen understand). But in fact, there is a long line of fudgers. For instance, in 1830, computer-science pioneer Charles Babbage published a book in which he distinguished "several species of impositions that have been practiced in science ... hoaxing, forging, trimming, and cooking." Babbage believed that forging was uncommon. Rarely are results completely counterfeited and pulled out of thin air.

For example, Gross classifies the Piltdown man as an example of hoaxing. This fossil combining parts of an ape and human skull was discovered in 1911 and not discredited until the 1950s. Most hoaxes are intended to poke fun at the public's credulousness, but the Piltdown hoax was undertaken by well-meaning British imperialists who hoped their construction would fill an awkward gap in the record. I suppose they were thinking, "If the missing link does not exist, we will just invent one."

"Trimming" is probably a form of scientific misconduct that few scientists confess to their most-exacting monitors such as the National Science Foundation but rather quietly cover up in bland hypocrisy. It consists of "eliminating outliers to make results look more accurate, while keeping the average the same." Who has not committed that little white sin? Let him who is without self-assurance cast the first chad.

"Cooking," on the other hand, the purposive selection and distortion of data, might be a real concern for all of us. Gross goes on to inspect the career of Harvard's "war crimes professor" Richard Herrnstein, who became a co-author after his death of the book *The Bell Curve* about racial differences in intelligence. It is not a very pretty kettle of fish.

But what of our scientific heroes? Like Darwin? It is thought that Charles Darwin either essentially stole the idea of natural selection from Alfred Russel Wallace, the father of biogeography, or at the least, failed to credit some of his predecessors in his rush to fame and self-glorification.

Personally, I prefer my line of real fudge makers to anyone stretching the truth or downright fudging the data any day. Real fudge is a lot tastier going down.

January 10, 2012

Autosomal Testing the New Standard

How fast does the molecular clock tick? Americans, especially, like most everything fast. We don't think too much about the word slow. But two scientists have changed our minds about that. As often happens in science, two research teams independently reached the same groundbreaking results. The breakthrough in the present case concerns the mutation rate of DNA and has profound implications for human evolution as well as for autosomal DNA ancestry analysis.

What is the DNA mutation rate? This is the rate at which a genetic marker mutates or changes over time. James X. Sun et al. in the recent article in *Nature Genetics*, "A Direct Characterization of Human Mutation Based on Microsatellites," and A. Kong et al. in the recent article, "Rate of de novo Mutations and the Importance of Father's Age to Disease Risk," in *Nature* both made an important, recent discovery. The speed of mutation in DNA is slower and more stable than previously thought. They discovered this how? By the magic of math.

How can math help us discover our ancestry? DNA is stable. Because it is so stable, we can calculate our way all the way back to when we swung in the trees and threw guavas at gorillas. (Even early man was in the trees longer than previously thought according to Charles Choi in his recent article in *Science*, "Early Human 'Lucy' Swung from the Trees.") And these calculations lead us to the stories of our ancestors.

What does this mean concerning autosomal DNA ancestry tests? They have even more scientific validity. Second-generation DNA ancestry testing is based on these very genetic markers, and that is confirmation that the alleles on your DNA that are examined using a statistical basis have been relatively unchanged for the past 20,000 years. That easily comprehends the period we know as world history and provides a time frame for valid inferences about population patterns and ancestry of individuals.

These common and not-so-common markers that everyone has are behind the method making it possible for anyone to take an autosomal ancestry test. Autosomal or non-sex-linked markers change at a much slower rate than the Y chromosome, for instance, which seems to be highly changeable, depending on the father's age (Kong 201). The Y chromosome is a marker only males have. It is used for other types of tests: male, haplotype, sex-linked DNA tests. Only males can take these tests, and it only provides information about that one male line.

Who knows what our DNA has yet to tell us? Or what DNA tests there will be in the future. This is an exciting new field. But what you can know now with a true and thorough autosomal DNA test is more than most realize is possible. The DNA Fingerprint Test is a simple test anyone can take that gives you a comprehensive snapshot of your cumulative ancestry.

September 15, 2013

14 DNA AND RELIGION

The Cohen Modal Haplotype

The whole field of genetic genealogy was launched, practically speaking, in 1997, when a Canadian physician noticed at temple that a fellow congregant who was a Sephardic Cohen looked nothing like an Ashkenazi Cohen. Theoretically, they should have the same pedigree and show a family resemblance.

He consulted with geneticists and wrote a letter to the journal *Nature* claiming that the most common set of Y chromosome microsatellite scores shared by men named Cohen (Hebrew for "priest") represented the genetic signature of Aaron, the patriarch who founded the Jewish priesthood: The inference was that all Jewish men with the surname Cohen were descended from Aaron, the brother of Moses.[6] A study published the next year proposed that Aaron's descendants had been found among carriers of the so-called Cohen Modal Haplotype, defined as DYS19=14, DYS388=16, DYS390=23, DYS391=10, DYS392=11 and DYS393=12.[7] A follow-up investigation of a South African tribe, the Lemba, who claim to be Jewish, showed that indeed they bore the Cohen Modal Haplotype, which they could only have received from Jewish

---

[6] Karl Skorecki *et al.*, "Y chromosomes of Jewish priests." *Nature* 385 (1997) 32.

[7] Thomas, M.G. *et al.*, "Origins of Old Testament priests." *Nature* 394 (1998) 138-40.

male founders, presumably far-traveling merchants from the Middle East.[8]

Over the following years, however, this theory began to unravel. The CMH was found in J1, as well as J2 and even J*. How could a single haplotype belong to different subclades of J? Was it a matter of "convergence," a hypothetical event by which the same set of scores, in time, could be produced from multiple unrelated parent haplotypes that just happen to coincide?

Such efforts to detect a genetic signature's survival in the Y chromosome unchanged over 3,500 years (approximately the amount of time since the time of the patriarchs) are misplaced. It should come as no surprise that the original "discovery" has never been proved. Here's where countless genetic genealogy enthusiasts miss the mark. The average mutation rate for Y-STRs, the "alleles" that create differences in haplotype scores, is 2.1×10^{-1}, or once every 476 generations, i.e., once every 9,520 years.[9] But this estimate is computed on the basis of *one* allele changing—either increasing by 1 or diminishing by 1; for instance, the number of repeated tags on the marker DYS14 might randomly mutate from a count of 14 to 15. Since the number on any given microsatellite can mutate in either direction, no matter how it changes over the centuries, the alleles tend to balance out and hover around a central score. In other words, the count does not keep going in either direction, up or down. Theoretically, then, a score of DYS19=14 could be produced by a mutation from either DYS19=13 or DYS19=15, in which case two different parent haplotypes might look the same and be indistinguishable.

Where most people make a mistake, however, is to forget that these

---

[8] Thomas, M.B. *et al.*,"Y Chromosomes Traveling South: The Cohen Modal Haplotype and the Origins of the Lemba the "Black Jews of Southern Africa." *American Journal of Human Genetics* 66 (2000) 674ff.

[9] See "Y-STR Haplotype Reference Database," created by Sascha Willuweit and Lutz Roewer, Institute of Legal Medicine, Humboldt-Universität Berlin, Germany, in cooperation with Michael Krawczak (Kiel), Manfred Kayser (Leipzig) and Peter de Knijff (Leiden), based on the continuous submission of Y-STR haplotypes by the International Forensic Y-User Group." Release 18 from January 31, 2006. Available online at http://www.yhrd.org. Also E. Heyer *et al.*, "Estimating Y chromosome specific microsatellite mutation frequencies using deep rooting pedigrees." *Human Molecular Genetics.* 6(1997): 799-803.

statistics apply to *one* microsatellite. Hence, when two haplotypes match (no matter how many markers are tested in common on them, whether it is 4, 8 or 12) it is conventionally reported that the "most remote common ancestor", the male ancestor they share, lived about 10,000 years ago. For multiple matching sites, however, one must *multiply* the probabilities, and if we do this, for nine markers (the searchable standard in the YHRD), we obtain a *combined* mutation rate of once every 952 years, i.e. a time depth of 500-1,500 years, about a millennium ago. In other words, matches on nine sites go back to about the year 1000.

And what about convergence—the creation of a single, ambiguous configuration from two different parent configurations? The probability of this randomly occurring could be computed, and the chances of such a fluke happening are more and more remote the greater the number of markers considered, but the possibility of it occurring and surviving is remote. The reason is a combination of genetic drift, founder effects, bottlenecks and what may be called political, or cultural (instead of natural) selection. Over time, a self-contained population has a tendency to emphasize certain genetic types while others die out or become insignificant.

It is unlikely that in the same relatively narrow time frame two distinct haplotypes formed by the rare process of convergence would become dominant, favored by political circumstances or the momentum of history. In Y chromosomal haplogroups, founder types like Genghis Khan (or what is more likely, his Mongol kinsmen as a whole) had many sons and were favored by military victories and wise governance, so that their genetic signature is evidently present in millions of Asian men.[10] But both Khan and the Viking Somerled (the progenitor of a sizable part of the Irish and Scottish population, according to Oxford Ancestors)[11] lived about 1,000 years ago, not 3,500 like Aaron. Even if Aaron's progeny did prosper and survive, it would have split into dozens of different genealogies (though these still could not cross subclade boundaries, since J1 and J2 went their different ways over 10,000 years ago, long before Aaron). The search for the ancient roots of the Jewish

---

[10] *Science News,* Feb. 8, 2003.
[11] *The Scotsman,* "DNA Shows Celtic Hero Somerled's Viking Roots," by Ian Johnston (April 26, 2005), available at www.scotsman.com.

priesthood is as misguided as the attempt to prove an unbroken genetic or cultural link between the inhabitants of the state of Israel and Biblical Israelites. If Skorecki, Thomas, Hammer and company found any modal Jewish haplotype at all, it is probably the genes of a prolific medieval rabbi who lived about 1,000 years ago and had numerous sons and grandsons. Whether his family was Old Palestinian or not is debatable. Certainly, six markers are not enough to judge.

We have included this tangent because among the Sephardic Jews there are several haplotypes which differ greatly from the Ashkenazi CMH but which are thought nonetheless to be Kohanim, including a Scottish J type associated with the name Gordon (claimed to be a corruption of Cohen), Iberian J2 type (often borne by males with the name Sanchez, a late Roman and Visigothic surname meaning "priest, or holy man") and even some R1bs (Cowan) and E3bs. Such diverse types prove, again, that medieval Jewish communities were formed as often from local convert populations as from a core of supposedly Old Palestinian males.[12]

"There are already many fine books on the history of the Jews in the Middle Ages," writes Theodore L. Steinberg, an English professor at State University of New York, in the preface to his *Jews and Judaism in the Middle Ages* (Wesport: Praeger, 2008). So why another one?

"All of those other studies, as excellent as they are, presume a certain degree of knowledge on the part of the reader—knowledge of Jewish customs and traditions and beliefs, as well as a general knowledge about the Middle Ages," he continues.

If you have never been to a synagogue service, don't know many observant Jews and, perhaps, just discovered an interest in Judaism after finding Jewish ancestry in your family tree, this is the book for you. Beginning with the first chapter, "Jews and Judaism, What Are They?" and continuing with "Talmud and Midrash," Steinberg skillfully guides

---

[12] Oddly, we did not find any Cohens, Katzes or related surnames among the Js in the Y-Haplogroup J Base at www.m410.net/yjdb/ and only one man specifically named Cohen claiming to be of the ancient priestly lineage in Family Tree DNA's Ysearch at www.ysearch.org. On the other hand, there were Stewart, Alexander, Forbes and Adair Js (all Scottish clan names), as well as numerous Cohens classified as R1b and E3b.

the reader through a crash course on Jewish history since the advent of Christianity. He introduces us to the rabbinical traditions of Judaism, Mishna, Gemara and all the flowering branches of halakah or Jewish law. We learn why Jews were blamed, and tolerated, by the Church. We learn about everyday life in cities where Jews, Christians and Muslims mixed, Jewish occupations, their literature, philosophy and the Cabala, all major areas of intersection with Christian society.

Appendix I has a good chronology of important events, from the life of Rabbi Akiva, which overlapped with that of Saul/Paul to the infamous date of 1492.

Joseph Jacobs

Another masterwork on Judaism intended primarily for non-Jewish readers is Joseph Jacobs' *Jewish Contributions to Civilization. An Estimate* (Philadelphia: Jewish Publication Society of America, 1919). This began life as the Australian Judaic scholar's *Studies in Jewish Statistics*, published in an anthropology journal in 1891, at the height of his fame. It was to be the first volume in a trilogy, the second book devoted to individual, rather than collective, contributions to European culture, the third a philosophical answer to anti-Semites about the value of Jews in the modern secular state. Alas, Jacobs died in 1916, leaving only notes for the second book and nothing at all of the third.

For Jacobs the watchword is always "judicious." He never exaggerates Jewish influence, emphasizing again and again that the number of Jews at no time, probably, rose above one-half of 0.1 percent in Western Europe, outside of countries where Jews were tolerated such as Moorish Spain and Poland/Lithuania under the Jagellon dynasty.

Did Jewish thinkers transform medieval philosophy from the stale dogmas of the theologians into more modern ideologies? Maybe, but they were only part of the movement.

His considered assessment of the Jewish contribution to medieval fables and folklore, one of his academic specialties, is "about one-tenth" of the material. There are those today, however, who would go so far as to say all of troubadour poetry and half of courtly love romances were inspired by the Judeo-Arabic tradition of Spain and southern France, with deeper roots in Arabia, Egypt and Babylon.

Did the Radanite ("From Persian *rah dan*, knowing the way") merchants plying the Silk Road in the early Middle Ages introduce all the choice import goods and enlightened ideas we associate with the East?

"Europe owes to the Jewish Radanites the introduction of oranges and apricots, sugar and rice, Jargonelle pears, and Gueldre roses, senna and borax, bdellium and asafoetida, sandalwood and aloes, cinnamon and galingale, mace and camphor, candy and julep, cubebs and tamarinds, slippers and tambours, mattress, sofa, and calabash, musk and jujube, jasmine and lilac " (p. 203).

But their influence was limited. Beginning in the 12th century, Venice took away the monopoly on the Levantine trade, just as Lombard merchants replaced Jews as bankers and moneylenders throughout most of Europe. The transition to a moneyed economy, according to Jacobs, was not due to Jews.

On the subject of Jews and capitalism, including stock exchanges and paper money, Jacobs takes a Ciceronian position. He denies that any Jews were involved with the South Sea Bubble, Mississippi Scheme of John Law or other experimental business models of the day. He does not even think that Jews fostered acceptance of bills of exchange or letters of credit. He points out that even with the Dutch East India Company, Jews had only very moderate ownership (although we wonder if, like the silent partner behind the usurious kings of medieval times, the partners named Coen and Hendricks and the like chose rather to have their names out of it), while the Jewish presence was absent at the first exchanges in Antwerp and London, and later minimal.

We think he might be too circumspect here. The Jewish role in the discovery, exploration and development of the American colonies is set forth in detail in Elizabeth C. Hirschman and Donald N. Yates, *Jews and Muslims in British Colonial America* (Jefferson: McFarland).

Jacobs is particularly harsh toward his contemporary Werner Sombart, who wrote an enthusiastic and eulogistic apology for the Jews in economic history (*Die Juden und das Wirtschaftsleben*, Leipzig, 1911, trans. by M. Epstein, 1913, *The Jews and Modern Capitalism).* Jacobs calls the European professor of economics' work a "farrago of fantasies about Jewish life and religion" (p. 259). The dour Victorian explicitly rejects Sombart's equation of the American way of business with Jewish practice and influence ("what we call Americanism is nothing else than the Jewish spirit distilled"). Again, however, we think Jacobs is bending over backward not to appear partisan. His objections to Sombart's arguments seem academic and petty if Sombart was on the right track, as most people concede today.

Prudent or Prudish?

In the same way, we feel that Jacobs' inability to detect any Jewish influence in the Puritans, or indeed the entire Reformation, is willfully blind.

"It cannot be by chance," he writes, "that the three most prominent voices among the Politiques (advisors to French king Henri IV), who laid down the principles which were to result in the Edict (of Nantes, granting toleration to Protestants, 1598)— Michel de l'Hopital, Jean Bodin and Michel de Montaigne—were all partly of Jewish race" (p. 281). But what he gives with one hand Jacobs takes away with the other, for he goes on to say that the influence of these men stopped with the borders of France, while "freethinkers" in other countries like Baruch Spinoza certainly did not meet with much success.

And of course, looking ahead to the eventual playing out of these policies, we cannot attribute either the American Bill of Rights or French Revolution to Jews, can we? Jacobs has a bowdlerized version of Jewish influence. He is willing to take modest credit for the good things Jews introduced, while letting their Christian imitators line up in the limelight and get the glory, but he doesn't want to admit that Jews might have been responsible for or at least complicit in some evils in modern society, such as communism, the militaristic state and free love ("the sexual vagaries of Enfantin," p. 308).

June 25, 2011

Plato Geneticist

Mary Settegast is described on the jacket simply as an archeological researcher, the 20-year-old book being *Plato Prehistorian: 10,000 to 5,000 B.C. Myth, Religion, Archaeology* (Hudson: Lindisfarne, 1990).

It's obvious she is not a member of the entrenched academic community of archeologists and prehistorians, for she spends most of the introduction to her fascinating study inveighing against the Old Model and New Archeology and defending the value of myth.

She then retells the Egyptian Priest's tale from Plato's *Timaeus* about how Solon's ancient Greek ancestors defeated an aggressive Atlantic sea-power situated on a now-lost continent beyond the Straits of Gilbraltar—the so-called Atlantis myth, which has no other source but the writings of Plato. Her thesis is that Plato is representing what he believed to be historical fact. Among other arguments, Settegast points out that it would have been impious for him to contrive a political fiction and put it in the mouth of Critias, who attributes the story to his grandfather, who received it from Solon himself, given the occasion of the dialogue, a celebration of Athena's festival day. She asks, "Would Socrates have Critias offer to the goddess as 'a just and truthful hymn of praise' (*Timaeus* 21) an intentional misrepresentation of Athena's own past history with the Greeks?"

Once Plato's word and intentions are vindicated, it is possible to study the scattered clues he gives us to prehistory of the Mediterranean world in a new light. Settegast makes a good case that the Magdalenian cave art of 17,000/15,000 to 9000 B.C.E. preserves the fading glory of an Atlantic culture of enormous power and sophistication that came to an abrupt end toward the end of the 10th millennium. She brackets the question of the location of a sunken continent and dwells instead on the blunders of modern prehistorians who fail to grasp the advanced picture of civilization left to us in Paleolithic remains like the Lascaux paintings. For instance, most anthropologists have explained the paintings as vehicles for sympathetic hunting magic without noting that it is the horse that is most commonly depicted while excavations of Magdlenian sites reveal almost exclusively the remains of reindeer as their principal animal food. The religious significance of the animals is lost on most analysts. Plato, as usual, provides the pertinent clue: the Atlantics worshipped Poseidon and regarded his sacred animal the horse with great awe. A revisionist look at the horses in cave paintings clarifies that the lines on horses' heads represent harnesses, not natural contours or anatomical details, proving that the Magdalenians or Atlantic peoples had tamed the horse by 12,000 B.C.E., some 8,000 years before the date assigned to the domestication of the horse in the conventional model.

I've just started to read the book and will conclude this preview for the blog by mentioning that one obstacle to accepting Plato's story at face value was that he describes the Atlantics as literate. The recent re-evaluation of the "magic signs" in Magdalenian caves as a writing system with heirs in many Old World alphabets seems to bear him out once again ... and make his detractors look stupid and full of hubris. It is the effect many Socratic dialogues were meant to have on their readers.

Addendum: One of the offshoots of Atlantic Culture according to *Plato Prehistorian* was the Çatal Hüyük civilization that flourished in Anatolia from 6200-5300 B.C.E. Only 2 percent to 3 percent of the 32-acre site has been excavated, but what has come to light so far includes amazing cyclopean walls, refined wall paintings and peculiar religious practices such as a vulture-bull rite, leopard shrine and Mistress of the Animals cult reminiscent of Venus figurines. It is conceivable that Atlantic Culture itself was spurred to life originally by admixture of Europeans with Neanderthals, since there are numerous signs of Neanderthal culture in archeological remains. Significantly, the Venus figures once associated with Gravettian Culture now appear to have had their origins with Neanderthals, who occupied Europe for 350,000 years before *H. sapiens sapiens*. Venus figurines were worn about the neck by Neanderthals, as proved in several excavations in Spain and elsewhere. In 1961, archeologists unearthed the skull of a Neanderthal man in the ancient site of Chalcedon on the east side of the Bosporus in Asia Minor, although the find is seldom mentioned today.

Our Neanderthal Index is based on affinities with archaic populations presumed to carry the highest rate of admixture with Neanderthals. These include many of the Atlantic and Mediterranean populations mentioned in *Plato Prehistorian*, including Greek, Turkish, Syrian, Arabian, Basque, Egyptian and Berber.

June 6, 2010

Now Come Jewish DNA Studies

Two major research articles on Jewish DNA appeared in June. As reported by Nicolas Wade in *The New York Times* in an article titled "Studies Show Jews' Genetic Similarity," they settle an old controversy.

One of the surveys of genomic or autosomal DNA was conducted by Gil Atzmon of the Albert Einstein College of Medicine and Harry Ostrer of New York University and appears in the *American Journal of Human Genetics*. The other, led by Doron M. Behar of the Rambam Health Care Campus in Haifa and Richard Villems of the University of Tartu in Estonia, is published in the journal *Nature*.

The two teams reached similar conclusions independently and simultaneously. Their findings refute the long-standing contention that Jews "have no common origin but are a miscellany of people in Europe and Central Asia who converted to Judaism at various times."

One of the articles, published online June 9, and in print July 8, was "The Genome-Wide Structure of the Jewish People," (*Nature* 466, 238-42). The editors' summary for it goes like this:

> A comparison of genomic data from 14 Jewish communities across the world with data from 69 non-Jewish populations reveals a close relationship between most of today's Jews and non-Jewish populations from the Levant. This fits in with the idea that most contemporary Jews are descended from ancient Hebrew and Israelite residents of the Levant. By contrast, the Ethiopian and Indian Jewish communities cluster with neighbouring non-Jewish populations in Ethiopia and western India, respectively. This may be partly because a greater degree of genetic, religious and cultural crossover took place when the Jewish communities in these areas became established.

An abstract for the other article mentioned by *The New York Times*, "Abraham's Children in the Genome Era," is given as follows by the publisher, *The American Journal of Human Genetics*:

> For more than a century, Jews and non-Jews alike have tried to define the relatedness of contemporary Jewish people. Previous genetic studies of blood group and serum markers suggested that Jewish groups had Middle Eastern origin with greater genetic similarity between paired Jewish populations. However, these and successor studies of monoallelic Y chromosomal and mitochondrial genetic markers did not resolve the issues of within and between-

group Jewish genetic identity. Here, genome-wide analysis of seven Jewish groups (Iranian, Iraqi, Syrian, Italian, Turkish, Greek, and Ashkenazi) and comparison with non-Jewish groups demonstrated distinctive Jewish population clusters, each with shared Middle Eastern ancestry, proximity to contemporary Middle Eastern populations, and variable degrees of European and North African admixture. Two major groups were identified by principal component, phylogenetic, and identity by descent (IBD) analysis: Middle Eastern Jews and European/Syrian Jews. The IBD segment sharing and the proximity of European Jews to each other and to southern European populations suggested similar origins for European Jewry and refuted large-scale genetic contributions of Central and Eastern European and Slavic populations to the formation of Ashkenazi Jewry. Rapid decay of IBD in Ashkenazi Jewish genomes was consistent with a severe bottleneck followed by large expansion, such as occurred with the so-called demographic miracle of population expansion from 50,000 people at the beginning of the 15th century to 5,000,000 people at the beginning of the 19th century. Thus, this study demonstrates that European/Syrian and Middle Eastern Jews represent a series of geographical isolates or clusters woven together by shared IBD genetic threads.

Critics of the new research findings point out that there are still no known markers for Jewish ancestry in genomic DNA. They obviously are not among our customers, however, since those who have purchased the 18 Marker Ethnic Panel available since last year are routinely screened for Jewish I, Jewish II and Jewish III. Their locations on genomic DNA were discovered by the company last August.

A Jewish genetic signature expressed in terms of autosomal DNA was predicted last year in a study titled, "A Genome-Wide Genetic Signature of Jewish Ancestry Perfectly Separates Individuals with and without Full Jewish Ancestry in a Large Random Sample of European Americans," by Ann C. Need et al. (*Genome Biology* 2009, vol. 10). That study spoke of "near perfect genetic inference of Ashkenazi Jewish ancestry." Interestingly, it also foresaw that "the genetic distinction between Jews and non-Jews may be more attributable to a Near-Eastern (i.e. Middle

Eastern) origin for Jewish populations than to population bottlenecks." (American usage favors "Middle East" over the British and outdated "Near East" still retained in academia.)

A final study of Jewish autosomal DNA that deserves mentioning is "Genomic Microsatellites Identify Shared Jewish Ancestry Intermediate between Middle Eastern and European Populations," published also last year in *BMC Genetics*, vol. 10, by Naama M. Kopelman et al. It used 678 autosomal microsatellite loci in 78 individuals, but what is proven on a large scale seems equally true on the small scale of our three Jewish markers based on microsatellites forming part of your DNA fingerprint.

August 20, 2010

Signs of Crypto-Judaism

As part of our series on Jewish ancestry, we reproduce below an appendix from the book, *Jews and Muslims in British Colonial America*, by Elizabeth C. Hirschman and Donald N. Yates.

Rituals and Practices of the Secret Jews of Portugal
The following is an incomplete list of practices that may be indicative of Jewish origin among Anusim or secret Jews or former Jewish families in the New World today. Adapted from Professor Eduardo Mayone Dias, Department of Spanish and Portuguese, UCLA. .

Told one is Jewish explicitly by parents, grandparents, or other relatives, a boy when he turns 13, a girl at 12.
Having Jewish family names: Duran, Lopez, etc.
Secret synagogues; secret prayer groups.
Avoiding church.
Churches without icons.
Lighting candles on Friday night when the first star appears.
Clean house and clothes for Shabbat.
Not allowed to do anything Friday night (not even wash hair).
El Dia Puro (Yom Kippur).
Celebrating a spring holiday.
Fasts: three days of Tanit Esther; every Monday and Thursday, fast of Gedalia.

Venerating Jewish saints, with celebrations: Santa Esterika, Santo Moises, etc.

Eight candles for Christmas.

Circumcision; consecration on eighth day (avoiding circumcision because that would bind child to the laws of Moses.

Biblical first names, like Esther.

Women taught Tanakh and ruled on questions.

Married under huppah/canopy.

Rending of garments; burial within one day; covering mirrors; spigots in cemeteries.

Seven days, then one year, of mourning.

Tombstones bearing Hebrew names, designations such as "daughter of Israel," and Jewish symbols (hand pointing to a star, open book of life, torah, star of David).

Possessing talit and tefillin, mezuzot, Tanakh, siddurim other Jewish objects.

Sweeping the floor away from the door (to avoid defiling mezuzah).

Having Cabalistic knowledge and practices.

Ritual slaughter (special knives, tested on hair or nails); covering blood with sand; removing sinew.

Purging, soaking, salting, boiling meat.

Avoiding pork and shellfish and other non-kosher foods (squirrel, rabbit).

Avoiding blood; throwing out eggs with bloodspots.

Avoiding red meat in general.

Waiting between meat and milk.

Eating only food prepared by mother or maternal grandmother.

Birth Rituals

To place a rooster's head over the door of the room where the birth will occur.

After the birth, the mother must not uncover herself or change clothes for 30 days.

To throw a silver coin into the baby's first bath water, especially a son's.

To say a prayer eight days after birth in which the baby's name is included.

Belief that the fairies (*hadas*) preside over a naming ceremony at birth

Wedding Rituals

Only home weddings.

To fast on the wedding day (both bride and groom, as well as two male friends of the groom and two female friends of the bride).

To bind the bride and groom's hands with a white cloth while a prayer is said.

To follow the wedding ceremony with a light meal consisting of a glass of wine, salt, bitter herbs, honey, an apple and unleavened bread.

At the wedding ceremony, bride and groom eat and drink out of the same plate and glass.

Marrying your brother's widow (Levirate law).

Funeral Rituals

To have ritual meals to which a beggar is invited and serve the food the deceased liked best.

To throw away all water in the home of the deceased.

To leave furniture overturned to show how a relative's death has upset the family.

To appear disheveled and careless about your own appearance during mourning.

To go to the deceased's room for eight days and say: May God give you a good night. You were once like us, we will be like you.

Not to shave for 30 days after the death of a relative.

Not to eat meat for one week after a death in the family, then fast on the anniversary.

Naming Rituals

Having two names, a private one in Hebrew (*kinnui*, e.g. Moses) and public one in the vernacular (Morris). Others: Jacob/James, Raphael/Ralph, Hannah/Johannah, Adina/Adelaide.

Allusions to mascots of Hebrew tribes like deer (Naphthali) and wolf (Levi).

Belief in being descended from the Biblical King David.

Naming after religious objects: Paschal, Menorah.

Translating Hebrew names, especially girls': Hannah into Grace, Esther into Myrtle, Peninah into Pearl, Roda into Rose, Shoshannah into Lillian, Lily. Simchah into Joy, Tikvah into Hope, Tzirrah into

Jewel, Golda into Goldie.

Allusions to Jacob's blessing of his sons and grandsons, e.g. Fishel for Ephraim because he was to multiply like the *fish* of the sea.

Use of Hebrew, but non-biblical names (e.g., Meir, Hayyim, Omar, Tamarah/Demarice).

Use of names from Jewish legend and folklore (e.g. Adinah, Edna, Adel, progenitress of the tribe of Levi).

Use of hypocoristic or pet names within the family alluding to Hebrew ones, for instance Zack or Ike for Isaac, Robin (Rueben) instead of Robert.

Adding the theophoric suffix -el to surnames, e.g., Lovell, Riddell, Tunnel.

Naming after a living relative, preferably the eldest born after the grandfather or grandmother, the next born after uncles and aunts and only after the father when these names are exhausted (Sephardic) or naming only after dead relatives (Ashkenazic).

Use of double names like Edward Charles and James Robert.

Changing the name of a child who becomes ill to foil the angel of death.

Giving a child an amuletic name like Vetula ("old woman") to bring long life.

Favoring names that begin with Lu- to remind the child that the family was once Portuguese (Lusitanian): Louise, Luanne, etc.

Belief in gematria (numerology of names, determined by Hebrew alphabet)

Avoiding saint's names (Paul, Peter, Barbara) and using Marianne or Mariah instead of Mary.

Jokes about the virgin birth of Jesus by Mary

Using names like Christopher or Christina to dispel doubts about conversion to Christianity.

Knowing whether your family belongs to the Kohanim (priestly caste), Levite (House of David) or Israelite (all the rest) division of Jews.

Other

Swearing an oath with your hat on.

Not mentioning the name of God. Writing it G*d.

Washing your hands before prayer.

A father blessing a son in public.

Saying grace after the meal.

Bowing and bobbing during religious service.

Jokes about the central tenets of Christianity (Immaculate Conception of Mary, rising from the dead of Jesus, etc.).

Deriding idolatry of saints and ornate decor of churches.

Hatred of the pope.

Preparing Saturday's meal (often a slow-cooking stew, for instance of eggplant) on Friday afternoon so no work is performed on the Sabbath.

Eating preferably fruits that grow in the land of Israel (dates, olives, oranges, grapes, peaches etc.).

Spreading sand from Israel on a grave or in a sanctuary.

Eating tongue on Rosh Hashanah to symbolize *head* of the year.

Having Bibles containing only the Old Testament and prayer books consisting only of the Psalms.

Having pictures of rabbis and scholars rather than saints in the sanctuary.

Performing tashlich, letting old clothes float away in running stream to mark a new year.

Forgiving a debt on Yom Kippur.

Facing Jerusalem during rituals.

Uttering brief blessings when you see lightning, mountains and other natural wonders.

Using only percussion instruments like the tambourine and hand clapping in services.

Silent prayer by congregation after prayers made out loud.

Worship services in the home.

Having 11 elders in a place of worship (minyam).

August 25, 2010

Abraham's Children: A Review

Implications for Jewish History and Genealogy

Article: Gil Atzmon et al., "Abraham's Children in the Genome Era: Major Jewish Diaspora Populations Comprise Distinct Genetic Clusters with Shared Middle Eastern Ancestry," *American Journal of Human Genetics* 86 (June 11, 2010) 850-859.

This blog post attempts to summarize this important article and translate its technical results into layman's language for the sake of our customers. From the genetic evidence, we hope to glean some useful information about Jewish history and genealogy, especially for those who find they have "some" Jewish ancestry in their family tree but who are non-Jewish in the way they identify.

The eleven authors represent a stellar team of international researchers specializing in population genetics. The institutions involved are leading centers of genetics and biomedicine, starting with the Albert Einstein College of Medicine in New York and including Tel Aviv University in Israel. Appearance in the *American Journal of Human Genetics*, a publication of the American Society of Human Genetics, assures prestige and finality of the highest order.

The study uses a new sample of 237 carefully qualified Jewish subjects, which it compares with data from the Human Genome Diversity Project started at Stanford in the 1990s, as well as the Population Reference Sample (PopRes) project, containing hundreds of thousands of SNPs (single nucleotide polymorphisms, used to differentiate genes that show heredity lines and disease linkage in populations).

The scale of the project is colossal, and it must have taken years to complete. Financial support came from private and public sources, including the U.S.-Israel Binational Science Foundation and NIH. It supersedes (while it confirms or clarifies) the two types of previous approaches to the problem: 1) genetics studies on blood groups from the 1970s, and 2) recent studies of Y chromosomal and mitochondrial haplotypes. The latter "uniparental" studies could only offer a limited view of the subject. This study uses autosomal DNA to its full advantage. The scientists had supercomputers and the latest tools in biostatistics at their disposal.

Material and Methods

The two cornerstones of any statistical study are reliability and validity. Experts would be challenged to find anything wrong with the reliability of "Abraham's Children." State-of-the-science genotyping, phylogeny, bootstrapping and GERMLINE algorithms are employed. But validity issues may be mentioned as likely to call some of the findings into slight question in some areas. Subjects were recruited in New York (Iranian, Iraqi, Syrian and Ashkenazi Jews), Seattle (Turkish Sephardic Jews), Greece (Greek Sephardic Jews) and Rome (Italian Jews). The main divisions into Italian, Ashkenazi, Syrian, Middle Eastern (termed Mizrahi, or Eastern, Jews here) can be seen in some instances to be question-begging, and there are equivocations in labels that might, conceivably, create "desired results." Moreover, subjects "were included only if all four grandparents came from the same Jewish community." Such a rule might have influenced one of the findings, namely, that "Jewish Communities Show High Levels of IBD," or Informative By Descent, shared segments of DNA (p. 855f.).

Notably, the conclusion that Ashkenazi Jews are highly inbred, with most being as close as third or fourth cousins to each other, could be an effect of the sample selection from New York Ashkenazi Jews, who often emigrated together and belonged to the same synagogue for several generations. A larger, more randomized selection would have been better. The total number of Ashkenazi Jews used was only 34 persons.

Jews now living in Rome are not the same as Roman Jews. Traditional schemes of Jewry speak of the Romaniot Jews, but these were traditionally at home in the Byzantine Empire. Are Syrian Jews mostly Middle Eastern or are they another amalgam of émigrés and remnants? Where are Central Asian Jews like the large Uzbekistani population my local barber represents? The absence of Caucasian and Central Asian Jews may have affected the study's rather resounding, overconfident statement that it "refutes" the contention that Ashkenazi Jews have little Middle Eastern heritage and rather more of a contribution to their genetics from Khazars and Slavs. The study, like many, seems at pains to prove Middle Eastern connections of Ashkenazi Jews, who are the leading force in present-day Israel.

There are no German, Dutch, Spanish, Portuguese or English Jews in the study. "Ashkenazi" seems to represent Jews living in New York with rather uniform ancestry in Poland, Hungary, Ukraine, Lithuania and Russia. All Jews in the study seem to be of the "go to temple" (at least on the High Holy Days) sort, although no mention is made of religious orientation. Subjects were chosen because of their ostensibly "pure" Jewish roots.

Seven Jewish population clusters are defined:

- Irani Jews
- Iraqi Jews
- Syrian Jews
- Ashkenazi Jews
- Italian Jews
- Greek Jews
- Turkish Jews

These were sorted out into three major groups of the Jewish Diaspora:

- Eastern European Ashkenazim
- Italian, Greek and Turkish Sephardim
- Iranian, Iraqi and Syrian Middle Easterners

How Different are Jews?

Table 1 on page 853 shows Fst values, which indicate the degree of projected inbreeding within a cluster and allow one to make comparisons between clusters. If we look at the table of genetic diversity among Jews, we can pick out some of the following points of interest:

- Ashkenazi Jews are very, very different genetically from Italian (Sephardic) Jews: genetic distance between the two populations is over 3.1, whereas that between Ashkenazi Jews and non-Jewish European populations such as Russian or French hardly ever goes above 1.0.

- Syrian and Iraqi Jews are also very different from each other; Syrian Jews group together with European (Ashkenazi and Sephardic) Jews, a major finding of the study.

- All Jewish populations, including Ashkenazim, resemble genetically the Druze, a Middle Eastern population that is still in its original place in Lebanon and Israel. Syrian and Turkish Jews (Sephardim) most resemble Druzes.

The study identifies two major groups of Jews, characterizing them as Middle Eastern Jews and European/Syrian Jews. Of the Middle Eastern Jews, the Irani and Iraqi were scarcely distinguishable. The Europeans have varying degrees of admixture with non-Jewish populations into which they have been dispersed. European Jews are 20 percent to 40 percent (or on another page, 30 percent to 60 percent) European, the study concludes, while Sephardic Jews have 8 percent to 11 percent North African (Berber) DNA. Italian, Syrian, Iranian and Iraqi Jews are the most inbred.

All these statistics "highlight the commonality of Jewish origin" and expose the Middle Eastern origins and genetic unity of Jewish people, even in dispersal from their homeland in ancient Israel.

The flipside of inbreeding is outbreeding or exogamy. If European Jews are highly admixed, by the same token, many Europeans must have some degree of Jewish admixture. Gene flow did not go just one way. What are the most outbred Jewish clusters according to the study? It is the Sephardic Jews from Italy, Greece and Turkey, with low rates of shared IBD and greater diversity than the Ashkenazim.

Lost Tribes Still Lost

The brief discussion section at the end of the article attempts to put the statistical findings together with an outline of Jewish history and resolve some of the mysteries of Jewish genetic makeup, chiefly the Khazar question. "Each of the Jewish populations formed its own cluster as part of the larger Jewish cluster," the authors conclude. "Each group demonstrated Middle Eastern ancestry and variable admixture with European populations."

Next, the authors attack the difficulty of determining when the major split between Eastern and Western Jewish groups occurred (with Syrians falling out with the Sephardim and Ashkenazim in the West, although intermediate in genetic distance with Middle Eastern Jews like the Iraqi and Irani). Based on IBD or calculations of shared genes on the front end or bottlenecks and back end or genetic drift, "a split between Middle Eastern Iraqi and Iranian Jews and European/Syrian Jews…is 100-150 generations [or] 2500 years ago." They correlate this population divide with the Babylonian and Persian periods of Jewish history. These disturbances began in the 900s-800s B.C.E. when the Kingdom of David fell apart into the Kingdom of Jerusalem and the Northern Kingdom (Samaria , Israel, Ephraim) and were finalized in the 6th century B.C.E. when King Cyrus of Persia granted permission to the exiles to be repatriated from Babylon to Jerusalem.

In the shuffle, the so-called Lost Tribes representing the Northern Kingdom had been dispersed "beyond the river" into Central Asia, leaving the Judeans in the ascendancy in Israel. Although the authors do not explicitly state it, the fact that Syrian Jews are several quantums' length closer in affinity with European Jews than Middle Eastern Jews seems to reflect the historic Return to Zion and inauguration of the Second Temple in Jerusalem in 516 B.C.E. and to explain why European Jews, especially Sephardim, emphasize the House of David and Tribes of Levi in their traditional ancestral accounts. So far, the autosomal clock of change seems to beat with the march of history.

The article is clear about another thing, "the idea of non-Semitic Mediterranean ancestry in the formation of the European/Syrian Jewish groups," attributing this to the mass conversions and proselytism during Greco-Roman times. It mentions that there were 6 million Jews in the Roman Empire, when Judaism accounted for 10 percent of the population. But here is where the authors' sense of history begins to get fuzzy. They do not speak of medieval conversions or mention, for instance, the Jews of Visigothic and Berber Spain and Septimanian France and gloss over the huge Khazar conversion and expansion event, 600-1000 C.E. Again, medieval history does not seem to be geneticists' strength.

"Abraham's Children" claims that the genetic composition of European/Syrian Jewish groups "is incompatible with theories that Ashkenazi Jews are for the most part the direct lineal descendants of converted Khazars or Slavs" (p. 857). In this brief sentence, are a number of fallacies, special pleading, misnomers, false assumptions, sleights of hand, straw arguments and equivocations. This blog posting can only touch on the controversy, but to begin with, as we have seen, Ashkenazi Jews (n=34, all from New York) are lumped together with Sephardim and Syrian Jews as having Middle Eastern core pedigrees, even though they have the highest amount of local population admixture. Are Khazars the same as Slavs? Not ethnographically. They are sometimes treated as

Middle Easterners and sometimes as Europeans or other interlopers in this article, but it doesn't seem to matter really, because there are no Khazars in the study sample. The authors are thus stalking a phantom, which they duly track down and slay.

Ghost in the Machine

Using haplogroup studies, the authors raise the possibility of 12.5 percent non-Middle Eastern admixture per generation in Ashkenazi Jews (based on Hammer et al.). They also draw attention to the 7.5 percent frequency of haplogroup R1a1 among Ashkenazi Jews, a non-Middle Eastern, rather Eastern European male lineage. But R1a1 is not diagnostic of Khazars; try maybe G. R1a1 descendants are typically fair haired and light-eyed; Khazars had a Middle Eastern, Turkic appearance, with dark features.

In a sense, geneticists always seem to find support for what they set out to prove in the first place. It is small wonder that "Abraham's Children" comes at the end of and completes a decades-long push by Big Science to legitimize Jewish claims to Middle Eastern roots. It is a splendid survey, the last in a long series, but it is not the final word on the subject. There are flaws both in the sampling and historical thinking.

The Thirteenth Tribe by Arthur Koestler, a Hungarian Ashkenazi Jewish author, advanced the thesis that the modern Jewish population in Central and Eastern Europe is not descended from the historical Israelites of antiquity, but from Khazars. This Turkic people of the Caucasus region converted to Judaism in the eighth century and moved into Russia, Hungary, Ukraine, Poland, Belarus, Lithuania and Germany during the 12th and 13th century when the Khazar Empire was collapsing. Koestler's work was founded on that of the French scholar Ernest Renan and set off a firestorm of controversy among Zionists.

August 26, 2010

Does He or Doesn't He? Only His Geneticist Knows for Sure

A visit to the temple by a Canadian doctor and one-page letter to the journal *Nature* in 1997 started it all, a frenzied hunt for the multi-study, double-blind, placebo-controlled proof that Jewish men of the surname

Cohen carried the same genes as the biblical patriarch Aaron, the mystical Cohen Modal Haplotype (Skorecki et al., 1997). Early testing showed that nearly half of men named Cohen—Cohanim in the Hebrew plural—had the same values at six locations on their Y chromosome. For the record, here are the magic numbers that first defined the "Y chromosome of Old Testament Priests":

DYS19 - 14
DYS388 - 16
DYS390 - 23
DYS391 - 10
DYS392 - 11
DYS393—12

But having those scores did not make you a member of the club. As the authors of an article on the "extended CMH" finally appearing last year write, the original research by Thomas et al. (1998) "produced a 'low resolution' CHM that was shared among many non-Jewish populations" (M.F. Hammer et al., "Extended Y Chromosome Haplotypes Resolve Multiple and Unique Lineages of the Jewish Priesthood," *Human Genetics* 126:707-17).

That would never do. Moreover, such sketchy data "did not provide the phylogenetic resolution needed to infer the geographic origin of the CHM lineage."

So, Hammer at the University of Arizona, partnering with the National Laboratory for the Genetics of Israeli Populations, spent 10 years and millions of dollars narrowing down the definition of the CMH and pinpointing its origin in history. Glossing over the statistics, subjects and methods—are you ready—the *extended* CMH is not much different from the *original* CMH, only it has no living matches.

Yes, you read that right. Whereas Skorecki's minimal six-locus haplotype might be proudly shared by thousands who eagerly sent in their DNA samples to genetic testing companies over the years, the new and improved CMH of Hammer et al. with twice the definition or 12 loci when "compared with the YHRD (Y-STR Haplotype Reference Database in Berlin) ... yielded zero out of 10,243 possible haplotypes in 66 populations" (p. 711).

Way to go, I hear from Marketing.

From a dybbuk the CMH has gone to being a dynosaur. It was once doubted whether it existed. It is now proven to be extinct.

No further research is needed on this subject as far as I'm concerned.

August 28, 2010

Heretical History

Ashkenazi Jews May Have Founded Israel, But Hebrew-Speaking Khazars Created Ashkenaz

Yes, the Khazars spoke Hebrew. It was the official language of the medieval kingdom, just as it is today in Israel. That is one of the surprising revelations of a book titled *The Invention of the Jewish People*, by Tel Aviv history professor Shlomo Sand, who devotes half a chapter to the "strange empire in the East" (London: Verso, 2009).

Judaism has not always been a closed society. During all but the last four or five centuries of its existence, it has been a proselytizing (and stabilizing) force in world events. Great states like the now-forgotten Khazaria played a role in the balance of powers and destiny of world civilization. The policies of the Kagan and his viceroy the Bey were praised everywhere for creating a prosperous, multiethnic, multifaith state that held sway over the lives of millions for five centuries (700-1200).

In an older book, *Catastrophe* (New York: Ballatine), David Keys has a chapter on Khazaria, "The Jewish Empire." In it, after reviewing what we know about the medieval state, he concludes: "The Khazar empire prevented the westward spread of Islam. If it had not been for the military might of the empire, Islam would likely have rolled west into pagan eastern Europe and possibly even into pagan Scandinavia in the eighth and ninth centuries A.D." He postulates that the Vikings could well have become Muslim, as well as Poland, Hungary, Romania, eastern Austria, the Czech and Slovak lands, Germany, Denmark, Sweden, Norway and Danelaw, the Viking state that emerged in eastern England.

"If the Khazar empire had not prevented Islamic expansion, it is even possible that the Normans (originally Vikings from Denmark) might have already been Muslims for two hundred years by the time they conquered England in 1066. What's more, if the Arabs had occupied what is now the Ukraine and Russia (rather than the Khazars, who founded Kiev, Russia's first capital), a Viking people known as the Rus would never have been able to push south and east from the Baltic to establish Russia" (p. 98).

An "investigative archeologist," Keys attributes the rise and spread of the Khazar state to a devastating volcanic explosion in Java in 535 that caused worldwide darkness, crop loss, famine, droughts, floods, migration of peoples and destabilization of regimes. Among the losers in what has passed into history as the Dark Ages were the Byzantines, Britons, French, Spanish, south Arabs, Tang Chinese, Teotihuacan and Peruvians. The winners were Avars, Huns, Koreans, Japanese, Toltecs, Incas, Mohammedans, Angles, Saxons, Jutes, Vikings and Visigoths. The global disaster had far-reaching effects that formed the modern world.

Sand is a professor in the department of European history (separated, curiously, from "Jewish" history in his country, Israel) and is concerned, as all historians are, with books. On the basis of records and written accounts, he reveals Khazaria's rise as no opportunistic accident to adorn a schoolboy's tale but as a historical phenomenon central to the formation of world Jewry. What Sand and Keys have in common is their iconoclastic, multidisciplinary approaches. Independently, they assign a prominent role to Khazaria in world history. Both attack taboo subjects with relish and finesse and are soundly ignored, no doubt, in the balkanized institutional world of learning.

Sand was apparently unaware of Keys or the global catastrophe of 535, but Keys has an excellent genetic summary of the heritage of the Khazars, one that is consistent with the latest research into the unity and diversity of the Jewish people(s).

"The Jewish empire's other legacy was the creation of a large pool of Jews of ethnically non-Jewish origin who subsequently became a major part—perhaps even the numerically dominant part—of northeast European Jewry and subsequently of world Jewry," Keys writes (p. 99). He goes further and substantiates and characterizes the convert origin of the Ashkenazi community in the ascendancy in modern Israel and the United States.

"This group, according to tradition, comprises the majority of the descendants of the ancient Israelite tribe of Levi—people who today still bear the name Levi or Levy. Significantly, it does not include a Levite subgroup—the Priests themselves—who often have the name Cohen. ... This genetic marker does not even show up among the Cohens (descendants of the ancient Israelite Chief Priests)—but only among the descendants of Assistant Priests (Levites). And then only within Ashkenazi (northern European) Jewry . ... If some top Khazars were adopting Cohenic Levitical status (i.e., Chief Priest status), then it is more than likely that others—a larger number—were adopting ordinary Levitical status (i.e., Assistant Priest status). Adoption of Cohenic or ordinary Levitical status by converts was and is expressly forbidden by rabbinical law, so the Khazars had to develop a mythic national history that gave them the right to Levitical status" (p. 100).

Kaftans and Yarmulkes

Both Keys and Sand write of the revealing role of Yiddish, the lingua franca of medieval European Jewry and first language of 80 percent of the settlers in the modern-day Land of Israel. They agree that Yiddish, even though it has a medieval German base, has more elements that are Slavic, Romance, Hebrew, Aramaic and even Turkic and is more accurately to be regarded as the result of a relexification process by originally mixed Sorbs, Magyars, Khazars and others, not as the dialect of west German or Rhineland Jews. In conventional Jewish teaching, Yiddish is regarded as a prestigious import from the Rhadanite and other Romano-Frankic Jews (who were Judean merchants under the Romans, according to most, hence retaining the desired link to ethnic origins in the Middle East). In the new view, Yiddish loses a lot of its special historical claims and becomes simply a language of convenience in the polyglot Khazar domain of former times.

The two books attribute also a superabundance of anthropological characteristics of Ashkenazi Jews to Khazar predominance, including shtetls (townlets), silk kaftans, fur headdresses, naming individuals after Jewish holidays (we have an Aunt Hanukkah and Uncle Pesach in our family tree) and innumerable place-names in Eastern Europe. Finally, Sand notes that the word yarmulke is derived from a Turkic word (p. 247).

Although the "Khazar thesis" is largely ignored or even silenced by

Israeli leaders and Jewish scholars, it is unlikely that the rulers of the modern state of Israel will make any constructive impact on the course of world events on the scale of Khazaria. Jews should reject the chauvinistic histories of 19th-century Zionists. As Sand says, "Why not begin to dream the future afresh, before it becomes a nightmare?"

September 4, 2010

Round Up the Usual Suspects

Book Review: *Abraham's Children. Race, Identity, and the DNA of the Chosen People*, by Jon Entine (New York: Grand Central, 2007).

Jewish DNA is full of controversies and the journalist Jon Entine shies away from none of them in this bestselling compendium. Interesting for our readers may be to note where he falls out on some of the more-vexed issues after taking the trouble to interview genetic news makers such as Karl Skorecki ("Genes of Old Testament Priests") and Father William Sanchez, the "poster boy" for New Mexico crypto-Jews.

Is there a Jewish "race"? That is probably not the right word, but yes, writes Entine, there is definitely a Jewish ethnicity that has been preserved in exile from the beginnings of Judaism in the Middle East over 3,000 years ago. Even Ashkenazi Jews are, genetically speaking, more similar to themselves and Middle Easterners than they are to Czechs, Poles and other Central and East European neighboring populations.

Is the Bible a true and accurate history of the Jews? No, for one thing, events in the Old Testament are corroborated by only a handful of contemporary records, including one Egyptian document and one Assyrian proclamation. The early books were rewritten several times, most famously by the patriarch scribe Ezra. The Jews returning from Babylonian Exile burnished the existing scriptures and introduced political themes that put the Northern Kingdom in a bad light. The word "Jew" was not actually used of the inhabitants of ancient Israel until around 520 B.C.E., when the "battered capital city, Jerusalem, surrounded by a scattering of towns" (p. 107) was called Yehud, the Aramaic name of a new province in the Persian Empire.

Do the Samaritans retain the closest genetic resemblance to Abraham's descendants? Here is what Entine writes:

The scientists speculate that not only are today's Samaritans likely descended from the Israelites, they may be the ancestral remnants of a breakaway group of Jewish priests that did not go into exile when the Assyrians conquered the northern kingdom in 721 B.C.E. Instead, these Cohanim may well have stayed, "but married Assyrian and female exiles relocated from other conquered lands, which was a typical Assyrian policy to obliterate national identities." It may just be that the tiny clan of Samaritans are a rare surviving branch of the ancient Israelites.

Among Entine's sources are Batsheva Bonne-Tamir, an Israeli geneticist who began studying Samaritan DNA when the population had dwindled to only a few families in the 1950s. Other authorities interviewed include: Karen Avraham (deafness in Jews), Doron Behar (Jewish founding mothers), Neil Bradman (Lemba Jews), Luca Luigi Cavalli-Sforza (Stanford's grand old man of DNA), Jared Diamond, David Goldstein (Jewish diseases), Michael Hammer (Y chromosomes), Mary-Claire King (disease studies), Jonathan Marks (critic), Tudor Parfitt (African and Indian Jews) and Mark Thomas (Cohanim).

What happened to the 10 Lost Tribes? Entine adopts a wry and caustic attitude toward this subject, beginning his chapter "Wandering Tribes" with a piece on the Worldwide Church of God, a Pasadena, California, sect founded by an ex-advertising agent, Herbert W. Armstrong, in the early 1930s. Armstrong was a proponent of British-Israelism, the belief that England is the heir to ancient Israel, the tribe of Ephraim having settled in Britain and the word "British" being derived from the ancient Hebrew word *beriyth*, which means covenant. "When Armstrong died in 1986, the WWCG claimed more than 150,000 members and an annual budget of $130 million" (p. 130).

Are Ashkenazi Jews really Khazars (i.e. non-Semitic)? Their genetic mix contains some Turkic elements from the Khazars, but even the Khazars were not 100 percent Turkic. Entine does not believe in the mass conversion portrayed in works such as Judah Halevi's medieval account *The Kuzari: A Book of Argument in Defense of a Despised Religion*. Following the author Kevin Brook ("not a formally trained historian but an impressively self-taught scholar ... creator of the Web site khazaria.com," p. 199), Entine says "the number of Khazarian Jews probably numbered no more than 30,000 out of a total population of 100,000, including a few thousand nobles and royalty" (p. 201).

Do you have to have a Jewish mother to be Jewish? Entine investigates this ruling very thoroughly and shows that Judaism was spread primarily by men with Middle Eastern roots who married local women of non-Semitic ancestry who converted to the husband's religion. The Jewish mother criterion came about in rabbinical times under the influence of Roman law. In Biblical times, Jewishness always came from a Jewish father.

Is there an "intelligence gene" among Ashkenazi Jews? Yes, it emerged in the age of the ghetto when survival selected for males who could earn a livelihood with their wits rather than hands or bodies.

Having touched on these questions, I would like to point out that many of the solutions seem rather superficial. Entine's research is not very deep or wide ranging. Nowhere in the chapter on mass conversions does he speak about the Babylonian principality of Narbonne in the South of France during Carolingian times. His treatment of Sephardic Jews is meager. There are other limitations in the scope of the work, but, in general, *Abraham's Children* is to be recommended as a solid, reliable, seemingly effortless account of a subject on which blood, sweat and tears have been spilled on every page in the past. That is no small feat.

October 7, 2010

Note on German Jews

Jewish II is characterized as the "strongest" marker, correlating very often with Ashkenazi Jewish parentage, especially in its double allele form (two checks under parents' columns on an 18 Marker Ethnic Panel or Jewish DNA Ancestry report). This description is validated by ENFSI results for 22 countries of Europe, where its strongest showings are in Czechoslovakia, Croatia and Slovenia (all Central European Slavic populations).

Its lowest frequency occurs in Germany, Denmark and the Netherlands, countries with few Ashkenazi Jews today. There are no data for Hungary in ENFSI. Czech is the closest population in geography and historical composition. Likewise, there are no data for Slovakia, the southern part of the former Czechoslovakia, as Slovakia chose not to participate in ENFSI.

In Czechoslovakia, nearly 1 in 26 people carry Jewish II in its double allelic form, suggesting both parents were Jewish. The lowest frequency of Jewish II occurs in Scotland, where only 1 in 65 have it.

Note that even though most Ashkenazi Jews spoke Yiddish, a Germanic dialect, they were not necessarily German in origin. More likely they belonged to the Slavic peoples. Yiddish is not linguistically descended from Rhinish or Low German but a High German language of convenience adopted by Slavic-speaking Jews during Germany's medieval Drang nach Osten (Push to the East). Germany itself was considered a predominantly Slavic nation by geographers until the 19th century, when it was first unified under the Prussian monarchy.

The reason Jewish II is not detected at a higher level in today's German population is because of the ethnic cleansing and genocide committed by the Nazi government in the Second World War. For various reasons, Czechoslovakia and Hungary's Jewish populations, many of them assimilated Jews, survived better than Germany's.

The best book about German Jewry is probably *The Pity of It All, A History of Jews in Germany, 1743-1933*, by Amos Elon (Macmillan, 2002), although many readers will find the author's approach to Jewish ethnicity Zionist and doctrinaire.

October 13, 2010

The Greatest Divide

You hear a lot of talk about the Neolithic Revolution—the gradual adoption and spread of agriculture, animal husbandry and town life by our prehistoric European ancestors—but the most important epoch in the course of civilization goes largely unnoticed in the history books. That was the abrupt shift from matriarchy and worship of the Great Goddess to the warrior-based governments and language stocks of the steppe-dwelling Indo-Aryan barbarians who invaded Old Europe beginning in the late fourth millennium B.C.E.

The roots of Europe's original female-oriented religion are lost in the mists of the early Stone Age, and may even precede the arrival of "modern humans" in Europe and be part of the heritage of Neanderthals. This substratum of a long-lasting peaceful hunter-gatherer society organized around the religion of the Great Goddess absorbed the spreading practice of agriculture from the Middle East beginning in the fifth millennium and reached its apogee of development in a pure form in the fourth millennium.

The cult of the Great Goddess, as depicted in an enthroned deity with flanking felines from Çatal Hüyük, an 8,000-year-old shrine in present-day Turkey (p. 107), was the lifelong object of study by Lithuanian-American archeologist Marija Gimbutas, whose most influential book is probably *The Language of the Goddess* (London: Thames & Hudson, 2006).

The ax fell on this ancient civilization—quite literally—around 3000 B.C.E. As confirmed in Jane McIntosh's *Handbook to Life in Prehistoric Europe* (New York: Facts on File, 2006), there was a clear line of demarcation between old and new Europe, from the Balkans to Britain, Spain and Scandinavia. The archeological record tells the story of a sweeping and abrupt end to things. The first metal weapons appear in the graves of elite males along with hoards of gold and jewels. Axes previously used to clear forests for agriculture are now battle-axes. Burials are single rather than family and clan-oriented. Whole villages were massacred and depopulated. Fortifications grew as violence escalated. The horse, venerated as just one of the totem animals of the Goddess since the early Stone Age, becomes the symbol of the warrior, along with the chariot and boat. Rock art features ithyphallic warriors wielding weapons or shooting arrows at each other. The transition can also be seen in the establishment of the Pharaohs in Egypt about 3500 B.C.E.

The invaders brought their male pantheon of war gods, Indo-European languages, aristocratic forms of government and Central Asian/Caucasian genes. The goddess cult underwent radical male adaptations, surviving in out-of-the-way places like Crete and Brittany.

So, rather than one transformation, European civilization first went through a Neolithic Revolution, then conversion to warrior-dominated patriarchal societies. It can be postulated that the matriarchal societies eagerly adopted agriculture but exhausted soils, destroyed vital forests and became weaker and smaller-bodied due to a changed diet, falling prey around 3000 B.C.E. to the barbarian warriors of the steppe, who found the accumulation of wealth and unprotected agrarian settlements of Old Europe easy pickings. Climate change could have been a contributing factor.

James Joyce called history "the nightmare from which one cannot wake." If we take a long view of human events, this nightmare began about 5,000 years ago. Otherworldly religions like Christianity introduced a further element of alienation and turning away from the sources of life. Before that, to judge from archeology, people were happily alive, awake and in tune with nature under the auspices of matriarchy.

October 28, 2010

Abraham Lincoln's Jewish Roots

The Abraham Lincoln Presidential Library and Museum is interested in their namesake's Jewish ancestry. Do a Google Search and you'll get numerous hits under "Abraham Lincoln Jewish Ancestry." They can all be traced back to the work of Rutgers University professor Elizabeth C. Hirschman, who published her findings in *The Melungeons: The Last Lost Tribe in America* in 2005 (Mercer University Press). Melungeons.com immediately reported the news in an article on Abraham Lincoln's mother, Nancy Hanks, in March 2005.

Actually, the revelation goes back to Abraham Lincoln's own lifetime. In 1909 a book was written by Isaac Markens titled *Abraham Lincoln and the Jews.* In it is the germ of the controversy. Rabbi Isaac Wise of Cincinnati wrote in 1865, "Abraham Lincoln believed himself to be bone of our bone and flesh of our flesh. He supposed himself to be of Hebrew parentage, *he said so in my presence,* and indeed he possessed the common features of the Hebrew race both in countenance and features."

Lincoln's son Robert later denied the admission of Hebrew ancestry. He told Markens that he had "never before heard that his father supposed he had any Jewish ancestry." Markens dismissed the exchange between Lincoln and the rabbi as a "pleasantry."

Pleasantry or not, the rumor seems to reflect historical truth.

"The Lincoln material in my book traces his family's arrival from England into the New Hingham Colony in Massachusetts and their migration down to the Appalachian area," Hirschman told the curator of the Lincoln Collection, James M. Cornelius, who contacted her in September. "I use genealogies, marriage practices, wills and cemetery inscriptions to build the case for his Jewish—and likely Sephardic—ancestry."

Now that Lincoln is beginning to be seen as having Jewish ancestry, what about his status as a Melungeon? That has now slipped into jeopardy, it seems. Jewish and Melungeon ancestry go hand in glove, except in the eyes of those who don't believe Melungeons are a distinct ethnic group at all. They are opposed to finding any unwanted ethnicity in people claiming to be Melungeon. We predict Lincoln's Jewishness will rapidly be disposed of by these devotees of the received standard version of American history.

September 12, 2010

Rabbinical Court Recognizes Majorcan Conversos as Jews
Ruling Applies to All Their Ancestors and Descendants

Majorcan Jews belong to the Sephardic (Spanish) division of world Jews. The name of the island is Majorca in Spanish, Mallorca in Catalan (in whose orbit historically it fell). In 2011, a rabbinical court recognized the Majorcan Chuetas and all their genealogically related families as Jews. It is possible that the Franciscan Majorcan Junipero Serra who founded California's Spanish mission system was Jewish (Converso), as was probably the family of Christopher Columbus.

DNA Consultants is fortunate to have population data on Majorcan Jews.

The Majorcan–Chueta (also spelled Mallorcan and Xueta) population data represent DNA samples from 102 unrelated Chuetan (Jewish) individuals on Majorca, one of the Balearic Islands, which mark the easternmost part of Spain.

Samples were obtained by the Genetics Laboratory of the Biology Department at the University of the Balearic Islands.

Source publication: Genetic Variability at Nine STR Loci in the Chueta (Majorcan Jews) and the Balearic Populations Investigated by a Single

Multiplex Reaction, *International Journal of Legal Medicine* (2000) 263-267.

March 9, 2012

The Thirteenth Tribe

Khazarian Hypothesis of European Jewish Origins Vindicated; New Genetic Study Shows Rhineland Hypothesis False

In "Heretical History" and numerous other posts, we have argued that the genetic and cultural contributions of the Turkic-Iranic Khazars deserve much more attention than the cosseted theories of European Zionist Jews and official views of the state of Israel on Jewish history. A new study by Eran Elhaik titled "The Missing Link of Jewish European Ancestry: Contrasting the Rhineland and the Khazarian Hypothesis," (*Genome Biology and Evolution* 5.1:61-74) bears out our thinking with hard evidence that seems capable of settling that rancorously-disputed question once and for all.

According to *Science Daily* (Jan. 16, 2013), "Despite being one of the most genetically analysed groups, the origin of European Jews has remained obscure ... but the new study ... sets to rest previous contradictory reports of Jewish ancestry." Elhaik's findings strongly support the Khazarian Hypothesis, as opposed to the Rhineland Hypothesis, of European Jewish origins.

Ashkenazi ("Germanic") Jews embraced a Western European origin myth not only because it presented Jews as very white, at the top of the race pyramid, but because of the prestige it brought them of being a spinoff of the Roman Empire.

The Khazarian thesis acknowledges that the most-important element is Middle Eastern among "brown" peoples, and that the period of efflorescence of Judaism in Europe began in the late Middle Ages under the influence of migrating Khazars.

That's an entirely different version of history, one much closer to Arthur Koestler's "Thirteenth Tribe" account, a theory for which he was castigated by fellow Jews and especially Zionists.

The new study was not possible until recently, when many of the gaps in Caucasian and Jewish genetics were filled for the first time, using autosomal approaches rather than sex-linked haplotype surveys.

Elhaik's masterwork examines a comprehensive dataset of 1,287 unrelated individuals in eight Jewish and 74 non-Jewish populations genotyped over a range of half a million single nucleotide polymorphisms (SNPs) or markers. These data were adapted from a study by Doron Behar and colleagues from three years ago.

The central role of Khazaria was also not wanted or wished for among Eurocentric scholars, who tended to denigrate *Ostjuden* or Eastern Jews. Few historians conceded even the fact that Khazaria was a Jewish state that lasted nearly a millennium, where Hebrew was spoken, preferring to think of it as a sort of travelers tale or land of religious fiction.

Elhaik used seven measures of ancestry, relatedness, admixture, allele-sharing distances, geographical origins and migration patterns to identify the Caucasus-Near Eastern and European ancestral signatures in European Jews' genome along with a smaller, but substantial Middle Eastern genome. "The results were consistent in depicting a Caucasus ancestry for all European Jews," according to *Science Daily*.

Elhaik wrote: "The most parsimonious explanation for our findings is that Eastern European Jews are of Judeo-Khazarian ancestry forged over many centuries in the Caucasus. Jewish presence in the Caucasus and later Khazaria (a Hebrew-speaking Central Asian empire) was recorded as early as the late centuries B.C.E. and reinforced due to the increase in trade along the Silk Road, the decline of Judah (1st-7th centuries), and the rise of Christianity and Islam. Greco-Roman and Mesopotamian Jews gravitating toward Khazaria were also common in the early centuries, and their migrations were intensified following the Khazars' conversion to Judaism ... The religious conversion of the Khazars encompassed most of the Empire's citizens and subordinate tribes and lasted for the next 400 years until the invasion of the Mongols. At the final collapse of their empire in the 13th century, many of the Judeo-Khazars fled to Eastern Europe and later migrated to Central Europe and admixed with the neighbouring populations."

According to *Science Daily*, Elhaik's findings explain otherwise conflicting results describing high heterogeneity among Jewish communities and relatedness to Middle Eastern, Southern European, and Caucasus populations not accounted for under the Rhineland Hypothesis. Although the study links European Jews to the Khazars, there are still questions to be answered. How substantial is the Iranian ancestry in modern-day Jews (Khazars were themselves mixed)? Since Eastern European Jews arrived from the Caucasus, where did Central and Western European Jews come from, those usually called Sephardic?

Finally, if there was no mass migration out of Palestine at the seventh century, what happened to the ancient Judeans? Shlomo Sand, author of *The Invention of the Jewish People*, has maintained there never were any expulsions or exoduses out of Palestine, only wholesale conversions to Islam. Thus, the true heirs of Judah are the persistent inhabitants who still occupy Jerusalem and the Holy Land, that is, Palestinians.

It is ironic, to say the least, that these ancient Judeans are dispossessed by a nationalist colonial power with roots no deeper than 19th century Europe, which exercises a force majeur based on mistaken notions of genetics and history.

March 22, 2013

Anasazi: Cannibals or Witch-Hunters?

Though having an exotic ancestry might be interesting, there are limits. You might not want to have cannibals for relatives. Luckily, you probably don't have to worry about that. Even if you are Navajo, or have some Navajo ancestry, they do not believe their "enemy ancestors," the Anasazi, were cannibals. And some anthropologists believe the violence is explained in other ways—like the executions of witches (or what the Navajo call "skinwalkers").

Whatever the explanation for the early violence of these people, it seems to have stopped around 1200 A.D. (Alexandra Witze). That is long enough ago that no one should be concerned about it. Why? First, scientists cannot determine exactly what happened. And many others of us probably had ancient ancestors that were violent in one manner or another that long ago as well.

We first visited Chaco Canyon in northern New Mexico, and the wind was cold and eerie as we walked along the deep, narrow canyon and gazed upward at the buttery apartment complexes made of stone and mud high above us. Tiny windows had been etched in them. An eagle flew high ahead.

Who were these people? The staff at the visitor's center was mostly Navajo and not descendants of the original inhabitants of the site, and either could not or would not tell us much. They said they were the Anasazi, a Navajo term meaning "enemy ancestors." What does that mean? When we pressed for more answers, I was told they belonged to the Chaco Culture. They were Chacoans. They created magnificent architecture. According to the *Smithsonian* article, "Riddles of the Anasazi," by David Roberts they were marvelous architects:

> The Anasazi built magnificent villages such as Chaco Canyon's Pueblo Bonito, a 10th-century complex that was as many as five stories tall and contained about 800 rooms. The people laid a 400-mile network of roads, some of them 30 feet wide, across deserts and canyons. And into their architecture, they built astronomical observatories.

But who were these people? According to Alexandra Witze's *National Geographic News* article, "Researchers Divided over Whether Anasazi Were Cannibals," there are some archeologists who think the Anasazi were cannibals because of "piles of butchered human bones, some of which were apparently roasted or boiled. In one instance, ancient human feces even seem to contain traces of digested human tissue." Archeologist Christy Turner has collected over decades what he calls "incontrovertible evidence of cannibalism and violence among the Anasazi" at sites dating "between about A.D. 900 and 1250."

According to Amelie A. Walker in her article, "Anasazi Cannibalism?" there was evidence of seven people "not from burials" who were "dismembered, defleshed, (with) their bones battered, and in some cases burned or stewed, (and left) in the same condition as animals for food." She suggests that there are three possibilities: "ritual" or "hunger" cannibalism or "something else altogether." Is the culture of the Anasazi one of cannibalism? Others disagree and think this is something else altogether.

What else could it be? Local tribes are "deeply offended" by the suggestion that the Anasazi were cannibals says Witze. However, most scientists agree that the Anasazi experienced "brutal violence" though this "drops off after about A.D. 1200." They just disagree about how to interpret that. J. Andrew Darling, an archeologist with the Gila River Community in southern Arizona thinks the answer is witch (or what the Navajo call "skinwalker") executions. Darling discovered "after researching the folklore of modern Pueblo tribes" that "dismembering a witch" was the "only way to prevent the witch from wreaking revenge after death." Debra Martin, an archeologist at Hampshire College in Amherst, Massachussets, argues that "the bone destruction is best explained by several different reasons, including witch execution, chewing by a carnivorous animal, or being reburied" (Witze).

"The best documented indication that the Basketmakers were headhunters is ... Kinboko Canyon, evidence discovered by archeologist Samuel J. Guernsey of the Peabody Museum of Harvard in 1915, and reported in a 1919 publication of the Smithsonian Institution's Bureau of American Ethnology," writes G. F. Feldman in the chapter on the Anasazi. He goes on to describe this and other early excavations in the Four Corners area that were quickly hushed up and reburied in horror, including Battle Cave in Canyon del Muerto, now part of the Canyon de Chelly National Monument inside the Navajo Indian Reservation. We read now of flesh-stripping, bone crushing, roasting pits, and sliced off mastoids.

According to this article, around A.D. 950, 11 persons, including women and children, were killed and butchered, cooked, and eaten on Burnt Mesa in New Mexico north of the San Juan River. At a site near the Hopi villages in Arizona, a group of 30 individuals, 40 percent under the age of 18, were slaughtered and eaten. In a Colorado rock shelter, a large jar was found filled with splintered human bones. ...

The grisly record goes on and on. Feldman writes that, by the year 2000, the number of such sites in the San Juan drainage where the Chaco Culture was centered had risen to 40 (p. 136).

So, witch hunters or cannibals ... or something else? We may never know the truth. According to Michael Adler, an archeologist at Southern Methodist University, ... "Scientists may never be able to prove whether the cannibalism or witch killing theory is correct. Or something else altogether."

April 18, 2013

Cave of Forgotten Rites

Cave of Forgotten Dreams is a mystical and breathtaking 3D journey into a winding and enormous cave in Southern France, the Chauvet Cave, where the earliest Paleolithic cave drawings are filmed by Herzog and his team. This is far older than Catalhoyuk, the Neolithic site in Turkey previously thought to be the oldest site for cave drawings.

Animals of all sorts (horses, rhinos, hyenas, and lions) seem to dance in shocking, realistic, and vivid detail along the walls "as if they had been done yesterday" one expert suggests. Moreover, this is not primitive art. The brush strokes of these artists are those of skilled masters who not only depict these animals in exquisite detail but show them running and charging by painting more legs and providing other details to expertly express fluidity and movement. This is Paleolithic animation! 3D seems to make absolute sense for this movie.

Despite the odd Herzog touches, like the New Age choir in the background, and the shot of albino crocodiles at the end, this is an excellent, must-see film about one of the greatest discoveries in our lifetime. I suppose the music is supposed to make you feel this is a mystical and otherworldly journey, but I don't think it was needed. Also, there is an inference that we could not possibly know anything about these people—hardly more than if we were albino crocodiles. I could not disagree more.

Though someone on the team did suggest the drawings "could have been made by a woman," I found it startling that no one suggested any connection to the Goddess religion or the Goddess symbols in the cave like the bull intertwined in a loving embrace with a female figure. These are ancient and well-known symbols since the Minoan civilization and from antiquity. Marija Gimbutas, in *Language of the Goddess*, discusses how the bull was both her son and her lover—in perhaps the first "trinity."

According to Michael Balter, author of *The Goddess and the Bull: Catalhoyuk: An Archeological Journey to the Dawn of Civilization*, the Catalhoyuk cave included both symbols of the horns of a bull and the goddess.

Moreover, it was in a cave, in the womb of the Mother Earth—a symbol of the Mother Goddess. Horses also were sacred to the Goddess, and they are everywhere. This was not an early museum of sorts. This was a very ancient sacred place.

July 9, 2014

15 DNA AND LAW

Wild, Wooly World of DNA

The first in a series of articles in the *New York Times*, titled "The DNA Age" presents case histories of people whose DNA tests are turning out to be mixed blessings, arousing more expectations than may be justified. From the adopted twins who are looking for financial aid after finding out they are part African and part Native American to the man raised a gentile attempting to invoke the law of return to Israel following the revelation his DNA matched Ashkenazi Jews, the series by Amy Harmon apparently intends to explore the two sides of DNA—the answers it brings, along with the new questions it raises.

On another front, Indian tribes routinely refuse to accept DNA evidence. According to the article, though, this has not deterred prospective new enrollees.

"It used to be 'someone said my grandmother was an Indian,' " says Joyce Walker, the enrollment clerk who regularly turns away DNA petitioners for the Mashantucket Pequot tribe, which operates the lucrative Foxwoods Resort Casino in Connecticut. "Now, it's 'my DNA says my grandmother was an Indian.'"

The title of the first of the series is "Seeking Ancestry in DNA Ties Uncovered by Tests." One of the featured DNA test takers was a customer of DNA Consultants.

April 13, 2006

Uncle Sam Wants Your DNA

U.S. Set to Begin a Vast Expansion of DNA Sampling
Julia Preston
New York Times
February 5, 2007

"The Justice Department is completing rules to allow the collection of DNA from most people arrested or detained by federal authorities, a vast expansion of DNA gathering that will include hundreds of thousands of illegal immigrants, by far the largest group affected.

"The new forensic DNA sampling was authorized by Congress in a little-noticed amendment to a January 2006 renewal of the Violence Against Women Act, which provides protections and assistance for victims of sexual crimes. The amendment permits DNA collecting from anyone under criminal arrest by federal authorities, and also from illegal immigrants detained by federal agents."

February 5, 2007

DNA and Prisons

Exoneration Using DNA Brings Change in Legal System
Solomon Moore
New York Times
October 1, 2007

"State lawmakers across the country are adopting broad changes to criminal-justice procedures as a response to the exoneration of more than 200 convicts through the use of DNA evidence.

"All but eight states now give inmates varying degrees of access to DNA evidence that might not have been available at the time of their convictions. Many states are also overhauling the way witnesses identify suspects, crime labs handle evidence and informants are used. ...

"Nationwide, misidentification by witnesses led to wrongful convictions in 75 percent of the 207 instances in which prisoners have been exonerated over the last decade, according to the Innocence Project, a group in New York that investigates wrongful convictions. …

"'The legislative reform movement as a result of these DNA exonerations is probably the single greatest criminal-justice reform effort in the last 40 years,' said Peter J. Neufeld, co-director of the Innocence Project."

October 1, 2007

When Objects Become Subjects

Review of Paul Brodwin, " 'Bioethics in Action' and Human Population Genetics Research"

Population-genetics experts who lecture in the groves of academe or trudge through the jungles of the Amazon are not immune to racist bombshells and political dynamite. In 1991, Stanford geneticist Luigi Luca Cavalli-Sforza announced a project to study human genetic diversity. The ponderous monograph that issued forth in 1994 became as revered as it was unreadable. His *History and Geography of Human Genes* posited two main limbs in the human DNA tree, the African and non-African, with the latter branching off into Europeans (Caucasians) and Northeast Asians. Included in Northeast Asians were the so-called Amerindians. Amerinds were closest in genetic distance to Northern Turkic, Chukchi and other Arctic and Mongolian peoples.

Little did Cavalli-Sforza and his team expect to encounter any opposition to their benign project, much less withdrawal of funding by the U.S. government and United Nations, but this is exactly what happened. The genial professor was surprised one day by a letter from a Canadian human-rights group called the Rural Advancement Foundation International. The group demanded he stop his work immediately. It accused the Human Genome Diversity Project of biopiracy, stealing DNA from unsuspecting indigenous people and mining it for valuable information pharmaceutical companies could use to make drugs Third World people could not afford.

Paul Brodwin's article published in 2005 in the journal *Culture, Medicine and Psychiatry* (29:145-78) reviewed this controversy, which had some positive repercussions in forcing researchers to rethink colonialist attitudes toward their subjects. But in the second case of "bioethics in action," Brodwin painted a much-more-ambiguous picture. It concerned the use of genetics by the ethnic group called Melungeons of Tennessee and Virginia to prove identity claims and press their ideas of special entitlements.

In the section of the article titled "The Reinvention of Melungeon Ethnicity," Brodwin chronicles the conflict between scientific genetics and the Melungeons' demand for collective recognition. Complicating this issue is that the academics were by no means certain among themselves about who or what Melungeons were from an anthropological perspective. A rancorous standoff between Virginia DeMarce and N. Brent Kennedy was matched by the tendentious nature of the Melungeons' own theories and assertions about themselves. Was there even such a thing as Melungeons or were they simply genealogical ghosts and lurid creations of popular journalism? Did they truly have some black and American Indian ancestry? Was the title only to apply to people in and around Newmans Ridge in Hancock County, Tennessee, or be extended to a wide range of persons of mixed ancestry like the Carolina Turks and Lumbee Indians? If the Melungeons went back before the arrival of Europeans, could they seek legal recognition as an indigenous American Indian tribe?

Questions abounded, and it seemed all of them were murky, emotionally charged and political. Unlike the Human Genome Diversity battle, neither party seemed to gain any advantages in the free-for-all. There were apparently no lessons to be learned on either side. At the end of the day, everyone just gave up and went home, exhausted.

Brodwin obviously sympathizes with the forces of the Academy in all this. He throws his lot in with the geneticist Kevin Jones, who found "he did not control the goals of research or the interpretation of findings." The Melungeon fracas illustrated "the political and conceptual vulnerabilities of human population genetics." In my opinion, however, Brodwin missed the point. Whom do university professors and academic researchers serve, if not the public? They should rejoice that so many of the great unwashed (even in the hills and hollers of Tennessee) are engaged by and even interested in their research.

And if they cannot achieve a satisfactory dialogue with their lay critics, whose fault is that? The debate should continue, not be swept under the rug of philosophical reflection. Whatever else they might be,

Melungeons are people. As such, they should not be dismissed when they become intractable.

September 1, 2009

Clearing out the Closets

Proposal Raises Bones of Contention
Anthropologists lobby to retain Native Indian skeletons for study
Rex Dalton
Nature 450/469 (November 22, 2007)

"Alarm is growing among anthropologists in the United States over a plan that could empty institutions of about 120,000 human skeletons currently stored for research purposes," according to this news item in *Nature.*"

A new proposal under the Native American Graves Protection and Repatriation Act (NAGPRA) would repatriate the bones at museums, universities and federal facilities across the nation to local Native American tribes "even if the skeletons are not culturally identifiable to the tribes."

Affected would be ancient skeletons similar to the disputed Kennewick Man specimen from Washington state.

"The rules would be disastrous," says Phillip Walker, an anthropologist at the University of California, Santa Barbara. He is a former member of NAGPRA's seven-person review committee and helped prepare the American Association of Physical Anthropologists (AAPA) comments.

Donald Yates comments: Even if the Federal Government decides to extend ownership of ancient remains and relics to locally visible American Indian groups for disposal as they see fit, authorities will have to take American Indian spokespersons' word on who they are, how old they are, where they came from and how long they have been where they are—in other words, the very questions science seeks to answer by study of those remains and relics. It's like giving students the chance to set course policies and grade themselves. I say this as a university professor and American Indian.

American Indian cultural continuity and collective identification is quite recent. The oldest Indian writing, the Walam Olum, says the last-but-one phase of Lenape history comprised 36 reigns of their chiefs from their settlement along the Ohio to a figure known as Lekhihiten the Author down to about 800 C.E. Before about the year 1 C.E., all these nations were still in Asia around the area of Lake Baikal. The Cherokee, Iroquois and Lenape lived together in the former territory of the Moundbuilder Indians (likely a mixture of Mexican Indians and peoples from the Old World) for an estimated 500 years. Another 36 generations passed until the Lenape arrived in their last dwelling place along the Delaware River, where they were encountered by the Dutch and English. A special black wampum belt was made to commemorate this event. When the beads were counted in historical times, the belt showed that the Lenape arrived in their easternmost homeland in 1396.

So much for ancient roots on the land, and the Algonquian Indians are regarded as the grandfathers by other Indian nations. There are more Indian nations of Algonquian speech today in North America than any other. Yet they, like all others, appear to be relative newcomers, historically, genetically and culturally.

Scholars such as Isabel Stewart have suggested that the Athabaskan Navajo and Apache Indians, today's largest and most-favored tribes in the eyes of the Federal Government, did not arrive in the American Southwest until shortly before the Spaniards.

It's time to stop throwing political sops and go on with the science of objective study of peoples, whether that involves historical accounts, genetics or whatever means. The truth is far more empowering than anything else.

November 22, 2007

Is Science Only for Scientists?

That's the import of a trio of opinions in this week's *Nature*. One of them, "Genetics without Borders," criticizes a "UK government scheme to establish nationality through DNA testing (as) scientifically flawed, ethically dubious and potentially damaging to science." The "scheme" is a peer-reviewed program of the UK Border Agency to test whether some 100 asylum-seekers are Somali nationals. The testing uses a combination of SNPs, mitochondrial DNA and Y chromosome, plus other forensic means, to determine whether they are actually Somali or not. (That is, within a high degree of probability, since all inductive conclusions are probabilistic.)

The editors of *Nature* fulminate against such methods. Yet, these are the tools of the trade used by law enforcement officials and academic geneticists, to say nothing of commercial DNA testing companies.

"The idea that genetic variability follows national boundaries is absurd," they scoff.

They are not impressed by the work of fellow scientists John Novembre et al., "Genes Mirror the Geography of Europe," in *Nature* 456, 98–101; 2008), saying that the idea that genetic variability follows man-made national boundaries is absurd. What is absurd is the idea that genetic variability is not molded and delineated by language, culture and historical events—the foundation of national boundaries. It seems to escape the opinion makers that Novembre et al. found that genetic patterns echoed linguistic divisions in Europe. This makes eminent sense in that courtship between most males and females is conducted in the same language. That means within the same nationalistic boundaries.

Random "mating" of an exogamous nature as envisaged by them is not in the nature of humans. It may be a generalization that can be formed of evolution, which is judged in sweeping retrospective, but it is not true of living people at any given time, in any given land or country. Until the 20th century (and perhaps even today) most people marry someone of the same rather narrowly defined ethnicity as themselves. In fact, until the modern period, an Englishman was most likely to marry a woman whose house was situated only an easy walk away. His horizons—and, thus, the eligible gene pool—was limited to a 24 mile square specifically labeled his "country."

Geneticists are wont to see human genetics in terms of geologic time, whereas the time depth and landscapes of history are more pertinent. The authors end by urging geneticists, "and indeed all scientists," to nip the government's "scheme" in the bud before the public finds out about it and an uprising ensues. This call to action seems to combine scientific cant with a patronizing view of the public.

Lay persons, and sometimes people outside one's narrow scientific specialty, just cannot be trusted to get anything quite right, can they?

Another day's blog will address the other two articles in this week's *Nature*, which exhibit similar mandarin attitudes.

October 6, 2009

US, EU Move to Regulate Direct-to-the-Consumer Genetic Testing

Discussion is accelerating in the United States and European Union to regulate private genomic testing that provides consumers medical information, according to *Science* magazine and the *European Journal of Human Genetics.* No mention is made in the reams of white papers about ancestry testing, but some of the pitfalls and bureaucratic morasses in the thinking about true genetic/medical testing are fairly ominous, if not silly.

"Although there has been speculation about the potential psychosocial harms of testing (that is, genomic medical testing), such as an increase in anxiety or encouragement of fatalistic behavior, there are, to date, few studies addressing these concerns," writes the reporters for Policy Forum in the Oct. 8 issue of *Science*. "The limited evidence tends to be reassuring, even for risk information associated with relatively serious ailments ... however, the scope for potential harm from unnecessary or unproven treatment after genetic risk assessment is an important unstudied question" (pp. 181f.).

We commend scientists and physicians for finding a new field of study divorced from reality but have to wonder what they will do about ancestry testing once they have conquered and tamed Frankenstein's elder monster. We suggest the following guidelines (just kidding, folks):

• Labeling on Internet sites and Zen Shopping Carts that explicitly states, "The claims for this ancestry product have not been evaluated by the U.S. Government Accountability Office (GAO), U.S. Federal Trade Commission (FTC), House Energy and Commerce Committee, Food and Drug Administration, National Institutes for Health or Department of Bioethics and Humanities, University of Washington School of Medicine, Seattle, WA 98195 USA."

- Predictive ancestry information may be hazardous to your progeny.
- No animal has been harmed in the production or clinical evaluation of this ancestry test.
- If you discover you have ancestry you did not expect, take a deep breath. Then take a healthy dose of skepticism, followed by two aspirins and a glass of water.

We're waiting for the next gambit from the genius bar in Washington!

November 28, 2010

Obstructionist Participants

In an article titled "Indian Tribe Wins Fight to Limit Research of Its DNA," Amy Harmon reports that Arizona State University has agreed to pay the Havasupai Indians of the Grand Canyon $700,000 and return blood samples collected from them for diabetes studies in the 1990s.

The university's Board of Regents apologized to the tribe for ... well, that part of the story is not clear. Not informing them that the samples might be used for "wider-ranging genetics"? Not informing the subjects that they reached negative conclusions and found no "diabetes gene" as they believed they had in a Pima Indian study? Not getting permission (no, that was done with simple-to-understand, signed consent forms, as was proper)? Coming to different conclusions about the Havasupai's origins than their myths and legends? Allowing people to "get degrees and grants" using "our blood"? Implying that the Havasupai are inbred? One Havasupai woman found that offensive.

Many tribal members were disgruntled because they were still suffering from diabetes after the university "took their blood."

Sorry, Havasupai Indians, a project participation consent form is not a treaty. But if you signed it, you should honor your word. You cannot go back now and require the researchers who use your samples to come to research conclusions that suit you and be silent about those that do not. Science (and society) don't work like that.

The tribe's dictates to the university were mercenary, and the university's decision to pay the tribe off, wrong. The case sets a bad precedent and places another barrier between Indian peoples in remote areas and the real world.

April 22, 2010

Rigged Genetics

If the facts don't fit the evidence, change them ...

We always suspected the genetics community of clinging to stale dogmas and being slow to acknowledge emerging new evidence about American Indians. But we did not dream that their officiousness extended to changing the information given by test subjects to bring it into conformity with preconceived conclusions.

Not until we heard Marcy's story.

"Over the years, I've heard complaints that (a DNA testing company) is not really responsive when you have questions about unexpected results," Marcy said. "They usually suggest further testing, which, of course, means more revenue to them.

"I've had some major disagreements with (a DNA testing company) over how they list results for mitochondrial haplogroup ancestral origins. ... I found out they were taking dozens of T2's who had listed their earliest-known female ancestor as being from America or the United States, changing this and placing them in the 'unknown' category. They claimed that because our haplogroup was designated European, our ancestors couldn't be from the United States!

"Now, this was nonsense, because at the same time, they allowed people to claim other similarly-colonized western countries, like Cuba. It's my opinion that if participants list a country of origin for their earliest-known female relative, that should be what is on the web page, not something assigned by (a DNA testing company) because as they told me, it may 'confuse people,' or contradict current scientific data.

"As a consequence (the DNA testing company's) publicly reported ancestral origins has nothing to do with our haplogroup's ancient Cherokee clan mother. The chips should fall where they may."

Now, this is not professional behavior on the part of a DNA testing company and it prevents new findings from coming to light.

In a study of 52 individuals claiming direct maternal descent from an American Indian woman, mostly Cherokee, we found that they were unmatched anywhere else except among other participants. Haplogroup T emerged as the largest lineage, followed by U, X, J and H. Similar proportions of these haplogroups were noted in the populations of Egypt, Israel and other parts of the East Mediterranean.

DNA testing companies do a disservice to their customers and to science by failing to call results as they appear without doctoring them. It is time geneticists stopped bringing all American Indians over the Bering Straits and forcing test subjects into the Procrustean bed of outmoded theory.

August 23, 2011

Jim Bentley, DNA Frontiersman

We interviewed one of Chromosomal Labs Bode Technology's senior staff members, Director of Sales and Marketing Jim Bentley, to get his perspective on industry changes over the past 35-plus years.

When did you first get interested in DNA?

JB: I'll have to preface my answer with a few remarks on "the early days." When I graduated from Arizona State University in the 1970s, DNA testing as we know it, was not really a field that was in existence. There was not a lot going on. The little work I did with chromosomes was using electron microscopy. I worked in the biochemistry department, however, and performed hundreds of assays using poly-acrylamide gel electrophoresis, mainly for separation of proteins. This technique, although improved and streamlined remains in use today for DNA-STR separation. The field we're in today where we can determine a person's profile and compare it with others for forensics for relationships, ancestry, missing persons, adoptions and the like, that technology hadn't been developed yet. It wasn't quite as easy as it is today.

Tell us more about the evolution of DNA testing.

JB: It basically began with blood groups and types. The first paternity test was done in a court case with Charlie Chaplin in the 1940s. He was excluded as the father, but the court said he could go ahead and pay child support anyway—probably, because he could afford it. Since that time, scientists started moving past groups and types into some other techniques. Human Leukocyte Testing (HLA), DQ-Alpha, and Restriction Enzyme STR testing (RFLP) are examples of the evolution of DNA testing.

The big breakthrough came when Dr. Alec Jeffreys at the University of Leicester discovered STR testing in England the late 1980s. He used STR profiling on the Colin Pitchfork case. Colin Pitchfork became the first criminal convicted on the basis of DNA evidence and as a result of a mass DNA screening operation. He was charged with raping and murdering two teenage girls. Since that time, the forensic community has really refined the techniques to perform STR testing. They've made it simpler and more accurate. It's really moved exponentially in the last 20 years. Today, competent biologists and chemists can produce excellent results, every time. Dr. Jeffreys has been knighted for his contributions.

So, what got you involved?

JB: I came out of college as a chemist, one interested in the medical field. I started out working in clinical chemistry and toxicology. The work we did with DNA was extremely limited and very costly. But I did stick with a career in clinical chemistry. Within four years after graduating from school, I was managing a clinical laboratory in Houston, Texas, called National Health Laboratories. It was a laboratory of about 100 scientists and support staff. After mergers, acquisitions and such, that company remains as Lab Corp. (It performs more than 1 million tests on more than 370,000 specimens each day.)

What opportunities for professional growth did you have over the years?

JB: Through taking a lot of continuing-education coursework, I became proficient and qualified as a general supervisor in clinical chemistry, toxicology, hematology, parasitology, microbiology, serology— everything except for tissue work like histology and cytology, which was done by certified medical experts in those specialties. My interests kept me in touch with the staff pathologists, however, as well as all the rest of the laboratory. Though my present-day field did not exist at the time I graduated, by staying current I was able to benefit from the changes and be part of an emerging valuable service provided not only to the medical community but also to the forensic one, and the general population at large.

What are some famous cases you've been involved with ... that you can talk about?

JB: Actually, that's my problem. We've been involved in a number of high-profile cases, but we're not allowed to talk about any of them. Most have been on the forensic side, serial-killer trials in Arizona, also in California, some that made the news in Florida … Texas ... Georgia.

Were you involved in catching the Grim Sleeper?

JB: Actually, that's an ongoing case in Los Angeles we are familiar with, but we didn't do the work on it, so we can talk about that one. The importance of the Grim Sleeper case has to do with familial testing and autosomal DNA. It was termed the Grim Sleeper case because there were a number of homicides that took place beginning in the mid-1980s, all with the same basic MO (modus operandi), and then the murderer went underground for 14 years. The victims were typically prostitutes shot with a firearm. In 2010, a suspect, Lonnie David Franklin Jr., 57, was arrested and charged with multiple counts of murder. He has not yet been convicted, nor the evidence against him tested in court.

How was DNA used to catch him?

JB: So, here were a number of cold cases, but they were being tracked, and the law enforcement authorities in Los Angeles continued to monitor progress. The sole survivor of one of the Grim Sleeper's attacks furnished a description of him as a black man in his 30s, along with other details. According to her story in the press, he lured her into an orange Ford Pinto, shot her in the chest with a pistol, took Polaroid's and raped her, leaving her for dead. In 2008, the body count was 13, and a $500,000 reward was put out for "America's Most Wanted."

It became the first use in California, and one of the first three cases in the United States to use familial DNA searching, that is, drawing on the FBI's CoDIS database to match one family member's profile with a suspect's profile. The L.A. police were able to provide a close partial match to Franklin's crime-scene profile with that of his son, whose CoDIS markers were on file for a minor crime. They then set up a kind of mini-sting operation at a pizza parlor in Buena Park, where they knew the family liked to eat.

Undercover detectives masqueraded as waiters and busboys. When the family left, they whisked away an unfinished pizza slice. The crust yielded DNA, which police linked on a more-solid basis to Lonnie

Franklin. It was the first high-profile case in which a family member's DNA had been used to catch a criminal. The ACLU and others had been critical of familial searching on grounds of privacy, and there is still a lot of debate over familial searching because it might open up the search and include those who hadn't committed any crime.

Did this help produce new commercial products like the "cousin finders"?
JB: Only a few states are doing familial searching, and they are pretty guarded about it. It's hard for me to make a connection. Certainly, these developments have been concentrated in the past three or four years, but the use of this technique is spreading.

Are people legitimately suspicious about DNA databases?
JB: Fears surface from time to time. There have been claims that keep popping up that someone's going to take everything that's in the database and use it to determine genetic deficiencies that could lead to medical issues down the road. Once it was speculated that if such information was released, insurance companies would begin denying people coverage based on their profiles.

This is the mother of conspiracy theories, isn't it?
JB: It really is. For the most part—not for everyone—the vast majority of the markers we are using are in the "junk DNA" area. That is, they don't by themselves "do" anything or give you genetic information on the face of things. There may be one or two markers that possibly could be construed as yielding some medical information—such as a trisomy at vWA or TPOX (a CoDIS locus). But by and large, you are not going to be able to do any medical diagnostics with the markers we run. Usually trisomies such as Down's syndrome would be physically expressed and not hidden. It's a little different with SNP panels (single nucleotide polymorphisms) such as those run by 23andMe. With a high number of those, it's entirely possible to predict medical predisposition. That's what they base their business on.

Let's talk some more about the CoDIS database.
JB: It's important to realize that even law enforcement doesn't provide

much access to the CoDIS (Combined DNA Identification System) databank. That's something I have to give the FBI credit for. They have developed a system that is secure. It's the DNA administrator at each facility who has undergone FBI training and uploads the data under very strict rules, and they are notified of any "hits" that involve them, but, otherwise, there is very little access, and the use of the database is very even across the country. There are not a large number of portals that can be used to access the CoDIS database. There are several hundred law enforcement laboratories that are running profiles across the country, and the database is best thought about on three different levels: LDIS, SDIS and NDIS, local, state and national versions. Between our labs in Phoenix and Virginia, we've tested over a million profiles for entry into CoDIS. That's about one-tenth of the entire number. I can tell you there is tight security. Hundreds of thousands of investigations have been aided by a DNA hit (we don't like to say "match" so much, because statistically nothing is 100 percent) generating a lead.

How did you get bitten by the genealogy bug?
JB: I've always been fascinated with ancestry. I think it came about because my father took an interest in discovering our family's roots and had to do so at the time by traveling to Salt Lake City, Utah, and poring over whatever records he could find there about our fathers, and great-grandfathers, and great-great-grandfathers, and so forth. He had tintypes of some of the relatives. We had various pieces of the puzzle. My father pretty much consolidated everything back to William Bentley, who settled in Rhode Island in the early 1700s and had come from Bedfordshire, England. He put together a book for family use. He glorified a few of them and left a few out that weren't ready for glorification. For the sensitivity of some of the relatives, he left a few details out, but it was a pretty solid piece of work. For me, it kind of fostered this interest in ancestry and its importance. Certainly, when I started at Chromosomal Labs • Bode Technology, we started looking at the various tools that could be used. Our history, to be sure, is passed down from generation to generation. Initially, we were using mitochondrial DNA, Y-SNP's and Y-STRs and then autosomal STRs to determine how we're connected to general and specific individuals back to the Revolutionary War days and how you are linked with the world

population, what your roots were. I have a particular Y haplogroup of G2a, which is not one of the more common ones.

Hmm ... you and Joseph Stalin.

JB: [Laughs]. Is that what his haplogroup was? Uh-oh! He was one of the worst. Well, I got interested in G2a and hooked up with about 50 other Bentleys, and we identified our founder patriarch haplotype. I get emails from them on a regular basis. The other thing we tried to find out was what in the world were all these G2a's doing in England. I don't know. But one of the things I find in the literature most often was that the Sarmatians were horsemen that gave the Romans a pretty rough time. Eventually, they were decimated. The Romans took their remaining cavalry and pressed them into service for 12 to 13 years or longer. Some were dispatched to Hadrian's Wall. Now, do I know for a hundred percent certainty that's where I came from? No, but it's fun to regard that as a hypothetical personal history.

You have a Scythian gene, don't you?

JB: Yes, I do, according to the analysis DNA Consultants did for my autosomal ancestry. The work Dr. Yates has done on the rare alleles supports a lot of the stuff the family has been putting together for years and years. I was very pleased to get my Rare Genes from History report back showing I had the Scythian gene. That seems to go along with the Sarmatian theory about the Bentleys.

How do you see the industry changing over the next few years?

JB: I can speak best about changes I am seeing in the field. They're getting closer to having rapid DNA testing on a chip. This gives flexibility to those who want to use DNA as "point of use" testing. The FBI, this past year, came out at the Promega conference and said that within the next two years they would like to see wide adoption of "point of use" testing. The IntegenX prototype allows you to put your swab into a cartridge, insert it into the instrument on the fly and get your STR results in a few hours. Previously, Rapid DNA testing was not only time-consuming and lab-bound, but it was very expensive. It cost several hundred dollars in reagents alone. As the technology improves to allow two-hour testing, in our lab or on a chip, reagent and personnel time continue to drop, Now, the FBI would like to see point-of-use testing in every booking station in the country. At the last show, I also saw an instrument from Illumina that would run Y-STRs, mtDNA and autosomal DNA profiles simultaneously on one sample. Another change that is coming is we will see an expanded profile becoming the standard, perhaps something similar to the GlobalFiler kit from Life Technologies with its 24 loci. With the new technology you can increase the speed for amplifying the specimen by five times and achieve nine times the discriminating power or resolution.

Any final remarks?
JB: The DNA testing field is on the threshold of even greater accolades of appreciation both from the scientific community and the public. If DNA wasn't even in anyone's mind 20 years ago, soon it will be part of everyone's daily lives.

January 26, 2013

Junk DNA

We are our DNA. It was not a surprise to find that our entire DNA is functional ("Junk DNA Isn't Junk, and That Isn't Really News"). The surprise is in the discovery of what we can do with what we once thought was junk.

According to that recent NPR article, "It is a massive control panel that regulates the activity of our genes." Our genes "would not work" without it. So, instead of being junk, they are critical and "control how cells, organs, and other tissues behave." But we can also now read the markers and mutations on this "panel" and discover much more information than knowing it is just working efficiently for our body. This knowledge is considered a "major medical and scientific breakthrough" (Ibid.). We just have to read it well.

But first, what is DNA exactly? John Wilwol, in his recent NPR article, "A 'Thumb' on the Pulse of What Makes Us Human," quotes Sam Kean, author of the book *The Violinist's Thumb And Other Lost Tales of Love, War, and Genius, As Written by Our Genetic Code*, as saying that DNA is what makes us who we are. Wilwol further quotes

Kean to help us understand what DNA is and how it differentiates from genes: "While DNA is a thing—a chemical that sticks to your fingers," he writes, genes are more conceptual in nature, "like a story with DNA as the language the story is written in."

So, if DNA is a language how are we able to read it? All parts of our genetic code are now readable and meaningful. Marker locations (loci) are spread across one's entire genome, not confined to one's male (Y chromosome) or female (mitochondrial) DNA. (This is how sex-linked, haplotype tests that follow one line at a time are analyzed). Different mutations are handed down genetically—different according to the region where one's ancestors lived.

Because of this new ability to read markers, consumers are now able to buy autosomal DNA tests that provide a complete analysis of where all one's ancestors' ethno-geographic origins may lie—reflecting the entire spectrum of all their ancestral lines. Not just one line at a time as in haplotype testing. This is next-generation ancestry DNA testing and the wave of the future. Anyone can take an autosomal DNA test because it does not rely on X or Y chromosomes. A female is unable to take the male Y- linked test and must entice a male in her line, if one is available, to take this test. The future is now in many ways.

What else can you learn from autosomal DNA testing? Anne Tergesen, in a recent article in the *Wall Street Journal,"* quotes Megan Molenyak, author of, *Hey America, Your Roots Are Showing,* as saying that this relatively new test deciphers the amount of DNA shared between those whose common ancestors lived within the last half-dozen or so generations. Tergersen explains it like this, "Y-DNA and mitochondrial DNA can connect people whose common ancestors lived recently or hundreds of years ago. But to find out how closely you are related—and to locate relatives besides those on your direct maternal or paternal lines—you will need an autosomal DNA test." (Of course, you would both need one to compare) and "in general, the more DNA two people share, the closer their connection".

But there are even more things on the horizon with autosomal DNA for the future. According to a recent *Smithsonian* article, "Fetal Genome Sequenced Without Help From Daddy," a fetus' entire genome can now be sequenced from the mother alone with "99.8% accuracy." How is that possible? It was just "last month clinicians announced that they could sequence a fetus's entire genome by taking samples from the pregnant mother's blood and that of the father to be" ("Fetal Genome"). Now, they have a "more difficult, but more complete method (that) uses DNA from the pregnant woman and the fetus to map out every last letter of the fetal genome …with the advantage that it can pick up mutations that a fetus has but its parents do not" (ibid.). Rob Stein quotes Dr. Alan Guttmacher, director of the National Institute for the Child Health and Human Development in a recent NPR article, "Genome Sequencing For Babies Brings Knowledge and Conflicts," as saying, "Instead of screening for currently something like 30 conditions, it would allow you to screen for hundreds if not thousands, of conditions) at birth." He goes on to say that "one could imagine a day where knowing someone's entire genome sequence at birth, you could really begin to think about structuring health care, their dietary choices, their exercise choices … early in life, in a way that would have an impact on truly lifelong health." Stein says that this gene sequencing could "spot babies that are prone to conditions such as obesity, diabetes, heart attacks or cancer" and that we may soon be "sequencing all babies when they're born." It could be a wonderful tool. But we are not there yet.

According to Rob Stein in another NPR article, "Perfection is Skin Deep: Everyone has Flawed Genes," scientists have determined we are all more flawed than they thought. "Researchers discovered that normal, healthy people are walking around with a surprisingly large number of mutations in their genes." Chris Tyler-Smith of the Wellcome Trust Sanger Institute in Cambridge, England, and his colleagues analyzed the DNA of 179 people from several countries who volunteered their genetic information to the 1,000 Genomes Project.

In a paper published in the *American Journal of Human Genetics*, researchers reported that though none of the people whose DNA was studied were sick, the average person has about 400 minor flaws and one or two that could contribute to disease. Tyler-Smith says, "It's a bit surprising that people should be walking around apparently healthy yet we're seeing known disease-causing mutations in their genomes," he says. "But the answer was that these tended to be for mild and very often late-onset conditions. Things like heart disease, an increased risk of disease or developing cancer."

On its website, the American Diabetes Association highlights the interaction of genetic and environmental factors: "You inherit a predisposition to the disease then something in your environment triggers it. Genes alone are not enough."

So, the problem is not so much with the analytical tool but rather the possibility of over-interpretation. Again, we just have to read it well, with the same critical eye for what is written *in* us as that written *by* us. And who knows what else we will soon be able to discover from reading our DNA?

January 14, 2013

Genes Are the Teachers

Sam Kean, author of the book *The Violinist's*, says, "While DNA is a thing—a chemical that sticks to your fingers," genes are more conceptual in nature. He says genes are the story while DNA is the words that make up the story.

Chromosomes are long, cylindrical-like blocks that contain the words allowing them to create the story, similar to children's brightly colored alphabet blocks. Alleles are any alternative forms of the story (read gene) that may occur at any given locus (point) on the chromosome, attaching themselves like people do on a train or bus or subway—grabbing on for safety and, perhaps, deciding to go a different route. Or tell a different story. Reminds me of the crazy, alternate endings that one sometimes finds on a DVD. For instance, there are Internet rumors of alternate endings for *The Titanic*—Jack surviving and swimming to shore and Rose dropping the diamonds in the ocean as her granddaughter watches. Personally, I don't like alternate endings. But alleles? Alleles we need. They make us unique.

What do alleles do? Alleles hang out together like cute little couples or a group of friends on the chromosomes. Just like people, some of them are dominant, and some of them are shy (recessive). If you get two dominant people in the party, you know who is having their way. It is the same with alleles though it usually takes more than two people to really get anything done just like it usually takes more than two alleles to determine a trait. It might take quite a crowd of them. They might call another party (interact with another site of alleles). All of our genetic traits are caused by alleles interacting. The genes we inherit are all the same as humans, but it is how the alleles express them that makes us unique (*Encyclopædia Britannica*).

What about genes and DNA? According to Ananya Mandal, M.D., in her article, "Genes—What Are Genes?", DNA is the "chemical information database." The DNA carries the "complete set of instructions for the cell" and genes are the "working sub-units" of DNA. You might also see DNA as a library of information and the genes as the teachers and librarians who use the library and tell the students (each cell) how to work (how to make proteins), as each gene has a set of instructions, roughly speaking, for making a certain protein, be it melanin or insulin. Mary Kugler, R.N., in her article, "What are Genes, DNA, and Chromosomes?" says a gene is a "distinct portion of a cell's DNA" and has "coded instructions for making everything the body needs." These genes are "packed in bundles" of chromosomes. Something like all the librarians packed into offices that are really two offices in one.

How many chromosomes do we have? Everyone has 23 pairs of chromosomes (46) but only one is sex-linked. The other 22 pairs are autosomal chromosomes determining the rest of the body's makeup (Kugler). What do chromosomes do? Keep DNA in its place.

Or if they were people, alleles would be the artists, genes the teachers, chromosomes the disciplinarians, and the DNA would be the great Wizard of OZ—except with real instructions.

October 13, 2013

Too Big to Feel

We don't often write editorials in this space. Normally, you will see nothing but news in the DNA Consultants Blog. Some sparse marketing messages may appear whenever we have a new product or study. But the FDA's "stop and desist" letter last Friday to personal genomics giant 23andMe has sent shockwaves through the industry. Although we are not in the business of providing medical information to customers, only ancestral background analyses, we feel compelled to weigh in on the FDA's warning, which we think is overdue.

First, it is important to note that the FDA took action against 23andMe because the company has been selling an unapproved diagnostic device and medical service. The FDA demanded 23andMe "immediately discontinue marketing the Personal Genome Service" after years of protracted and unsuccessful requests for proof of safety and efficacy to back up the company's marketing claims.

A check of 23andMe's website on November 26 showed little change in the promises it makes to consumers. Splashed across the welcome page in large letters was "Get to know you." The site boasted having "Reports on 240+ health conditions." In language that sure sounds medical to us, it speaks of carrier status, health risks and drug response.

An appeal to parents suggests, "Find out if your children are at risk for inherited conditions, so you can plan for the health of your family." Under "drug response," the company advises that you can take "information on how you might respond to certain medications" from your $99 DNA test to your next doctor's visit.

The FDA found "some of the uses for which Personal Genomic Service is intended are particularly concerning." A false positive result for the BRCA gene, for instance, could cause a patient to remove breasts or ovaries to avoid getting cancer. A false negative could lead patients into an unfounded sense of security and make them ignore that an actual risk exists. An inaccurate result for warfarin drug response could lead a patient to self-manage their dosage or skip it altogether, leading to "illness, injury, or death." Some pretty dire concerns.

DNA Consultants simply does not do DNA testing for medical information. If a customer calls and asks for such a service, we explain that is something they should discuss with their healthcare provider. Neither our marketing nor fulfillment of tests contains any medical language.

We specialize in ancestry analysis exclusively and believe we do that better than anybody else.

Of course, the field of genetic screening has made enormous progress over the past 10 years. We monitor many of those advances in our blog. But we do not believe the field is by any means "there" yet, certainly not ready to be packaged and hawked to consumers. We doubt it ever will be. Or at least we hope not.

Despite popular anthems of genetic determinism, your health is not all in your genes.

But your ancestry certainly is. Find out what your ancestry is and you will be able to form an idea of what your ancestral medical history might look like, going beyond the two generations of family medical history covered by standard questionnaires at the doctor's office. But that part is entirely up to you and your healthcare providers.

No company is too big to fail. Even the largest are subject to oversight by the government as well as consumer pressure and the natural forces of the market. Nor is any company too big to feel. We hope 23andMe will respond both to the FDA and the public with understanding and responsibility, not indifference and arrogance.

November 26, 2013

Top Stories of 2013

What were the top genomics and DNA testing stories of 2013? It has definitely been the year that DNA from genetic testing has been rocking in the news. I have chosen what I think are the top three.

First, last June, there came the Supreme Court decision that police can now legally take DNA from anyone they arrest. They then enter this into a database where they can match it with existing samples (Dan Noswolitiz, "It's Now Legal for the Police to Collect DNA," *Popular Science).* Since we live in a rather scary world since 9/11, I thought that might be constructive, at first glance, until I realized that the key word is "arrest" not "charged." What is the difference? People are falsely framed and arrested every day as well as arrested for minor offenses. Consequently, there any number of ways this could be abused. Even if someone is declared innocent, guess what? They still have your DNA.

(Do you want others to have your personal genomic data? That is a question you might want to ask any DNA testing company you use as well. What do you do with my DNA? Do you keep it and put it in a database or share it with others?)

Also, in the same month, the Supreme Court ruled that human genes cannot be patented, except for synthetic genes. This was in response to a lawsuit between "Myriad Genetics, a medical diagnostics company, and the Association for Molecular Pathology" (Young). Many saw this as a win for women with an elevated genetic risk of breast and ovarian cancers as well as researchers and scientists (Richard Wolf, "Justices Rule Human Genes Cannot Be Patented," *USA Today*) Why? Myriad had a monopoly on the gene; as a result, no one else could produce or manufacture it, and the genetic test was inordinately expensive. Not every woman has a bank account matching Angelina Jolie's. Her decision to first take the genetic test and then have a preventive double mastectomy because of her high risk of breast cancer brought this case to the forefront.

But it also played into the biggest and most-controversial story of the year concerning genetics testing for disease and the battle between the FDA and 23andMe. The FDA told them to stop selling health- and medical-related information with their genetic tests. Some see this as the FDA stealing their right to their own personal genetic information. Since there is no genetic destiny for disease because of other lifestyle and epigenetic factors (Carl Zimmer, "Hope, Hype, and Genetic Breakthroughs" *Wall Street Journal*), others see this as part of a process to ensure that these tests are accurate and not misinterpreted. However, knowing you have a high genetic risk for a disease might mean you make better lifestyle choices. I think in the end it will not spell the end for direct-to-consumer genetic health services, but, hopefully, the industry will not only be resurrected, but there will also be better guidelines for the field as well as the consumer spurring a wider market so consumers have more choices. I am looking for that silver lining in the New Year.

January 2, 2014

Mired in Miracles

This past summer, the Supreme Court decided biotech companies could not patent human genes extracted through genetic testing. What are the implications of such a decision for businesses and consumers? This decision is advantageous for both. Why? It keeps prices down for the consumer assuring most anyone can afford needed genetic testing and prevents a monopoly on any genetic testing done for both research and medical purposes. That means more than one company can offer a particular type of genetic testing, and it is more affordable and accessible to the public.

The brouhaha all started with the company Myriad Genetics. First, this company discovered with genetic testing the "precise location and sequence" of two genetic mutations that are a significant risk factor for breast cancer, BRAC1 and BRAC2 genes. After making this discovery, they created their own form of genetic testing to find these mutations and decided to patent it. Angelina Jolie made the decision to have a double mastectomy after discovering she had these genes based on a very expensive test. It cost some $3,000 because of Myriad Genetics' monopoly. ("Biotech Companies No Longer Have the Right to Patent Human Genes").

Of course, the fight ended with the Supreme Court's decision, but our vision of it was of Angelina Jolie cast in the limelight once again. However, instead of being a hell-on-wheels wife doing her best to use every means to kill her husband in *Mr. & Mrs. Smith,* she sounded more like any woman afraid of getting breast cancer. The only difference was she could foot any bill they handed her.

Perhaps Angelina Jolie did not realize it, but the feeling many probably had when reading articles concerning her genetically inherited condition was not one of empathy alone but fear—fear upon the realization that most could ill afford genetic testing of this kind. Where did that leave the majority of women? Many, no doubt, were left wondering if they had the same genetic mutations especially if they had a high risk factor like Angelina Jolie's. But Angelina Jolie was not left to wonder. She got tested and made a decision.

The Supreme Court made the right call. No one should have a monopoly on genetic testing. Genetic testing should not just be for the Angelina Jolies of the world but affordable for anyone who needs it.

June 18, 2014

Forensic Fantasies

If DNA testing is to be done correctly, one has to follow certain procedures and protocols. Otherwise, a criminal investigation will misinterpret the results from contaminated samples. If the DNA testing results are not analyzed properly, the consequences could be devastating. It could even mean life or death. Amanda Knox's trial didn't follow correct forensic procedures.

Amanda Knox was a tourist in Italy in 2007 where she and her boyfriend were convicted of the murder of her British roommate, Meredith Kercher. Many Italian papers at the time suggested that they had wild orgies. It made for tabloid heaven. Where are the DNA testing clues? There is not any evidence. According to Claudia Dempsey, author of *Murder in Italy: Amanda Knox, Meredith Kercher, and the Murder Trial That Shook the World,* in her *Seattle Post-Intelligencer* article, "The Carnival Ride Continues," the judge who acquitted them and released the pair in 2011, Judge Claudio Patrillo Hellman, said there was *nessuna prova* (no evidence) against them. The prosecutor, Giuliano Mignini, believes, according to Dempsey, in a "four-way orgy" led by Knox that ended with Kercher's murder.

It is one thing to have fantasies and use them to create fiction for the screen or novels. It is quite another to use one's fantasies and botched DNA testing results to ruin lives. The acquittal of the murder trial, however, has since been overturned as Italy does not have a double indemnity clause as the United States does (one cannot be tried twice for the same crime). Consequently, Amanda Knox, is not attending the second trial and is not expected to be extradited whatever the outcome.

For one thing, recent DNA testing evidence shows that there was no blood of the victim on the knife. (Dempsey, "Good News for Amanda Knox: Victim's DNA Not Found on Knife"). Has anyone seen a horror movie like *Psycho* or any zombie movie where someone was killed with a knife, and they managed to *not* get blood on it? It would not have taken much DNA to be able to get a DNA sample of Meredith Kercher if it was used as the murder weapon. It was not in my opinion.

What was wrong with the DNA testing results? According to the Huffington Post article, "Amanda Knox Case: DNA Experts Cast Doubt on DNA Evidence," it was not up to international standards, according to DNA and forensic specialists. They used "a dirty glove and did not wear caps" and the bra clasp of Kercher that the prosecution insisted had Knox's DNA on it was not collected for 46 days. Even if Knox's DNA was on the bra clasp what does that prove? That she asked her girlfriend to help her fix her bra. Maybe bra-less men do not realize that women sometimes do that kind of thing? Also, the crime scene was not secured and handled properly. People moved around, and objects were moved ("Amanda Knox Case").

Also, most people do not realize that they already have someone in jail for this crime. Rudy Guede, by his own admission, decided to burglarize her place and then afterward "fled to Germany with his bloody shoes and pants." He is now in jail though soon to be released.

DNA testing is a wonderful tool. However, like any other tool, it must be used correctly following strict guidelines and procedures, or it is useless if not damaging.

October 11, 2013

Preserving Your DNA

We might have gotten a laugh from the movie *Identity Thief,* but it isn't that funny if it happens to you. Living in a fast-paced world full of online spammers, credit-card skimmers and petty thieves, most of us know to be careful with our finances, Social Security number and credit cards. But what about a more personal identity? What about your DNA? Your DNA is unique to you alone. Even identical twins show slight genetic differences with very expensive genetic testing costing thousands of dollars (Barry Starr, "Genetic Sleuthing"). Your DNA is you.

Why should I control my genetic information? Because of epigenetics and other factors, it is doubtful that prejudice will ever be used from your DNA testing results as it is portrayed in the film *Gattaca.* Nevertheless, in the near future, if a thief has your genetic profile, that would be the ultimate in identity theft. Why? It could well be how we soon identify replacing the now state-of-the art idea of using your voice or fingerprint. Your genetic information is like owning a mirror-image of you.

How is genetic information used? Forensics already uses genetics to catch criminals, exonerate suspects or find missing persons. And DNA testing companies that analyze DNA with autosomal DNA testing (non-cousin related) can give you a comprehensive analysis of ancestry based on and reflecting your unique genetic information.

How can I control my genetic information? First, if you have done autosomal DNA testing and received your genetic profile with your unique alleles, you should put it in a secure place. Many have done DNA testing but not as many have thought too much about how their DNA is being used. In order to control our genetic information, we need to become as careful with it as we are with our credit cards. You might wish to think twice about using a DNA testing company that shares your genetic information with third parties or stores it in a large genomics database for medical research without the individual's express permission to do so. Even if you have read all the fine print or given your permission for them to do this, you might want to rethink that option.

Why? Not all DNA testing companies share your DNA testing results, or gather it in a large database, so why not choose one that does not and keeps it secure?

How can I save my genetic information and keep it the most secure? Your DNA dies with you. Not only should you save your unique genetic profile, you should save a healthy copy of your and your loved ones' genomes now to take advantages of advances in medicine and gene therapies in the future. And there are other reasons. If someone goes missing, this will greatly help an investigation. Just a handful of DNA testing companies offer a way for you to securely and affordably save your genetic information. Take advantage of one that does and be ahead of the crowd.

March 26, 2013

Posts in Alphabetical Order

Index

By Judith K. Jarvis

www.ingramcontent.com/pod-product-compliance
Lightning Source LLC
Chambersburg PA
CBHW060114200326
41518CB00008B/821